Lecture Notes in Mathematics

A collection of informal reports and seminars
Edited by A. Dold, Heidelberg and B. Eckmann, Zürich

T0216478

333

Numerische, insbesondere approximationstheoretische Behandlung von Funktionalgleichungen

Vorträge einer Tagung im Mathematischen Forschungsinstitut Oberwolfach, 4.–8. 12. 1972

Herausgegeben von R. Ansorge, Universität Hamburg, Hamburg/BRD
W. Törnig, Technische Hochschule Darmstadt, Darmstadt/BRD

Springer-Verlag
Berlin · Heidelberg · New York 1973

AMS Subject Classifications (1970): 65-02, 65D30, 65J05, 65L10, 65Mxx, 65Nxx, 65Q05, 65R05

ISBN 3-540-06378-1 Springer-Verlag Berlin · Heidelberg · New York
ISBN 0-387-06378-1 Springer-Verlag New York · Heidelberg · Berlin

Offsetdruck: Julius Beltz, Hemsbach/Bergstr.

VORWORT

Im Mathematischen Forschungsinstitut Oberwolfach fand in der Zeit vom 4. 12. bis 8. 12. 1972 eine Tagung über die numerische Behandlung von Funktionalgleichungen statt. Die dort gehaltenen Vorträge befaßten sich mit verschiedenen Methoden zur numerischen Behandlung von linearen und nichtlinearen gewöhnlichen und partiellen Differentialgleichungen, Integralgleichungen, Integro-Differentialgleichungen sowie allgemeinen Funktionalgleichungen. Besonderes Gewicht kam der Diskussion von Fragen der numerischen Praxis, der Herleitung von Fehlerabschätzungen und der Gewinnung von Konvergenzaussagen zu. Dabei wurden teilweise approximationstheoretische und verwandte Hilfsmittel benutzt.

Die Veranstalter der Tagung freuen sich, daß in dem vorliegenden Band der größere Teil der Vorträge veröffentlicht werden kann. Den Autoren, den Herausgebern der Lecture-Notes und dem Springer-Verlag sei hierfür herzlich gedankt.

Dem Leiter und dem wissenschaftlichen Beirat des Mathematischen Forschungsinstituts Oberwolfach danken wir dafür, daß uns die Durchführung der Tagung ermöglicht wurde. Nicht zuletzt gebührt unser Dank den Herren Dipl.-Math. W. Gentzsch und Dipl.-Math. K. Merten für die Koordinierung und Überwachung der Druckvorlage.

Hamburg und Darmstadt im März 1973

R. Ansorge W. Törnig

INHALTSVERZEICHNIS

THE APPROXIMATION OF SOLUTIONS OF NONLINEAR ELLIPTIC BOUNDARY VALUE PROBLEMS HAVING SEVERAL SOLUTIONS

By E.L.Allgower and M.M.Jeppson

§1 In this paper we give an application of a recent extended constructive version of the Brouwer fixed point theorem [16] to obtain numerical approximations to the solutions to

$$(1.1) \qquad \begin{aligned} -\Delta u &= f(P,u,\nabla u) \quad \text{on} \quad D \\ u(P) &= 0 \quad \text{for} \quad P \in \partial D \end{aligned}$$

where D is a bounded domain in R^2 and ∂D is piecewise analytic. We also assume that f is bounded $(|f| \leq M)$ and continuous on $\bar{D} \times R^3$. Additional smoothness assumptions on f will be needed to assure existence of solutions to (1.1) in the context of appropriate Banach spaces and these will be mentioned below.

Of particular interest here are the cases in which (1.1) has several solutions. Recently several authors have treated the problem of obtaining numerical approximations to the solutions to (1.1) when f is independent of derivatives of u (see [15], [21], [22], [24], [25] and the references there). The approach in general is to formulate a consistent discrete analogue of (1.1) e.g.

$$(1.2) \qquad \begin{aligned} -\Delta_h u &= f(P,u,\nabla_h u) \quad \text{on} \quad D_h + C_h^* \\ u(P) &= 0 \quad \text{for} \quad P \in C_h \end{aligned}$$

where Δ_h is the usual five point approximation to Δ and ∇_h approximates ∇ by central differences after a square grid R_h has been placed on R^2, $C_h = \partial D \cap R_h$, D_h is the set of grid points in $R_h \cap D$

whose four nearest neighbors lie in $\bar{D}$ and C_h^* is the set of grid points in $R_h \cap D$ having at least one nearest neighbor outside $\bar{D}$. For $P \in C_h^*$, $\Delta_h u(P)$ is defined by the Shortley-Weller approximation. One then attempts to approximate solutions to (1.1) by an iterative procedure for solving (1.2), (see e.g. [22]).

There seems however, to be a general difficulty which is inherent to the iterative methods and in fact also with certain Newton-type algorithms when several solutions are present. The difficulty lies in the fact that some of the solutions to (1.2) will in general be strongly attractive when such algorithms are applied.

It is at this point that we make our departure and apply a topological fixed point algorithm which is impervious to the attractive or repulsive properties of any solution to (1.2). In general, for properly chosen mesh sizes our algorithm will approximate in one application as many solutions to (1.2) (and hence to (1.1)) as are present. Indeed, the solutions which have been defined as unstable, [13], [17], [26], are approximated with the same ease as any other solution.

At the end of the paper we give several numerical examples. The examples which we treat also indicate that our approach may be applied to a wider class of problems. For example, the boundedness condition on f may be relaxed somewhat, the Laplacian may be replaced by a uniformly elliptic operator and problems with mixed boundary data may also be treated.

Also elliptic boundary value problems in higher dimensions may be treated and certain other finite difference approximations may be used in place of the five point star.

§2 We first show that (1.2) will have solutions for each N under the simple assumption that f is continuous and bounded on D. To do this we assign an ordering to the points $P \in D_h + C_h^*$, e.g. the "natural ordering" [29] from left to right, top to bottom. When the boundary data has been incorporated in the divided differences, (1.2) may be rewritten as

$$(2.1) \qquad A\bar{u} = h^2 \bar{f}(\bar{u}) \quad \text{on} \quad D_h + C_h^*$$

where $\bar{u} = (u(P_1), \ldots, u(P_N))'$, $\bar{f}(\bar{u}) =$ $= (f(P_1, u(P_1), {}_h u(P_1)), \ldots, f(P_n, u(P_n), {}_h u(P_N))'$, N , is the number of grid points in $D_h + C_h^*$ and A is an $N \times M$ matrix having several specific properties [29] p.187. In particular, $A^{-1} \geq 0$. Hence the solutions to (1.2) are precisely the solutions to

$$(2.2) \qquad \bar{u} = h^2 A^{-1} \bar{f}(\bar{u})$$

and conversely. Now define the continuous mapping $T_N : R^N \to R^N$ by

$$(2.3) \qquad T_N \bar{u} = h^2 A^{-1} \bar{f}(\bar{u}) .$$

The fixed points of T_N are the solutions to (1.2) and conversely. From (2.3) we have

$$|T_N \bar{u}(\bar{P})| \leq h^2 M A^{-1} \bar{e}$$

where $\bar{e}$ is the N-vector whose components are all unity. Hence an a priori bound on any fixed point $\bar{u}^N$ of T_N is given by

$$|\bar{u}_i^N| = |u^N(P_i)| \leq h^2 M \sum_{j=1}^{N} a_{ij}^{-1} , \quad i = 1, \ldots, N$$

where a_{ij}^{-1} denotes the ij-th element in A^{-1}.

If we let

$$B_N = \{\bar{v} \in R^N \mid |\bar{v}| \leq h^2 MA^{-1}\bar{e}\}$$

then $T_N : R^N \to B_N$. Furthermore T_N is a continuous map since $\bar{f}(\bar{u})$ is continuous. In particular $T_N : B_N \to B_N$ and hence T_N has fixed points by the Brouwer fixed point theorem. Thus we have

Theorem 1. (1.2) has solutions for every square net on D .

In a later section we will take up the question of whether these solutions $u^N(P)$, $P \in D_h + C_h^*$ yield approximations which converge to solutions to (1.1).

§3 The algorithm we use for approximating the fixed points of T_N has been thoroughly described elsewhere [16] (also see [5], [6]) and hence we will only sketch it briefly here and state the results which we will use. Our discussions in this section will pertain to the case $B_N = C^N = \prod_1^N [0,1]$. This represents no restriction since in general the mapping $\bar{F} : C^N \to C^N$ whose fixed points we will first approximate is obtained by composing T_N with the linear homeomorphism $\bar{H}$ between C^N and B_N .

The continuous mapping $\bar{F} : C^N \to C^N$ induces a proper labeling (see [2]) of the points $Y \in C^N$ as follows:
Let $\bar{L}(Y) = (L_1(Y), \ldots, L_N(Y))$, where

$$L_i(Y) = \begin{cases} 0 & \text{if } f_i(Y) \geq Y_i \neq 1 \\ 1 & \text{if } f_i(Y) < Y_i \\ 1 & \text{if } Y_i = 1 . \end{cases}$$

Then set

$$\ell(Y) = \text{the number of initial 1's in the vector } \bar{L}(Y) .$$

Define

$$A_i = cl\{Y \in C^N \mid \ell(Y)=i\}\ ,\quad i = 1,\dots,N$$

where cl denotes the closure.

Theorem 2. $\bigcap_{i=0}^{N} A_i \neq \emptyset$ and any point belonging to this set is a fixed point of $\bar{F}$.

Points in $\bigcap_{i=0}^{N} A_i$ are approximated by N-dimensional simplexes which have a vertex in each of the sets A_i as follows.

We subdivide C^N into J^N small N-cubes by the hyperplanes $Y_i = k/J$ where $k = 1,\dots,J-1$ and $i = 1,\dots,N$ for a given positive integer J. Each small N-cube C_m^N is further subdivided into $N!$ simplexes of dimension N:

$$S_m(j_1,\dots,j_N) = \{(Y_1,Y_2,\dots,Y_N) \in C_m^N \mid Y_{j_1} \geq Y_{j_2} \geq \dots \geq Y_{j_N}\}$$

where $j_1,j_2,\dots,j_N$ is a permutation of $1,2,\dots,N$. Thus a simplicial subdivision of C^N into $N!J^N$ simplexes is described. By Sperner's Lemma there exists an odd number of N-simplexes in the subdivision whose vertices have the set of labels $\{0,1,\dots,N\}$ and any such simplex in our subdivision will be called a Sperner simplex.

In [16] an algorithm and a computer program is given for locating a Sperner simplex. It is also shown that points of a Sperner simplex are approximately fixed in the following sense.

Theorem 3. Suppose $\bar{F} : C^N \to C^N$ is continuous and that $\|X-Y\| \leq \delta$ implies $\|\bar{F}(X)-\bar{F}(Y)\| \leq \varepsilon$. Suppose that S^N is Sperner simplex with diameter $\leq \delta$. Then for any $X \in S^N$, $\|\bar{F}(X)-X\| \leq \varepsilon+\delta$. (Here we are using the norm $\|X\| = \max_{1\leq i\leq N} |X_i|$.)

From Theorem 3 we have a crude measure of the rate of convergence of Sperner simplexes to a fixed point of $\bar{F}$. It should be noted that it is never certain that a Sperner simplex actually contains a fixed point of $\bar{F}$ and hence it is not evident that one may obtain precise error estimates for the fixed points of $\bar{F}$ without making very much stronger assumptions on $\bar{F}$. However, from considerations outlined in [6] it appears that if $\bar{F}$ admits a linearization near a fixed point Y^N and $\{S^N(Y^N,J)\}$ is a sequence of Sperner simplexes which converge to Y^N, then the distance $d(Y^N,S^N(Y^N,J)) = O(1/J)$ as $J \to \infty$.

It is natural to ask which of the points in $\bigcap_{i=0}^{N} A_i$ can be approximated by Sperner simplexes. A sufficient condition is that for $Y \in \bigcap_{i=0}^{N} A_i$ and every neighborhood $N(Y)$, the sets $A_i \cap N(Y)$, $i = 0,1,\ldots,N$ are all open subsets of R^N. For then if J is sufficiently large, there will be a Sperner simplex $S_N(J)$ having a vertex in each of the sets $A_i \cap N(Y)$, $i = 0,1,\ldots,N$. That this is not a necessary condition may be seen in the case that a $Y \in \bigcap_{i=0}^{N} A_i$ is an isolated point of some A_i. Now Y may be approximated or not approximated by some Sperner simplex according as Y is or is not vertex point in the simplicial subdivision. In the application which we treat in this paper such points are likely to correspond to the maximal solutions u_h^* mentioned in [25].

A sufficient condition that the fixed points of $\bar{F}$ which may be approximated by Sperner simplexes can eventually be put into correspondence with a sequence of Sperner simplexes $\{S_N(J)\}$ which converge to a fixed point of $\bar{F}$ is given by

Theorem 4. Let $\bar{F} : C^N \to C^N$ be continuous, $\bar{F}(Y) = Y$ and $N(Y) \subset C^N$ a closed neighborhood of Y which contains no other fixed point of $\bar{F}$. Let $\{S_N(i)\}_{i=1}^{\infty}$ be a sequence of Sperner simplexes contained in $N(Y)$ such that diameter $\delta(S_N(i)) \to 0$ as $i \to \infty$. Then for every $Y_i \in S_N(i)$ we have $Y_i \to Y$ as $i \to \infty$.

§4 Let $G(P,Q)$ be the Green's function for the Laplacian operator on D and G_h be the corresponding discrete Green's function for the grid described in connection with (1.2). The assumptions which we have made on ∂D are sufficient to ensure the existence of the Green's function. By Green's identity, a solution to (1.1) will satisfy

$$u(P) = \int_D G(P,Q)f(Q),u(Q),\nabla u(Q))dz_Q \quad (4.1)$$

Conversely, a solution $u \in \mathring{W}_2^1(\bar{D})$ (the Sobolev space [27]) to (4.1) (or $u \in L_2(\bar{D})$ if f does not depend on derivatives of u) may be shown to be a classical solution to (1.1) under assumptions that f has sufficient smoothness to guarantee the existence of a classical solution to (1.1). For example, it suffices to assume that $f \in C^{(o,\lambda)}$ or $f \in C^{(1,\lambda)}$ (see e.g. [1], [11], [20] and the references there). Naturally, for stronger smoothness assumptions on f one may obtain better convergence rates between the solutions to (1.1) and (1.2) on the set $D_h+C_h^*$ as $h \to 0$.

We will now concentrate on approximating solutions to (4.1) in an appropriate Banach space X which may be taken to be $C(\bar{D}), L_2(\bar{D})$ or $C^{(1,\lambda)}(\bar{D})$, $\mathring{W}_2^1(\bar{D})$ according as f depends on derivatives or not or f is continuous or only in L_2 . It is easily seen that (4.1) must have solutions in X under the assumption that f is bounded by considering the operator

$$Tu(P) = \int_D G(P,Q)f(Q,u(Q),\nabla u(Q))dz_Q \quad . \quad (4.2)$$

It follows from the boundedness of f that

$$|Tu(P)| \leq M \int_D G(P,Q)dz_Q \quad .$$

If we let

$$B = \{w \in X \,|\, |w(P)| \le M \int_D G(P,Q)\,dz_Q\} ,$$

then $T : B \to B$. (Recall that $\int_D G(P,Q)\,dz_Q$ is the unique positive solution to Poisson's equation on D with $f \equiv 1$).

It is now routine to see that $T(B)$ is a compact, convex subset of X . Since T is also a continuous mapping, it follows from the Schauder fixed point theorem that (4.1) has a fixed point in B i.e. (4.1) has a solution in B .

We may rewrite (2.2) in terms of its components $u(P_i)$ $i = 1,2,\ldots,N$ by the formula

$$(4.3) \qquad \begin{aligned} u(P_i) &= h^2 \sum_{j=1}^{N} a_{ij}^{-1} f(Q_j, u(Q_j), \nabla_h u(Q_j)) \\ &= h^2 \sum_{j=1}^{N} G_h(P_i,Q_j) f(Q_j, u(Q_j), \nabla_h u(Q_j)) \end{aligned}$$

where G_h is the Green's matrix (under the particular ordering of the points) for Δ_h on $D_h + C_h^*$. This is also often referred to as the discrete Green's function for Δ_h on $D_h + C_h^* + C_h$, [12].

The discrete Green's function has several properties which are discrete analogues of the properties of G . Of particular use to us is the property that

$$(4.4) \qquad \int_D G(P,Q)\,dz_Q - h^2 \sum_{Q \in D_h + C_h^*} G_h(P,Q) = O(h^s)$$

where $s > 0$ is a number which is determined by the angles of ∂D , [30]. This result also holds for boundary value problems involving a mixture of Dirichlet and Neumann data on ∂D . For the case that ∂D has no acute angles, $s = 1$ and if $\partial D, f$ are sufficiently smooth, $s = 2$, [8]. Related results also appear in [7], [9], [10], [18].

We denote the fixed points of T_N defined in (2.3) by $\bar{u}^N$ and the corresponding solutions to (4.3) by u^N . We now define a continuous injection $\Phi_N : B_N \to X$. For the case that f is independent of derivatives, it suffices to extend u^N by performing a simple linear interpolation. Otherwise we define $\Phi_N u^N(P) = 0$ for $P \in \partial D$, interpolate linearly along grid lines and extend $\Phi_N u^N$ to a Hölder continuous function of exponent λ on all of D by an application of a theorem of McShane [19]. This extension may be performed so that

$$\|\Phi_N u^N\|_X = \|u^N\|_{R^N} = \max_{i=1,\dots,N} |u^N(P_i)| . \tag{4.5}$$

Theorem 5. There exists a subsequence of $\{\Phi_N u^N\}$ which converges to a function $u^\infty \in X$.

Proof. Since u^N satisfies (4.3) and f is uniformly bounded,

$$\|u^N\| = M\|\int_D G(P,Q)dz_Q\| + O(h^s) \quad \text{as} \quad h \to 0$$

by (4.4). Hence it may be seen by (4.5) that $\{\Phi_N u^N\}$ is uniformly bounded. Similarly, for $P,P' \in D_h + C_h^* + C_h$, there exists a $k>0$ such that

$$\begin{aligned} |u^N(P)-u^N(P')| &\le M|\int_D [G(P,Q)-G(P',Q)]dz_Q| + O(h^s) \quad &\text{as} \quad h \to 0 \\ &= M|(\nabla \int_D G(P,Q)dz_Q)_{P=P^*} \cdot (P-P')| + O(h^s) \quad &\text{as} \quad h \to 0 \end{aligned}$$

where P^* is some point lying on the line between P and P' . It is not necessary to assume that D is convex since $\int G(P,Q)dz_Q$ can also be defined outside of $\bar{D}$. Hence

$|u(P)-u(P')| \le (M\|\nabla(\int_D G(P,Q)dz_Q)+k_1)\|P-P'\|$ for some $k_1 > 0$.

Thus there exists a constant $c > 0$ such that

$|u^N(P)-u^N(P')| \le c\|P-P'\|$ for any $P,P' \in D_h + C_h^* + C_h$.

By the properties of Φ_N , we have

$|\Phi_N u^N(P) - \Phi_N u^N(P')| \le c\|P-P'\|$ for $P,P' \in D$ and each N .

If we choose $P,P' \in D$ such that $\|P-P'\| < \varepsilon/c$, then

$|\Phi_N u^N(P) - \Phi_N u^N(P')| < \varepsilon$ for each N.

Now by the Arzela-Ascoli theorem there exists a subsequence $\{\Phi_{N_k} u^{N_k}\}$ which converges uniformly in $C(\bar{D})$ and hence, it converges to a function in X, [11].

For convenience we will hereafter write $\{\Phi_N u^N\} \to u^\infty$.

Theorem 6. $Tu^\infty = u^\infty$. Hence by regularity arguments u^∞ is a solution to (1.1), [1], [11], [20].

Proof. Define $P_N : X \to R^N$ by $(P_N v)(P_i) = v(P_i)$ for $P_i \in D_h + C_h^*$ and any fixed ordering of the points of $D_h + C_h^*$. We have

$$\|Tu^\infty - u^\infty\| \leq \|Tu^\infty - \Phi_N P_N Tu^\infty\| + \|\Phi_N P_N Tu^\infty - \Phi_N T_N P_N u^\infty\| +$$
$$\|\Phi_N T_N P_N u^\infty - \Phi_N T_N u^N\| + \|\Phi_N T_N u^N - \Phi_N u^N\| + \|\Phi_N u^N - u^\infty\| .$$

There exists N_1, N_2, N_3, N_4 such that:

(i) $\|Tu^\infty - \Phi_N P_N Tu^\infty\| < \varepsilon/4$ for $N \geq N_1$ since Tu^∞ is smooth.

(ii) $\|\Phi_N P_N Tu^\infty - \Phi_N T_N P_N u^\infty\| < \varepsilon/4$ for $N \geq N_2$ since Φ_N is continuous and for $P \in D_h + C_h^*$,

$$P_N Tu^\infty(P) - T_N P_N u^\infty(P) = \int_D G(P,Q) f(Q, u^\infty(Q), \nabla u^\infty(Q)) dz_Q -$$
$$h^2 \sum_{Q \in D_h + C_h^*} G_h(P,Q) f(Q, u^\infty(Q), \nabla_h u^\infty(Q)) \to 0 \quad \text{as} \quad N \to \infty$$

by (4.4) and the boundedness and continuity of f.

(iii) $\|\Phi_N T_N P_N u^\infty - \Phi_N T_N u^N\| < \varepsilon/4$ for $N \geq N_3$ since $u^N \to P_N u^\infty$ on $D_h + C_h^*$ and $\Phi_N T_N$ is a continuous mapping for each N.

(iv) $\|\Phi_N T_N u^N - \Phi_N u^N\| = 0$ since $T_N u^N = u^N$.

(v) $\|\Phi_N u^N - u^\infty\| < \varepsilon/4$ for $N \geq N_4$ since $\Phi_N u^N \to u^\infty$.

The result now follows from (i)-(v).

Remarks 1. Using (4.4) and the uniform boundedness of f it may be seen that $u^{\infty}-u^{N} = O(h^{s})$ as $h \to 0$ on $D_h+C_h^*$. If (1.1) has a finite set of solutions $\{u_k^{\infty}\}_{k=1}^{j}$, the numerical algorithm we use to approximate solutions to $T_N u^N$ generally enables us to select without difficulty convergent sequences $\{u_k^N\}_{k=1}^{j}$ (see Theorem 4). Once this is done we have

$$(4.6) \qquad u_k^{\infty}-u_k^{N} = O(h^{s}) \quad \text{as} \quad h \to 0 \quad \text{on} \quad D_h+C_h^*$$

if $\lim_{N\to\infty} u_k^N = u_k^{\infty}$, $k = 1,\ldots,j$.

Results of this type were recently established by Simpson [25] under somewhat stronger smoothness assumptions on f which yield $s = 2$. Entirely similar arguments are applicable here to establish (4.6) and hence will be omitted.

2. The approach we use only relies on the facts that A has a positive inverse and that f has sufficient smoothness. Hence we may replace the five point difference scheme by certain higher order schemes subject to appropriate assumptions on ∂D, [10]. One may also assume only that $f \in L^2$ and bounded. In this case a more general convergence argument [3] or the theory of discrete convergence of mappings may be applied [28].

§5 In this section we treat the example

$$(5.1) \qquad \Delta u = f(u) \quad \text{on} \quad (-1,1)\times(-1,1)$$
$$u(x,y) = 0 \quad \text{if} \quad x \quad \text{or} \quad y = -1 \quad \text{or} \quad 1 .$$

Since this problem has lines of symmetry, $x = 0$, $y = 0$, $y = \pm x$, we will approximate solutions to (5.1) which are symmetric with respect to these lines. Hence we treat in our numerical calculations the mixed problem

$$\text{(5.2)} \qquad \begin{aligned} &\Delta u = f(u) \quad \text{on} \quad D = \{(x,y)\,|\,0 < y < x, 0 < x < 1\} \\ &u = 0 \quad \text{on} \quad \{(x,y)\,|\,x = 1, 0 \leq y \leq 1\} \\ &\frac{\partial u}{\partial u} = 0 \quad \text{on} \quad \{(x,y)\,|\,y = 0, 0 \leq x \leq 1\} \cup \{(x,y)\,|\,y=x, 0 \leq x \leq 1\} . \end{aligned}$$

We take square grids on $\bar{D}$ corresponding to $h = 1/n+1$. The truncation error for (5.1) is $O(h)$, [30]. The number of interior grid points for the finite difference analogue of (5.2) is $N = n(n-1)/2$. Rather than the "natural ordering" of the grid points we order the points by increasing distance from O . The motivation for this choice arises from a technical aspect of the search algorithm we employ which makes it preferable to have the sharpest a priori bounds on the components which occur nearest the end in the vectors $\bar{U}$. We exhibit in Figure 2 the matrices A in (2.1) for this boundary value problem after the boundary conditions have been incorporated. The principal submatrices of orders 3, 6, 10, 15 correspond to the grids $h = 1/4, 1/5, 1/6, 1/7$; i.e., $n = 3, 4, 5, 6$ respectively.

From Figure 2 it is evident how the sequence of matrices for each n may be formulated.

Figure 2

$$\begin{matrix}
2 & -2 & 0 & 0 & 0 & 0 & 0 & 0 & 0 & 0 & 0 & 0 & 0 & 0 & 0 \\
-1 & 3 & -1 & -1 & 0 & 0 & 0 & 0 & 0 & 0 & 0 & 0 & 0 & 0 & 0 \\
0 & -2 & 4 & 0 & -2 & 0 & 0 & 0 & 0 & 0 & 0 & 0 & 0 & 0 & 0 \\
0 & -1 & 0 & 3 & -1 & 0 & -1 & 0 & 0 & 0 & 0 & 0 & 0 & 0 & 0 \\
0 & 0 & -1 & -1 & 4 & -1 & 0 & -1 & 0 & 0 & 0 & 0 & 0 & 0 & 0 \\
0 & 0 & 0 & 0 & -2 & 4 & 0 & 0 & -2 & 0 & 0 & 0 & 0 & 0 & 0 \\
0 & 0 & 0 & -1 & 0 & 0 & 3 & -1 & 0 & 0 & -1 & 0 & 0 & 0 & 0 \\
0 & 0 & 0 & 0 & -1 & 0 & -1 & 4 & -1 & 0 & 0 & -1 & 0 & 0 & 0 \\
0 & 0 & 0 & 0 & 0 & -1 & 0 & -1 & 4 & -1 & 0 & 0 & -1 & 0 & 0 \\
0 & 0 & 0 & 0 & 0 & 0 & 0 & 0 & -2 & 4 & 0 & 0 & 0 & -2 & 0 \\
0 & 0 & 0 & 0 & 0 & 0 & -1 & 0 & 0 & 0 & 3 & -1 & 0 & 0 & 0 \\
0 & 0 & 0 & 0 & 0 & 0 & 0 & -1 & 0 & 0 & -1 & 4 & -1 & 0 & 0 \\
0 & 0 & 0 & 0 & 0 & 0 & 0 & 0 & -1 & 0 & 0 & -1 & 4 & -1 & 0 \\
0 & 0 & 0 & 0 & 0 & 0 & 0 & 0 & 0 & -1 & 0 & 0 & -1 & 4 & -1 \\
0 & 0 & 0 & 0 & 0 & 0 & 0 & 0 & 0 & 0 & 0 & 0 & 0 & -2 & 4
\end{matrix}$$

It is plain that for each n, the matrix A is irreducibly diagonally dominant (see [29], p.187) and $a_{ij} \leq 0$ for $i \neq j$. Hence, A is nonsingular and $A^{-1} > 0$. Thus T_N as defined in (2.3) is a continuous mapping of

$$B_N = \{\bar{v} \epsilon R^N \mid |\bar{v}| \leq h^2 M A^{-1}\bar{e}\}$$

into itself. From the standpoint of computational efficiency it seems preferable to use the mapping T'_N defined below since in general it involves fewer function evaluations. Let

$$(5.3) \qquad T'_N\bar{u} = (I-D_A^{-1}A)\bar{u} + h^2 D_A^{-1}\bar{f}(\bar{P},\bar{u},v_h\bar{u})$$

where $D_A = \text{diag } A$ and I is the $N \times N$ identity matrix. It is evident that any solution to (2.3) is a fixed point of T'_N and conversely. Also $D_A^{-1} > 0$ and $I-D_A^{-1}A \geq 0$. Hence, for any $\bar{u} \in B_N$,

$$\begin{aligned} |T'_N\bar{u}| &\leq |(I-D_A^{-1}A)\bar{u}| + h^2|D_A^{-1}\bar{f}(\bar{P},\bar{u},\nabla_h\bar{u})| \\ &\leq h^2 M(I-D_A^{-1}A)A^{-1}\bar{e} + h^2 M D_A^{-1}\bar{e} \\ &= h^2 M A^{-1}\bar{e} . \end{aligned}$$

Hence, $T'_N : B_N \to B_N$ and the fixed point algorithm may be applied for T'_N. The truncation errors arising from the mapping T'_N are the same as those for T_N.

Example 1. In (5.2), let $f(u) = \lambda^2 \sin u$. This problem corresponds to a 2-dimensional analogue of the pendulum equation. The trivial solution $u \equiv 0$ satisfies (5.2) for this case and any λ. If u satisfies (5.2), then so does $-u$. Thus the number of solutions to (5.2) is $2n+1$ if λ lies between the n-th and n+1-st eigenvalues of the Laplacian operator on the unit square $S = [-1,1]\times[-1,1]$. The eigenvalues in this case are $\lambda_n^2 = \frac{(2n+1)^2\pi^2}{2}$ and hence the problem

$$u+16 \sin u = 0 \quad \text{on} \quad D$$

$$u = 0 \quad \text{on} \quad \{(x,y)\,|\,x=1, 0 \leq y \leq 1\}$$

$$\frac{\partial u}{\partial n} = 0 \quad \text{on} \quad (x,y)\,|\,\{0 \leq x \leq 1 \text{ and } y=0 \text{ or } y=x\}$$

will have three solutions. In each of the applications of the algorithm, the expected three solutions were approximated in a single run. The symmetry in each case was nearly exact and the trivial solution was always exactly obtained. Hence, Table 1 exhibits just the upper solution. In each case $J = 256$ and the U_i approximates $U^N(P_i)$ where the P_i are ordered by increasing distance from 0.

N	U_1	U_2	U_3	U_4	U_5	U_6	U_7	U_8	U_9	U_{10}	$\sum_{i=1}^{N}[(T_N U)_i-(U)_i]^2$
3	1.51042	1.01172	.66406								3.57E-5
6	2.02431	1.72931	1.47259	1.06078	.88991	.52586					7.16E-5
10	2.21028	2.03766	1.88208	1.64663	1.51369	1.20646	.95500	.87590	.69597	.41719	1.75E-5

Table 1

Example 2. In (5.2) let $f(u) = \lambda e^u$. This equation arises in chemical reactor theory [13], [14]. There is a critical value λ_c such that the number of solutions to (5.2) is 2, 1, 0 according as λ is less than, equal to or greater than λ_c, [13]. (This is the 2-dimensional analogue of the Bratu equation). The mappings T_N and T'_N are not self mappings on any N-cell B_N which will include the upper solution when two solutions are present. However, the algorithm will nevertheless approximate the upper solution for a sufficiently large N-cell B_N. In particular, for

$$\Delta u+1/4\, e^u = 0 \quad \text{on} \quad D$$

$$u = 0 \quad \text{on} \quad \{(x,y)\,|\,x=1, 0\leq y\leq 1\}$$

$$\frac{\partial u}{\partial n} = 0 \quad \text{on} \quad \{(x,y)\,|\,0\leq x\leq 1 \text{ and } y=0 \text{ or } y=x\}$$

there are two solutions. The various approximations are given in Table 2. The two solutions must be positive and we take $B_N = \prod_{i=1}^{N} [0,8(1-x(P_i)^2)]$ where $x(P_i)$ is the x-co-ordinate of the i-th grid point.

N	P	$\sum_{i=1}^{N}[(T_N u)_i-(u)_i]^2$	U_1	U_2	U_3	U_4	U_5	U_6	U_7	U_8	U_9	U_{10}
3	128	1.75E-4	.05859	.04688	.02734							
3	128	2.64E-3	6.15234	2.53125	1.28516							
3	512	3.05E-5	.04028	.02930	.02222							
3	512	1.07E-4	6.12671	2.54004	1.28857							
6	128	8.86E-4	.13994	.13940	.13120	.09020	.08309	.04738				
6	128	1.88E-4	6.34187	3.55047	2.38338	1.58145	1.14577	.58509				
6	512	1.72E-4	.07653	.07175	.05102	.04208	.03061	.02073				
6	512	6.04E-5	6.32143	3.54432	2.37245	1.57844	1.14031	.57621				
10	128	8.19E-4	.18032	.17031	.15778	.14298	.12622	.10777	.08790	.06691	.06574	.03353
10	128	3.73E-3	6.44628	4.23037	3.23967	2.54907	2.08264	1.40496	1.22727	1.03048	.70248	.35795
10	512	3.00E-4	.12397	.12087	.10124	.09491	.08678	.06586	.04649	.04416	.03616	.02169
10	512	1.41E-4	6.49277	4.20015	3.13843	2.48128	1.99587	1.99587	1.16219	.96423	.66632	.33897

Table 2

The U_i are the approximations to $u^N(P_i)$ where the P_i are ordered by increasing distance from 0 .

Remark 1. For this problem the lower solutions may easily be approximated by simple linear iterations $u^{i+1} = h^2 A^{-1} e^{u_i}$. Under this iteration for $N = 3$, the point $[.026623300, .018600012, .013258392]'$ is fixed for at least the first nine places. Thus, our approximations in Table 2 are correct to only one decimal place. This is all we may expect for our choice of J. The above iteration procedure can only converge to the lower solution for any h. This phenomenon also occurs for certain Newton-type algorithms. A recent hybrid Newton algorithm [23] appears often (but not always) to be able to overcome the attractive or repulsive properties of solutions if the starting points are chosen sufficiently near a solution. Hence, it would appear that in general it would be most efficient to use the search algorithm to obtain initial approximations and subsequently refine the solutions by applying a fast iterative method such as [23]. The topological algorithm seems to be quite well suited to determine the number of solutions an elliptic boundary value problem and also to locate branch points.

2. In [4] the problem (5.1) on the unit disk has been treated under the assumption of radial symmetry of the solutions.

References

[1] Agmon,S., Lectures on Elliptic Boundary Value Problems, van Nostrand, Princeton, 1965.

[2] Alexandroff,P.S., Combinatorial Topology, vol.1, Graylock Press, Rochester, New York, 1956.

[3] Allgower,E.L., Guenther,R.B., A Functional Analytic Approach to the Numerical Solution of Nonlinear Elliptic Equations, Comp. 2, 25-33, (1967).

[4] Allgower,E.L., Jeppson,M.M., Numerical Solution of Nonlinear Boundary Value Problems with Several Solutions, (to appear).

[5] Allgower,E.L., Keller,C.L., A Search for a Sperner Simplex, Computing 8, 157-165, (1971).

[6] Allgower,E.L., Keller,C.L., Reeves,T.E., A Program for the Numerical Approximation of a Fixed Point of an Arbitrary Continuous Mapping of the n-Cube or n-Simplex into Itself, Aerospace Research Laboratories Report 71-0257, Wright Patterson Air Force Base, Ohio, November 1971.

[7] Batschelet,E., Ueber die numerische Auflösung von Randwertproblemen bei elliptischen partiellen Differentialgleichungen, Z. Angew. Math. Phys. 3, 165-193 (1952).

[8] Bramble,J.H., Hubbard,B.E., On the Formulation of Finite Difference Analogues of the Dirichlet Problem for Poisson's Equation, Numerische Math. 4, 313-327, (1962).

[9] Bramble,J.H., Hubbard,B.E., A Priori Bounds on the Discretization Error in the Numerical Solution of the Dirichlet Problem, Contributions to Diff.Eqs. 2, 229-252, (1963).

[10] Bramble,J.H., Hubbard,B.E., New Monotone Type Approximations for Elliptic Problems, Math. Comp. 18, 349-367, (1964).

[11] Courant,R., Hilbert,D., Methods of Mathematical Physics, vol.II, Wiley Interscience, New York, 1962.

[12] Forsythe,G.E., Wasow,W.R., Finite Difference Methods for Partial Differential Equations, Wiley and Sons, 1960.

[13] Fujita,H., On the Nonlinear Equations $\Delta u+e^u=0$ and $\partial v/\partial t=\Delta v+e^v$, Bull. Amer. Math. Soc. 75, 132-135, (1969).

[14] Gelfand,I.M., Some Problems in the Theory of Quasi-linear Equations, Uspehi Mat. Nauk. 14, (1959), Engl. transl., Amer. Math. Soc. Transl. (2), 29, (1963), 295-381.

[15] Greenspan,D., Parter,S.V., Mildly Nonlinear Elliptic Partial Differential Equations and their Numerical Solution II, Numerische Math. 7, 129-146, (1965).

[16] Jeppson,M.M., A Search for the Fixed Points of a Continuous Mapping, SIAM (1972), Mathematical Topics in Economic Theory and Computation.

[17] Keller,H.B., Cohen,D.S., Some Positone Problems Suggested by Nonlinear Heat Generation, J. Math. Mech. 16, 1361-1376, (1967).

[18] Laasonen,P., On the Solution of Poisson's Difference Equation, ACM Jour. 5, 370-382 (1958).

[19] McShane,E.J., Extension of Range of Functions, Bull. AMS, 40, 837-842, (1934).

[20] Miranda,C., Partial Differential Equations of Elliptic Type, Springer, Berlin, 1970.

[21] Parter,S.V., Mildly Nonlinear Elliptic Partial Differential Equations and their Numerical Solution I, Numerische Math. 7, 113-128, (1965).

[22] Parter,S.V., Maximal Solutions of Mildly Nonlinear Elliptic Equations, Numerical Solution of Nonlinear Differential Equations, ed. D.Greenspan, Wiley and Sons, (1966), New York.

[23] Powell,M.J.D., A Fortran Subroutine for Solving Systems of Nonlinear Algebraic Equations, Numerical Methods for Nonlinear Algebraic Equations, ed. P.Rabinowitz, Gordon and Breach, (1970), New York.

[24] Simpson,R.B., Finite Difference Methods for Mildly Nonlinear Eigenvalue Problems, SIAM J. Num. Anal. 8, 190-211, (1971).

[25] Simpson,R.B., Existence and Error Estimates for Solutions of a Discrete Analog of Nonlinear Eigenvalue Problems, Mathematics of Computation 26, 359-375 (1972).

[26] Simpson,R.B., Cohen,D.S., Positive Solutions of Nonlinear Elliptic Eigenvalue Problems, SIAM J. Numer. Anal. 8, 190-211, (1971).

[27] Smirnov,W.I., Lehrgang der Höheren Mathematik, Teil V, VEB Deutscher Verlag der Wissenschaften, 1961.

[28] Stummel,F., Discrete Convergence of Differentiable Mappings, (these proceedings).

[29] Varga,R.S., Matrix Interative Analysis, Prentice Hall, (1962), Englewood Cliffs.

[30] Wigley,N.M., On the Convergence of Discrete Approximations to Solutions of Mixed Boundary Value Problems, SIAM J. Num. Anal. 3, 372-382, (1966).

FEHLERABSCHÄTZUNGEN BEI AUFGABEN MIT SCHWACH STRUKTURIERTEN AUSGANGSDATEN

Von R.Ansorge und C.Geiger

1. Einleitung

Die Approximations- und Interpolationstheorie spielen innerhalb der numerischen Mathematik eine mehrfache Rolle:

Zum einen werden (und dies ist wohl die vordergründigste Bedeutung) Hilfsmittel bereitgestellt, die es gestatten, eine explizit vorgegebene Funktion komplizierter Struktur näherungsweise zu berechnen (etwa in Form einer Tafel) durch Berechnung einfacher strukturierter Funktionen einer *vorgegebenen* Funktionenmenge.

Daneben ist man jedoch (insbesondere in den letzten 10 Jahren) dazu übergegangen, auch andere Informationen als die einer expliziten Darstellung zur Charakterisierung der zu approximierenden Funktion heranzuziehen, etwa eine definierende gewöhnliche Differentialgleichung, Integralgleichung, partielle Anfangs- oder Randwertaufgabe oder allgemeiner eine definierende Funktionalgleichung. Hier ist wohl zunächst eine Arbeit von Meinardus und Strauer [1] aus dem Jahre 1963 zu nennen, bei der die gesuchte Lösung einer gegebenen linearen Differential- oder Integralgleichung vorerst mittels irgendeines Verfahrens durch eine kontinuierliche Näherungsfunktion ersetzt wird, die anschließend ihrerseits mit den Mitteln der Approximationstheorie durch eine Funktion des *vorgegebenen* Funktionenraums approximiert wird in der Hoffnung, damit zugleich eine gute Approximation der gesuchten Lösung bezüglich dieses Funktionenraums zu gewinnen. Dieses Vorgehen wurde 1966 [2] auf den Fall nichtlinearer gewöhnlicher oder partieller Differentialgleichungen übertragen mit der Abänderung, daß es sich bei dem zunächst benutzten Näherungsverfahren um einen Diskretisierungsprozeß handelt.

In anderen Untersuchungen wird versucht, für gewisse Problemklassen den Umweg über eine Zwischennäherung zu vermeiden und so direkt zu einer guten Approximation der gesuchten Lösung einer gegebenen Funktionalgleichung bezüglich eines gegebenen Funktionenraums zu gelangen. Hier sei exemplarisch auf Arbeiten von Collatz [3], [4] im Zusammenhang mit der Lösung von Randwertaufgaben aus den Jahren 1969 und 1970 verwiesen.

Zum anderen haben sich approximationstheoretische Ergebnisse auch als rein methodisches Hilfsmittel bewährt, z.B. zur Gewinnung von Schranken für die Fehler gegebener Näherungsverfahren. Die Anwendbarkeit derartiger Hilfsmittel ergibt sich insbesondere dann, wenn mit dem betrachteten Verfahren Näherungen für nicht-glatte Lösungen einer gegebenen Funktionalgleichung gesucht werden, d.h. für Lösungen, die den bei der Konstruktion des Näherungsverfahrens oder der Herleitung der Fehlerabschätzung unterstellten Regularitätsvoraussetzungen nicht genügen.

Diese Situation tritt beispielsweise bei der numerischen Bestimmung verallgemeinerter Lösungen von Anfangswertaufgaben mittels Differenzenverfahren auf. Entsprechende Fehlerabschätzungen wurden unter Benutzung der Theorie intermediärer Räume im Falle linearer Anfangswertaufgaben erstmals 1967 durch Peetre und Thomée [5] angegeben (vgl. auch [6]). Ähnliche Überlegungen, z. T. Verallgemeinerungen, finden sich bei Hedstrom [7]. Nachdem Thompson [8] 1964 die Existenz verallgemeinerter Lösungen auch im Falle gewisser halblinearer Anfangswertaufgaben nachgewiesen hatte, wurden die Ergebnisse von Peetre und Thomée unter Verwendung der in [6] benutzten Hilfsmittel auch auf solche (z.T. allgemeineren) halblinearen Fälle übertragen (vgl. [9]).

Dieser methodische Aspekt der Anwendung approximationstheoretischer Hilfsmittel im Zusammenhang mit Fehlerabschätzungen soll hier in einer konstruktiven Zwecken etwas angepaßteren Weise und losgelöst von den speziellen Fehlerabschätzungen linearer oder halblinearer Anfangswertaufgaben erneut vorgestellt werden, wobei sich die Grundidee lediglich auf die Hintereinanderschaltung der einmaligen Anwendung der Dreiecksungleichung, der Approximation bezüglich eines Mengensystems und auf einen Minimisierungsprozeß reduziert.

2. Rückführung der Fehlerabschätzung auf eine solche für stark strukturierte Ausgangsdaten

Gegeben sei eine Abbildung E des metrischen Raumes $(\mathfrak{M},\sigma)$ in den metrischen Raum $(\mathfrak{K},\rho)$.

Gegeben sei weiter eine Indexmenge $\mathfrak{H}$ sowie eine Schar von Abbildungen $\{Q_h\}$ $(h \in \mathfrak{H})$, die $\mathfrak{M}$ in $\mathfrak{K}$ abbilden.

Etwa ist Eu_o $(u_o \in \mathfrak{M})$ die (eventuell verallgemeinerte) Lösung einer Anfangs- oder Randwertaufgabe mit den Anfangs- (bzw. Rand-) Daten u_o. Eu_o kann aber auch den Wert eines Integrals mit einem von u_o abhängenden Integranden darstellen oder die Lösung einer Integralgleichung sein usw. Q_h sei eine Näherung für E mit einem das Näherungsverfahren beschreibenden Parameter h (etwa einer Schrittweite).

$\mathbb{R}_o^+$ sei die Menge der nichtnegativen reellen Zahlen.

Es gebe ein auf $\mathfrak{H} \times \mathbb{R}_o^+$ definiertes (nichtnegatives) Funktional $\psi = \psi(h,x)$, das bezüglich $x \in \mathbb{R}_o^+$ isoton sei, mit der Eigenschaft

$$\rho(Q_h u_o, Q_h v_o) \leq \psi(h, \sigma(u_o,v_o)) \quad \forall\, u_o,\, v_o \in \mathfrak{M} \tag{1}$$

sowie eine auf $\mathbb{R}_o^+$ definierte isotone Funktion ψ^* mit der Eigenschaft

$$\rho(Eu_o, Ev_o) \leq \psi^*(\sigma(u_o,v_o)) \quad \forall\, u_o,\, v_o \in \mathfrak{M}. \tag{1a}$$

(1) und (1a) stellen also verallgemeinerte Hölder-Stetigkeiten sowohl der *Näherungsoperatoren* Q_h wie des *Lösungsoperators* E dar.

Für alle hinreichend stark strukturierten Elemente $v \in \mathfrak{M}$, etwa für alle $v \in \mathfrak{J} \subset \mathfrak{M}$ (z.B. alle hinreichend glatten Anfangswerte einer Anfangswertaufgabe oder alle hinreichend glatten Integranden eines zu berechnenden Integrals usw.), existiere eine *Fehlerabschätzung*, d.h. es existiere ein auf $\mathfrak{h} \times \mathfrak{g}$ definiertes und explizit angebbares Funktional φ mit der Eigenschaft

$$\rho(Q_h v, Ev) \leq \varphi(h,v) \quad (\forall h \in \mathfrak{h}, \forall v \in \mathfrak{g}). \tag{2}$$

Wir fragen nun nach einer Fehlerabschätzung für $u_o \in \mathfrak{M}-\mathfrak{g}$, also für schwächer strukturierte Ausgangsdaten u_o.

Sei nun $\mathfrak{R}$ eine weitere Indexmenge und

$$\tilde{\mathfrak{g}} := \{ \mathfrak{g}_r \mid r \in \mathfrak{R}, \ \mathfrak{g}_r \subset \mathfrak{g} \}$$

ein Mengensystem.

Seien $v_r \in \mathfrak{g}_r$ "gute Approximationen" für $u_o \in \mathfrak{M}-\mathfrak{g}$ bezüglich $\mathfrak{g}_r$ (d.h.: $\sigma(u_o,v_r)$ soll nach Möglichkeit klein sein), die nach einem gewissen Prinzip bestimmt werden (etwa: v_r Minimallösung für u_o bezüglich $\mathfrak{g}_r$, falls solche Minimallösungen existieren); infolge des gewählten Approximationsprinzips wird $v_r = v_r(u_o)$.

Die Dreiecksungleichung ergibt

$$\rho(Q_h u_o, Eu_o) \leq \rho(Q_h u_o, Q_h v_r) + \rho(Ev_r, Eu_o) + \rho(Q_h v_r, Ev_r),$$

so daß mit (1), (1a) und (2) folgt

$$\rho(Q_h u_o, Eu_o) \leq \psi(h, \sigma(u_o,v_r)) + \psi^*(\sigma(u_o,v_r)) + \varphi(h,v_r). \tag{3}$$

Die explizite Bestimmung der v_r wird in der Regel schwierig sein, so daß (3) noch nicht befriedigend verwendet werden kann. Wir setzen deshalb voraus, daß hinsichtlich des Prinzips, gemäß dem die Elemente aus $\mathfrak{M}-\mathfrak{g}$ bezüglich der Elemente des Systems $\tilde{\mathfrak{g}}$ approximiert werden, *Jackson-Sätze* bekannt seien. Dies soll zunächst nur bedeuten, daß für jedes $r \in \mathfrak{R}$ ein auf $\mathfrak{M}-\mathfrak{g}$ definiertes Funktional $\sigma_r^*(u_o)$ explizit angegeben werden kann mit der Eigenschaft

$$\sigma(u_o, v_r(u_o)) \leq \sigma_r^*(u_o) \quad (\forall r \in \mathfrak{R}, \ \forall u_o \in \mathfrak{M}-\mathfrak{g}). \tag{4}$$

Ferner mögen auf $\mathfrak{M}-\mathfrak{g}$ bezüglich $\tilde{\mathfrak{g}}$ "methodenabhängige Bernstein-

Zamansky-Ungleichungen" gelten: Damit soll zunächst nur ausgedrückt werden, daß das durch die Fehlerabschätzung (2) (und damit durch die Methode zur Bestimmung der Näherungen Q_h) sowie durch das verwendete Approximationsprinzip definierte $\varphi(h,v_r(u_o))$ für $u_o \in \mathfrak{M}-\mathfrak{g}$ durch explizit angebbare Funktionale $\chi_r(h,u_o)$ abgeschätzt werden kann:

$$\forall r \in \mathfrak{R} \quad \forall u_o \in \mathfrak{M}-\mathfrak{g} \; : \; \varphi(h,v_r(u_o)) \leq \chi_r(h,u_o). \tag{5}$$

Aufgrund der vorausgesetzten Monotonieeigenschaften von ψ und ψ^* geht (3) mit (4) und (5) über in die Fehlerabschätzungen

$$\rho(Q_h u_o, Eu_o) \leq \psi(h,\sigma_r^*(u_o)) + \psi^*(\sigma_r^*(u_o)) + \chi_r(h,u_o) =: F(r,h,u_o), \tag{6}$$

die wegen der Existenz von $F^*(h,u_o) := \inf_{r \in \mathfrak{R}} F(r,h,u_o)$ bei jeweils festem $h \in \mathfrak{h}$ und jeweils festem $u_o \in \mathfrak{M} - \mathfrak{g}$ zu

$$\rho(Q_h u_o, Eu_o) \leq F^*(h,u_o) \tag{7}$$

verbessert werden können (evtl. wird man hier anstelle von F^* eine explizit angebbare obere Schranke $F^{**}(h,u_o)$, etwa ein hinreichend kleines $F(r,h,u_o)$ bei geeignetem $r \in \mathfrak{R}$ verwenden).

Die Güte der so gewonnenen erstrebten Abschätzung (7) für die schwach strukturierten Ausgangsdaten $u_o \in \mathfrak{M}-\mathfrak{g}$ im Vergleich zu der auf $\mathfrak{g}$ gültigen Abschätzung (2) (von der wir ausgingen) wird abhängen von der Reichhaltigkeit des Systems $\tilde{\mathfrak{g}}$, von der Güte des verwendeten Approximationsprinzips (bestmöglich wäre die Wahl von v_r als Minimallösung für u_o bezüglich $\mathfrak{g}_r$, sofern solche existieren) sowie insbesondere von der Güte der Abschätzungen (4) und (5).

3. Spezialisierung auf konkrete Bernstein-Zamansky-Aussagen und konkrete Jackson-Sätze

Es sei $\mathfrak{N}$ ein normierter Raum mit der Norm $\|\cdot\|_{\mathfrak{N}}$, und es gelte $\mathfrak{g} \subset \mathfrak{M} \subset \mathfrak{N}$. Auf der linearen Hülle $\mathfrak{g}^*$ von $\mathfrak{g}$ [1)] seien darüberhinaus subadditive Funktionale (z.B. Halbnormen) $\alpha_{\mathfrak{g}^*}^{(\mu)}$ $(\mu = 1,\ldots,m)$

---

1) Ist $\mathfrak{g}$ (wie bei vielen konkreten linearen Aufgaben) selbst ein linearer Raum, so kann die Hüllenbildung natürlich entfallen ($\mathfrak{g}^* = \mathfrak{g}$).

definiert, wobei für das in (2) auftretende Funktional φ eine Abschätzung der Form

$$\forall \; v \in \mathcal{G} \; : \varphi(h,v) \leq \varphi^*(h, \alpha^{(1)}_{\mathcal{G}^*}(v), \; \alpha^{(2)}_{\mathcal{G}^*}(v), \; \ldots, \; \alpha^{(m)}_{\mathcal{G}^*}(v)) \qquad (8)$$

mit einer in allen Variablen (evtl. mit Ausnahme der ersten) isotonen Funktion φ^* existiere.

Es sei $\mathcal{R} = \mathbb{N}$ oder auch nur $\mathcal{R} = \{1,2,\ldots,r^*\}$. Es gebe eine Kette linearer Teilräume $\mathcal{G}^*_r < \mathcal{G}^*$ $(\mathcal{G}^*_1 < \mathcal{G}^*_2 < \ldots < \mathcal{G}^*)$ mit

$$\mathcal{G}_r := \mathcal{G}^*_r \cap \mathcal{G} \neq \phi \qquad \forall \, r \in \mathcal{R} \, ,$$

wobei $\mathcal{G}_r$ die früher angegebene Bedeutung habe. Offenbar bildet damit auch $\tilde{\mathcal{G}}$ eine Kette

$$\mathcal{G}_1 < \mathcal{G}_2 < \ldots < \mathcal{G} .^{2)}$$

Es gebe Folgen $\{k_1,k_2,\ldots,k_m\} \subset \mathbb{R}^+_o$ und $\{c_1,c_2,\ldots,c_m\} \subset \mathbb{R}^+_o$ mit der Eigenschaft

$$\forall \; v \in \mathcal{G}^*_r \; : \alpha^{(\mu)}_{\mathcal{G}^*}(v) \leq c_\mu \; r^{k_\mu} \; \|v\|_{\mathcal{N}} \quad (r = 1,2,\ldots) \; (\mu = 1,\ldots,m); (9)$$

auf den Räumen $\mathcal{G}^*_r$ wird also die Gültigkeit solcher *Bernstein-Ungleichungen* (vgl. z.B. [10], S.60 ff.) vorausgesetzt.

Das früher genannte Approximationsprinzip bestehe in der Lösung der durch

$$E_r(u_o) := \inf_{v \in \mathcal{G}_r} \|u_o - v\|_{\mathcal{N}} \quad (r = 1,2,\ldots) \; (u_o \in \mathcal{M}-\mathcal{G}) \qquad (10)$$

beschriebenen Approximationsaufgabe.

Nunmehr existiere ein $\theta > 0$ derart, daß

---

2) Ist $\mathcal{G}$ ein linearer Raum, so folgt unmittelbar $\mathcal{G}_r = \mathcal{G}^*_r$, so daß auch die Mengen $\mathcal{G}_r$ lineare Räume sind.

$$\forall\ u_o \in \mathfrak{M} : \sup_{r \in \mathbb{R}} r^\theta E_r(u_o) < \infty . \qquad (11)$$

Somit gibt es eine Folge $\{v_r\}$ mit $v_r = v_r(u_o) \in \mathfrak{G}_r$ $(r = 1,2,\ldots)$ und ein auf $\mathfrak{M}$ definiertes Funktional $\omega_{\mathfrak{M}}$, so daß ein *Jackson-Satz*

$$\forall\ u_o \in \mathfrak{M} : \| u_o - v_r(u_o) \|_{\mathfrak{N}} \leq \frac{\omega_{\mathfrak{M}}(u_o)}{r^\theta} \quad (r = 1,2,\ldots) \qquad (11a)$$

gilt.

Besitzt (10) eine Lösung, so wähle man als v_r diese Minimallösung: $E_r(u_o) = \| u_o - v_r(u_o) \|_{\mathfrak{N}}$. Als Funktional $\omega_{\mathfrak{M}}(\cdot)$ ist dann jede (explizit angebbare) Majorante des Funktionals $\sup_{r \in \mathbb{R}} r^\theta E_r(\cdot)$ verwendbar. Existieren Minimallösungen nicht für alle $u_o \in \mathfrak{M}$, so gibt es bei beliebig fest vorgegebenem $\varepsilon > 0$ gewiß ein $v_r \in \mathfrak{G}_r$ mit $\| u_o - v_r \|_{\mathfrak{N}} \leq E_r(u_o) + \frac{\varepsilon}{r^\theta}$. Als $\omega_{\mathfrak{M}}(\cdot)$ ist dann jede Majorante des Funktionals $\sup_{r \in \mathbb{R}} r^\theta E_r(\cdot) + \varepsilon$ wählbar [3)].

Es mag zunächst als unrealistisch erscheinen, einen Jackson-Satz (11) für ein im allgemeinen nichtlineares Approximationsproblem (10) als erfüllt anzusehen. In praktischen Fällen (vgl. Abschnitt 5) findet sich bei nichtlinearen Problemen (lineare Probleme mit linearen Räumen $\mathfrak{M}$ und $\mathfrak{G}$ sind hier nicht tangiert) häufig folgende Situation:

$$\mathfrak{M} \subset \mathfrak{X}_\kappa := \{ u \in \mathfrak{N} \mid \| u \|_{\mathfrak{N}} \leq \kappa \}, \qquad \mathfrak{G}^* \cap \mathfrak{X}_\kappa = \mathfrak{G}^* \cap \mathfrak{M} = \mathfrak{G} .$$

Damit folgt wegen $\mathfrak{G}_r = \mathfrak{G}_r^* \cap \mathfrak{G}$: $\mathfrak{G}_r = \mathfrak{G}_r^* \cap \mathfrak{M} = \mathfrak{G}_r^* \cap \mathfrak{X}_\kappa (r = 1,2,\ldots)$.

---

3) Ist $\mathfrak{N}$ ein Banachraum, sind $\mathfrak{M}$ und $\mathfrak{G}$ lineare Teilräume von $\mathfrak{N}$ und wählt man $\omega_{\mathfrak{M}}(u_o) := \| u_o \|_{\mathfrak{N}} + \sup_{r \in \mathbb{R}} r^\theta E_r(u_o)$, so ist $\omega_{\mathfrak{M}}$ eine Norm auf $\mathfrak{M}$, in der $\mathfrak{M}$ zu einem Banachraum, sogar hinsichtlich $\tilde{\mathfrak{G}}$ zu einem Approximationsraum der Klasse $D_\theta(\mathfrak{N})$ (vgl. [10], S.60) wird, denn neben (11) gilt dann auch eine Bernstein-Ungleichung $\omega_{\mathfrak{M}}(v) \leq 2r^\theta \| v \|_{\mathfrak{N}}$ für alle $v \in \mathfrak{G}_r$ $(r = 1,2,\ldots)$.

Ist dann $u_o \in \mathfrak{M}$, so betrachte man vorerst die *linearen* Approximationsaufgaben

$$E_r^*(u_o) = \inf_{v^* \in \mathcal{G}_r^*} \| u_o - v^* \|_{\mathfrak{N}} \quad (r = 1,2,\ldots),$$

für die mit gewissen $v_r^* = v_r^*(u_o)$ und Funktionalen $\omega_{\mathfrak{M}}^*(u_o)$ ein Jackson-Satz $\| u_o - v_r^*(u_o) \|_{\mathfrak{N}} \leq \frac{\omega_{\mathfrak{M}}^*(u_o)}{r^\theta}$ gelten möge.

Wähle nun $v_r = v_r(u_o) := \dfrac{\| u_o \|_{\mathfrak{N}}}{\frac{\omega_{\mathfrak{M}}^*(u_o)}{r^\theta} + \| u_o \|_{\mathfrak{N}}} v_r^*(u_o)$. Da $\mathcal{G}_r^*$ ein linearer Raum ist, gilt $v_r \in \mathcal{G}_r^*$. Zugleich gilt

$\| v_r^* \|_{\mathfrak{N}} \leq \| v_r^* - u_o \|_{\mathfrak{N}} + \| u_o \|_{\mathfrak{N}} \leq \dfrac{\omega_{\mathfrak{M}}^*(u_o)}{r^\theta} + \| u_o \|_{\mathfrak{N}}$, mithin

$\| v_r \|_{\mathfrak{N}} \leq \| u_o \|_{\mathfrak{N}} \leq \varkappa$, d.h. $v_r \in \gamma_\varkappa$. Also folgt $v_r \in \mathcal{G}_r^* \cap \gamma_\varkappa = \mathcal{G}_r$,

und man erhält

$$\| v_r - u_o \|_{\mathfrak{N}} = \left\| \frac{\| u_o \|_{\mathfrak{N}}}{\frac{\omega_{\mathfrak{M}}^*(u_o)}{r^\theta} + \| u_o \|_{\mathfrak{N}}} v_r^*(u_o) - u_o \right\|_{\mathfrak{N}} =$$

$$= \frac{1}{\frac{\omega_{\mathfrak{M}}(u_o)}{r^\theta} + \| u_o \|_{\mathfrak{N}}} \left\| \| u_o \|_{\mathfrak{N}} (v_r^*(u_o) - u_o) - \frac{\omega_{\mathfrak{M}}^*(u_o)}{r^\theta} u_o \right\|_{\mathfrak{N}}$$

$$\leq \frac{\| u_o \|_{\mathfrak{N}} \| v_r^* - u_o \|_{\mathfrak{N}} + \| u_o \|_{\mathfrak{N}} \frac{\omega_{\mathfrak{M}}^*(u_o)}{r^\theta}}{\| u_o \|_{\mathfrak{N}}} \leq \frac{2\,\omega_{\mathfrak{M}}^*(u_o)}{r^\theta},$$

so daß in der Tat auch im Falle dieser nichtlinearen Approximation mit $\omega_{\mathfrak{M}}(u_o) := 2\,\omega_{\mathfrak{M}}^*(u_o)$ ein Jackson-Satz (11a) gilt.

Die Abschätzung (4) realisieren wir nun gemäß (11a) durch

$$\forall\ u_o \in \mathfrak{M} : \sigma_r^*(u_o) := \frac{\omega_{\mathfrak{M}}(u_o)}{r^\theta} \quad (r = 1,2,\ldots). \tag{12}$$

Weiterhin gilt das auf Zamansky [11] zurückgehende

Lemma:
Die Approximierenden $v_r = v_r(u_o)$ genügen für jedes $u_o \in \mathfrak{M}$ der Ungleichung

$$\alpha_{\mathfrak{g}^*}^{(\mu)}(v_r) \le f_\mu(r,u_o) \quad (\mu = 1,2,\ldots,m) \quad (r = 1,2,\ldots) \tag{13}$$

mit

$$f_\mu(r,u_o) := \begin{cases} \gamma_\mu r^{k_\mu - \theta}\, \tilde{\omega}_{\mathfrak{M}}(u_o) & \text{für } k_\mu > \theta \\ \gamma_\mu^*\left[2 + \frac{\log r}{\log 2}\right] \tilde{\omega}_{\mathfrak{M}}(u_o) & \text{für } k_\mu = \theta \\ \gamma_\mu^{**}\, \tilde{\omega}_{\mathfrak{M}}(u_o) & \text{für } k_\mu < \theta \end{cases} \tag{14}$$

bei gewissen angebbaren Konstanten γ_μ, γ_μ^*, γ_μ^{**} [4)] und mit $\tilde{\omega}_{\mathfrak{M}}(u_o) = \max\{\omega_{\mathfrak{M}}(u_o), \|u_o\|_{\mathfrak{N}}\}$.

Beweis:
Sei $p \in \mathbb{N}$ so bestimmt, daß $2^p \le r < 2^{p+1}$. (15)

Wegen der Subadditivität der $\alpha_{\mathfrak{g}^*}^{(\mu)}$ folgt

$$\alpha_{\mathfrak{g}^*}^{(\mu)}(v_r) \le \alpha_{\mathfrak{g}^*}^{(\mu)}(v_1) + \sum_{\tau=1}^{p} \alpha_{\mathfrak{g}^*}^{(\mu)}(v_{2^\tau} - v_{2^{\tau-1}}) + \alpha_{\mathfrak{g}^*}^{(\mu)}(v_r - v_{2^p}). \tag{16}$$

Da die $\mathfrak{g}_r^*$ eine Kette bilden und lineare Räume sind, gilt $v_{2^\tau} - v_{2^{\tau-1}} \in \mathfrak{g}_{2^\tau}^*$, so daß für $u_o \in \mathfrak{M}$ mit (9) und anschließend mit (11) die Ungleichung

---

4) In praktischen Fällen gilt gelegentlich sogar $f_\mu(r,u_o) = \gamma_\mu r^{k_\mu - \theta}\,\omega_{\mathfrak{M}}(u_o)$ für $k_\mu \ge \theta$ (vgl. z.B. [6]), so daß log r gar nicht auftritt. U.a. tritt dieser Fall ein, wenn $\mathfrak{M}$ und $\mathfrak{g}^* = \mathfrak{g}$ lineare Räume darstellen, $\omega_{\mathfrak{M}}$ gemäß Fußnote [3)] gewählt und $\mathfrak{g}$ jeweils mittels $\alpha_{\mathfrak{g}^*}^{(\mu)}(\cdot) := \|\cdot\|_{\mathfrak{N}} + \sup_{r \in \mathbb{R}} r^{k_\mu} E_r(\cdot)$ ($\mu = 1,\ldots,m$) normiert wird.

$$\alpha^{(\mu)}_{g^*}(v_{2^\tau}-v_{2^{\tau-1}}) \leq c_\mu (2^\tau)^{k_\mu} \| v_{2^\tau}-v_{2^{\tau-1}}\|_{\mathfrak{n}}$$

$$\leq c_\mu\, 2^{\tau k_\mu}\{\| u_o - v_{2^\tau}\|_{\mathfrak{n}} + \| u_o - v_{2^{\tau-1}}\|_{\mathfrak{n}}\}$$

$$\leq c_\mu\, 2^{\tau k_\mu}\left\{ \frac{\omega_{\mathfrak{m}}(u_o)}{(2^\tau)^\theta} + \frac{\omega_{\mathfrak{m}}(u_o)}{(2^{\tau-1})^\theta} \right\}$$

$$\leq c_\mu\, 2^{\tau k_\mu} \frac{2}{2^{\theta(\tau-1)}}\, \omega_{\mathfrak{m}}(u_o) = c_\mu 2^{1+\theta}\, 2^{\tau(k_\mu-\theta)}\, \omega_{\mathfrak{m}}(u_o)$$

resultiert.

Analog erhält man

$$\alpha^{(\mu)}_{g^*}(v_r - v_{2^p}) \leq c_\mu\, r^{k_\mu} \frac{2}{(2^p)^\theta}\, \omega_{\mathfrak{m}}(u_o) \leq c_\mu\, 2^{1+\theta}\, 2^{(p+1)(k_\mu-\theta)}\, \omega_{\mathfrak{m}}(u_o).$$

Damit liefert (16)

$$\alpha^{(\mu)}_{g^*}(v_r) \leq \alpha^{(\mu)}_{g^*}(v_1) + c_\mu\, 2^{1+\theta} \sum_{\tau=1}^{p+1} 2^{\tau(k_\mu-\theta)}\, \omega_{\mathfrak{m}}(u_o).$$

Berücksichtigt man mit (9) noch

$$\alpha^{(\mu)}_{g^*}(v_1) \leq c_\mu \| v_1\|_{\mathfrak{n}} \leq c_\mu\{\|v_1 - u_o\|_{\mathfrak{n}} + \|u_o\|_{\mathfrak{n}}\} \leq c_\mu\{\omega_{\mathfrak{m}}(u_o) + \|u_o\|_{\mathfrak{n}}\}$$

$$\leq 2c_\mu\, \widetilde{\omega}_{\mathfrak{m}}(u_o) \leq 2^{1+\theta} c_\mu\, \widetilde{\omega}_{\mathfrak{m}}(u_o),$$

so erhält man

$$\alpha^{(\mu)}_{g^*}(v_r) \leq c_\mu 2^{1+\theta} \sum_{\tau=0}^{p+1} 2^{\tau(k_\mu-\theta)}\, \widetilde{\omega}_{\mathfrak{m}}(u_o) = c_\mu 2^{1+\theta}\, \widetilde{\omega}_{\mathfrak{m}}(u_o) \cdot \begin{cases} (p+2) & \text{für } k_\mu = \theta \\ \dfrac{2^{(p+2)(k_\mu-\theta)}-1}{2^{k_\mu-\theta}-1} & \text{für } k \neq \theta \end{cases}$$

woraus mit der Einschließung (15) unmittelbar die Behauptung folgt.

Damit ergibt sich für die in (8) auftretende Funktion φ^* aufgrund der vorausgesetzten Monotonieeigenschaften wegen (13) die Aussage

$$\varphi(h, v_r) \leq \varphi^*(h, f_1(r,u_o), f_2(r,u_o), \ldots, f_m(r,u_o)) =: \chi_r(h,u_o), \quad (17)$$

womit (5) eine Realisierung erfahren hätte.

Um die zur Abschätzung (7) führende Minimisierung (über r) der rechten Seite der aus (6) mit (12) und (17) hervorgehenden Abschätzung

$$\rho(Q_h u_o, Eu_o) \le \psi\left(h, \frac{\omega_{\mathfrak{M}}(u_o)}{r^\theta}\right) + \psi^*\left(\frac{\omega_{\mathfrak{M}}(u_o)}{r^\theta}\right)$$
$$+ \varphi^*(h, f_1(r,u_o), f_2(r,u_o), \ldots, f_m(r,u_o)) \tag{18}$$
$$=: F(r,h,u_o)$$

vornehmen zu können, bedarf es weiterer Konkretisierungen.

4. Der Fall gleichgradig hölderstetiger Operatoren Q_h

Wir treffen nun weitere, auf praktische Anwendungen zugeschnittene Voraussetzungen über die Abschätzungen (1) und (8). Dabei können wir andererseits auf die Vorgabe einer Ungleichung (1a) verzichten. Der folgende Satz stellt für den betrachteten Fall eine Zusammenfassung der bisherigen Überlegungen dar; sein 2. Teil kann als eine unter den angegebenen speziellen Voraussetzungen mögliche quantitative Ergänzung eines Satzes von Rinow (vgl. [12], S.49) aufgefaßt werden.

Satz:
Seien $(\mathfrak{N}, \|\cdot\|_{\mathfrak{N}})$ ein normierter Raum, $(\tilde{\mathfrak{K}}, \rho)$ ein *vollständiger* metrischer Raum, $\mathfrak{M}$ eine Teilmenge von $\mathfrak{N}$ und $\mathfrak{G}$ eine in $\mathfrak{M}$ *dichte* Menge. Seien weiter $\{Q_h\}$ $(h \in (0,h_o])$ eine Schar *stetiger* Abbildungen von $\mathfrak{M}$ in $\tilde{\mathfrak{K}}$, E_o eine *stetige* Abbildung von $\mathfrak{G}$ in $\tilde{\mathfrak{K}}$ mit den Eigenschaften (vgl. (1),(2),(8))

$$\rho(Q_h u_o, Q_h v_o) \le C\|u_o - v_o\|_{\mathfrak{N}}^{\delta} \quad (0 < \delta \le 1) \quad \forall u_o, v_o \in \mathfrak{M},\ h \in (0,h_o], \tag{19}$$

$$\rho(Q_h v, E_o v) \le h^{\sigma} \hat{\varphi}(\alpha^{(1)}_{\mathfrak{G}^*}(v), \alpha^{(2)}_{\mathfrak{G}^*}(v), \ldots, \alpha^{(m)}_{\mathfrak{G}^*}(v)) \tag{20}$$

($\sigma > 0$, $\hat{\varphi}$ isoton in allen Variablen, $\alpha^{(\mu)}_{\mathfrak{G}^*}$ subadditive

Funktionale auf der linearen Hülle $\mathcal{G}^*$ von $\mathcal{G}$)
$\forall\ v \in \mathcal{G}$. [5)]

Es gebe eine Kette $\{\mathcal{G}_r^*\}_{r \in \mathcal{R}}$ linearer Teilräume von $\mathcal{G}^*$ ($\mathcal{G}_r^* \subset \mathcal{G}_{r'}^* \subset \mathcal{G}^*$ für $r < r'$, $\mathcal{R} = \mathbb{N}$), derart, daß mit reellen Zahlen $k_\mu \geq 0$ ($\mu = 1,\ldots,m$), $\theta > 0$ die Bernstein-Ungleichungen (9) und ein Jackson-Satz (11) gelten. Dann kann man aussagen:

1. Es existiert eine stetige Fortsetzung E von E_o auf $\mathcal{M}$, und $\{Q_h\}$ ist für $h \to 0$ auf $\mathcal{M}$ stetig-konvergent gegen E.

2. Gibt es ein $\beta \geq 0$ mit

$$\sup_{r \in \mathcal{R}} \{r^{-\beta}\ \hat{\varphi}\ (f_1(r,u_o),\ \ldots,\ f_m(r,u_o)) < \infty \qquad \forall\ u_o \in \mathcal{M}, \tag{21}$$

so ist mit $\vartheta = \frac{\theta\delta}{\theta\delta + \beta}$ und einem auf $\mathcal{M}$ definierten, explizit angebbaren Funktional κ

$$\varrho\,(Q_h u_o, Eu_o) \leq \kappa(u_o)\ h^{\sigma\vartheta} \qquad \forall\ u_o \in \mathcal{M},\ h \in (0,h_o] \ . \tag{22}$$

<u>Beweis:</u>
a) Die 1. Aussage folgt unmittelbar aus einem Satz von Rinow (vgl. [12], S.49); dabei wird nur ein Teil der Voraussetzungen benötigt.

b) Zu $\varepsilon > 0$ und u_o, $v_o \in \mathcal{M}$ gibt es Zahlen $h_1(u_o)$, $h_2(v_o) \in (0,h_o]$ mit (vgl. a))

$$\varrho\,(Q_h u_o, Eu_o) < \varepsilon \qquad \text{für} \quad h \leq h_1(u_o) \ ,$$

$$\varrho\,(Q_h v_o, Ev_o) < \varepsilon \qquad \text{für} \quad h \leq h_2(v_o) \ .$$

Mit $h_3 = \min\,(h_1,h_2)$ ist also wegen (19) für $h \leq h_3$

---

5) Eine Abänderung auf Abschätzungen durch Summen von Termen der angegebenen Art (vgl. [9], Ungleichung (26)) bereitet keine Schwierigkeiten.

$$\varrho(Eu_o, Ev_o) \leq \varrho(Eu_o, Q_h u_o) + \varrho(Q_h u_o, Q_h v_o)$$

$$+ \varrho(Q_h v_o, Ev_o) \leq 2\varepsilon + C \| u_o - v_o \|_{\mathfrak{N}}^{\delta} ,$$

folglich gilt

$$\varrho(Eu_o, Ev_o) \leq C \| u_o - v_o \|_{\mathfrak{N}}^{\delta} \; \forall \, u_o, v_o \in \mathfrak{M} . \tag{23}$$

c) Vergleicht man (1), (1a) mit (19), (23) sowie (2) und (8) mit (20), so gewinnt man durch das im vorigen Abschnitt geschilderte Vorgehen die Abschätzung (18)

$$\varrho(Q_h u_o, Eu_o) \leq 2\, C \left(\frac{\omega_{\mathfrak{m}}(u_o)}{r^{\theta}} \right)^{\delta}$$

$$+ h^{\sigma} \hat{\varphi}(f_1(r, u_o), \ldots, f_m(r, u_o))$$

$$\leq 2\, C \, (\omega_{\mathfrak{m}}(u_o))^{\delta} \; r^{-\theta\delta} + \tilde{\kappa}(u_o) \, r^{\beta} \, h^{\sigma} \quad \forall u_o \in \mathfrak{M}$$

und damit für optimale Wahl von r die Behauptung (22).

5. Beispiele

1. Als erstes Beispiel behandeln wir die numerische Lösung der ersten Randwertaufgabe der ebenen Potentialtheorie

$$\begin{aligned} &\Delta u = 0 \quad \text{in} \quad R \quad \text{(einfachzusammenhängendes beschränktes Gebiet der x-y-Ebene)} \\ &u = u_o \quad \text{auf } \Gamma \quad \text{(doppelpunktfreier Rand von R)} \end{aligned} \tag{24}$$

für den Fall, daß als Näherungsverfahren die Differenzengleichung

$$U(x+h,y) + U(x-h,y) + U(x,y+h) + U(x,y-h) - 4\, U(x,y) = 0 \tag{25}$$

($h \in [0,h_o] =: \mathfrak{H}$, $h_o > 0$) benutzt wird.

Dabei sei etwa $u_o \in C^o(\Gamma)$, so daß u_o nach Einführung des die Randkurve beschreibenden Bogenlängen-Parameters $s (0 \leq s \leq L)$ als stetige periodische Funktion dieses Parameters, also $u_o \in C_L^o =: \mathfrak{N}$ (wobei unter $\|\cdot\|_{\mathfrak{N}}$ die Tschebyscheff-Norm verstanden werde), gedeutet werden kann (C_L^k = Raum der L-periodischen k-mal stetig differenzierbaren Funktionen).

Wir denken uns den Definitionsbereich der Näherungsfunktion $U(x,y)$ von den Gitterpunkten des über $\bar{R} := R \cup \Gamma$ gelegten Quadrat-Gitters durch folgende Manipulationen auch auf die Zwischengitterpunkte (also auf $\bar{R}$) ausgedehnt (vgl. z.B. [13], S.284 ff.):
Setze $U(x,y) = u_o(x,y)$ für alle $(x,y) \in \Gamma$; (25) werde als gültig erklärt für alle $(x,y) \in \bar{R}$, deren in (25) auftretende Nachbarpunkte $(x+h,y)$ usw. ebenfalls noch zu $\bar{R}$ gehören; durch lineare Interpolation in Randnähe gelangt man dann schließlich zu einer auf $\bar{R}$ stetigen Funktion U:

$$U \in C^o(\bar{R}) =: \mathfrak{k} .$$

Auf $\mathfrak{k}$ führen wir ebenfalls eine Tschebyscheff-Norm ein und induzieren damit die früher auf $\mathfrak{k}$ als existent vorausgesetzte Metrik ρ (womit übrigens $\mathfrak{k}$ vollständig ist).

Es sei $\mathfrak{g}$ der Raum der auf Γ dreimal stetig differenzierbaren Funktionen:

$$\mathfrak{g} := C_L^3 \,{}^{6)} .$$

Als Glieder der Kette $\tilde{\mathfrak{g}}$ wählen wir die Räume $\mathfrak{g}_r$ der trigonometrischen Polynome der Periode L von Höchstordnung r (r = 1,2,...), so daß in der Tat Bernstein-Ungleichungen der Form (9) gelten (vgl. z.B. [14], S.56 ff.), wenn wir $\alpha_{\mathfrak{g}}^{(\mu)}(v) := \|v^{(\mu)}\|_{\mathfrak{N}}$ (für alle $v \in \mathfrak{g}$; $\mu = 1,2,3$) setzen:

$$\|v^{(\mu)}\|_{\mathfrak{N}} \leq \left(\frac{2\pi}{L}\right)^{\mu} r^{\mu} \|v\|_{\mathfrak{N}} \quad \text{für alle } v \in \mathfrak{g}_r \tag{26}$$

$$\left(c_\mu = \left(\frac{2\pi}{L}\right)^{\mu},\ k_\mu = \mu,\ m = 3\right).$$

Für den Fall eines achsenparallelen Quadrats Γ der Kantenlänge 1, auf

---

6) Da $\mathfrak{g}$ ein linearer Raum ist, entfällt hier die Hüllenbildung (vgl. Fußnote 2)).

den wir uns jetzt spezialisieren, zeigte Wasow 1952 ([15] ; vgl. auch [13] , S.309 ff.):

$$\forall u_o \in \mathcal{G} \; : \|Q_h u_o - E u_o\|_{\mathfrak{k}} = \|U - u\|_{\mathfrak{k}} \leq h^2 (1{,}4 \|u_o''\|_{\mathfrak{n}} + 0{,}43 \|u_o'''\|_{\mathfrak{n}}) \quad (27)$$

$$=: \varphi^*(h, \alpha_{\mathcal{G}}^{(1)}(u_o), \alpha_{\mathcal{G}}^{(2)}(u_o), \alpha_{\mathcal{G}}^{(3)}(u_o)),$$

womit (2) in Verbindung mit (8) für diesen Fall realisiert wäre.

Mit $\sigma = 2$ und $\hat{\varphi}(\alpha_{\mathcal{G}}^{(1)}(u_o), \alpha_{\mathcal{G}}^{(2)}(u_o), \alpha_{\mathcal{G}}^{(3)}(u_o)) = 1{,}4 \|u_o^{(2)}\|_{\mathfrak{n}} + 0{,}43 \|u_o^{(3)}\|_{\mathfrak{n}}$ ergibt (27) eine Realisierung von (20), wobei übrigens auch die im Satz des Abschnitts 4 vorausgesetzten Dichtheitsbeziehungen für jeden zwischen $\mathcal{G}$ und $\mathfrak{N}$ liegenden Raum $\mathfrak{M}$ erfüllt sind, da bereits $\mathcal{G}$ dicht in $\mathfrak{N}$ ist (Weierstraß'scher Approximationssatz). Auch die Bedingung (19) des in Abschnitt 4 bewiesenen Satzes ist mit $C = \delta = 1$ erfüllt, da nicht nur die Lösung des kontinuierlichen Problems (24) sondern auch die Lösung von (25) dem Maximum-Prinzip genügt (vgl. z.B. [13] , S.285):

$$\|Q_h u_o - Q_h v_o\|_{\mathfrak{k}} = \|Q_h(u_o - v_o)\|_{\mathfrak{k}} \leq \|u_o - v_o\|_{\mathfrak{n}}. \quad (28)$$

Sei nun etwa

$$\mathfrak{M} = \{ u_o \in \mathfrak{N} \mid u_o^{(\nu)} \in \text{Lip}\,\alpha,\ 0 < \alpha \leq 1 \} \; (\nu = 0 \text{ oder } \nu = 1 \text{ oder } \nu = 2).$$

Dann gilt bekanntlich in der Tat ein Jackson-Satz (vgl. z.B. [14] , S.52 ff.) der Form (11):

$$\|u_o - v_r(u_o)\|_{\mathfrak{n}} \leq \frac{\hat{c}\, M_\nu(u_o)}{r^{\nu+\alpha}},$$

wobei $\hat{c}$ eine von u_o und ν unabhängige, angebbare Konstante und $M_\nu(u_o)$ die Lipschitzkonstante von $u_o^{(\nu)}$ darstellen (somit: $\tilde{\omega}_{\mathfrak{M}}(u_o) = \text{Max}\{\|u_o\|_{\mathfrak{n}}, \hat{c}\, M_\nu(u_o)\}$, $\theta = \nu + \alpha$).

Schließlich ist auch noch die im Satz des Abschnitts 4 genannte Voraussetzung (21) erfüllt [7], denn es gilt mit (14) (wobei hier der Hinweis der Fußnote [4] gemäß [6] zutrifft)

---

[7] s. nächste Seite

$$\hat{\varphi}(f_1(r,u_o),\ f_2(r,u_o),\ f_3(r,u_o)) = 1{,}4\ f_2(r,u_o) + 0{,}43\ f_3(r,u_o)$$

$$< \tilde{\omega}_{\mathfrak{M}}(u_o)\cdot \begin{cases} (1{,}4\gamma_2\ r^{2-\theta} + 0{,}43\gamma_3\ r^{3-\theta}) & \text{für } 0 < \theta = \nu+\alpha \leq 2 \\ (1{,}4\gamma_2^{**} + 0{,}43\gamma_3\ r^{3-\theta}) & \text{für } 2 < \theta = \nu+\alpha \leq 3 \end{cases};$$

man wähle also $\beta = 3 - \theta$.

Damit liefert (22) für $u_o \in \mathfrak{M}$ die Fehlerabschätzung

$$\| Q_h u_o - Eu_o \|_{\mathfrak{k}} = \max_{(x,y)\in\bar{R}} |U(x,y) - u(x,y)| \leq \kappa(u_o)\ h^{2/3(\nu+\alpha)}$$

mit angebbarem $\kappa(u_o)$.

Für lipschitzbeschränkte Randwerte u_o ($\nu = 0$, $\alpha = 1$) sichert man somit z.B. (für $h \to 0$) die Konvergenzordnung $\frac{2}{3}$, für die Walsh und Young [16] (auf anderem Wege) lediglich den Wert $\frac{2}{7}$ angeben konnten (bei Beschränkung auf abgeschlossene Teilmengen von R konnten sie in [17] diesen Wert allerdings auf 1 verbessern).

Es sei noch bemerkt, daß im Falle glatter Ränder auch die in [5] benutzten Methoden auf Differenzapproximation linearer Randwertaufgaben übertragen worden sind (vgl. [18]).

2. a) Die Überlegungen der vorigen Abschnitte lassen sich unmittelbar auf Fehlerabschätzungen bei Quadraturformeln anwenden.

Verwendet man zur näherungsweisen Berechnung eines Integrals $\int_a^b u_o(x)dx$ Ausdrücke der Form

$$Q_h u_o := \sum_{i=1}^{n} A_i^{(n)}\ u_o(x_i) \quad \text{mit} \quad A_i^{(n)} \geq 0,\ x_i = a + ih,\ h = \frac{b-a}{n},$$

so kennt man nämlich häufig für hinreichend oft differenzierbare

---

7) Die Existenzaussage dieses Satzes ist im Falle der vorliegenden Randwertaufgabe allerdings ohne Relevanz, mithin auch nicht die Vollständigkeits- und Dichtheitsannahmen.

Funktionen gültige Fehlerabschätzungen der Art

$$\left| Q_h u_o - \int_a^b u_o(x)\,dx \right| \leq c\, h^{\sigma} \max_{x \in [a,b]} \left| u_o^{(k)}(x) \right|$$

(vgl. z.B. [19]). Damit liegt eine ähnliche Situation vor wie in Beispiel 1 (vgl.(27)). Als approximierende Funktionen v_r bieten sich hier Polynome vom Höchstgrad r an. Da für Polynome jedoch nur eine "schwache" Bernstein-Ungleichung (vgl. [14], S.60)

$$|v_r'(x)| \leq \frac{r}{\sqrt{(b-x)(x-a)}} \max_{x \in [a,b]} |v_r(x)| \tag{30}$$

gilt, muß man noch eine Zusatzüberlegung anstellen. Man kann etwa u_o unter Erhaltung seiner Stetigkeits- und Differenzierbarkeitseigenschaften in ein Intervall $[a - \varepsilon, b + \varepsilon]$ ($\varepsilon > 0$) fortsetzen, die Approximation dort ausführen und aus (30) eine auf $[a,b]$ gültige Bernstein-Ungleichung der gewünschten Art gewinnen.

b) Ein besonders leicht zu behandelnder Fall liegt dann vor, wenn man eine Kette $\{\mathcal{G}_r\}_{r=1,2,\ldots}$ finden kann, derart, daß für ein (2) genügendes Funktional φ mit einem $r^* = r^*(h)$ gilt

$$\varphi(h,v) = 0 \quad \forall h \in \mathfrak{Z}, \ \forall v_r \in \mathcal{G}_r, \quad r = 1,\ldots,r^*. \tag{31}$$

Dabei kann auf die Gültigkeit von Bernstein-Ungleichungen (9) verzichtet werden. Denn bestimmt man $v_r = v_r(u_o)$ gemäß (11a), so erhält man aus (3) unter der weiteren Voraussetzung, daß $\mathcal{G}_1$ konstante Funktionen enthält ($\|\cdot\|$ bedeute die Maximum-Norm in $C[a,b]$)

$$\begin{aligned} \left| Q_h u_o - \int_a^b u_o(x)\,dx \right| &\leq \sum_{i=1}^{n} A_i^{(n)} \|u_o - v_r\| + (b-a)\|u_o - v_r\| \\ &\leq 2\,(b-a)\,\frac{\omega_{m}(u_o)}{r^{\theta}} \qquad r = 1,\ldots,r^*, \end{aligned}$$

also insbesondere

$$\left| Q_h u_o - \int_a^b u_o(x)\,dx \right| \leq 2\,(b-a)\,\frac{\omega_{m}(u_o)}{(r^*)^{\theta}}. \tag{32}$$

Ist beispielsweise $u_o \in C^\nu(-\infty, +\infty)$ 2π-periodisch, $u_o^{(\nu)} \in \mathrm{Lip}\,\alpha$ $(0 < \alpha \leq 1)$ mit einer Lipschitzkonstanten $M_\nu(u_o)$ und

$$Q_h\, u_o := \frac{h}{2}\left[u_o(0) + 2\sum_{i=1}^{n-1} u_o(ih) + u_o(2\pi)\right]$$

die bekannte Trapezregel, so wähle man als $\mathcal{G}_r$ die Menge der trigonometrischen Polynome vom Grad $\leq r$; dann gilt

$$Q_h\, v - \int_0^{2\pi} v(x)dx = \frac{2\pi}{n}\sum_{i=0}^{n-1} v(i\,\frac{2\pi}{n}) - \int_0^{2\pi} v(x)dx = 0$$

für $v(x) = 1$, $\cos x$, $\sin x$, ..., $\cos(n-1)x$, $\sin(n-1)x$, d.h.

(31) ist für $r^* = n - 1$ erfüllt; realisiert man (11a) wieder mittels des klassischen Jackson-Satzes, so erhält man schließlich aus (32) für die betrachteten Funktionen u_o

$$\left|Q_h u_o - \int_0^{2\pi} u_o(x)dx\right| \leq 4\pi\, c_1 \frac{M_\nu(u_o)}{n^{\nu+\alpha}} = c_2\, M_\nu(u_o)\, h^{\nu+\alpha}.$$

Einen mit anderen Methoden geführten Beweis für dieses Ergebnis findet man in [19], S.56.

Analog kann man Fehlerabschätzungen für die Gauß'schen Quadraturformeln (Approximation mit Polynomen, vgl. [19], S.122 ff.) oder für die zusammengesetzten Newton-Cotes-Formeln (Approximation mit geeigneten Spline-Funktionen) gewinnen.

3. Als letztes Beispiel diskutieren wir die (in [9] ausführlicher dargestellte) Behandlung halblinearer Anfangswertaufgaben mittels Differenzenverfahren.

Sei $C(B)$ der Raum der auf einem Bereich $B \subset \mathbb{R}^d$ beschränkten und stetigen Funktionen und $\mathfrak{N} = \mathfrak{k} \subset C(B)$ ein Banachraum mit der sup-Norm. Die Anfangswertaufgabe

$$u_t = \sum_{|\nu|=0}^{m} a_\nu\, D^\nu u(t) + f(.,t,u(t)) \quad (0 \leq t \leq T), \quad u(0) = u_o \qquad (31)$$

($a_\nu \in C(B)$) sei für alle u_o aus einer in einer beschränkten Menge $\mathfrak{U} \subset \mathfrak{N}$ enthaltenen Teilmenge eindeutig lösbar und definiere stetige Lösungsoperatoren $E_o(t)$ durch

$$u(t) = E_o(t)u_o.$$

Die Aufgabe werde numerisch durch ein Differenzenverfahren (explizites k-Schritt-Verfahren) behandelt; es liefere mit einer in t-Richtung äquidistanten Schrittweite h zu $u_o \in \mathfrak{N}$ Näherungen $Q_h(t)u_o$ für $E_o(t)u_o$ ($t \in \{0,h,2h,\dots\} \subset [0,T]$). Unter geeigneten Voraussetzungen (u.a. der *L-Stabilität*, vgl. [9]) sind die Operatoren Q_h auf $\mathfrak{U}$ gleichgradig lipschitzstetig, d.h. es gilt auf $\mathfrak{U}$ die Ungleichung (19) mit $\delta = 1$. Ist nun weiterhin die Differenzapproximation auf einer in $\mathfrak{U}$ dichten Menge mit der Anfangswertaufgabe *konsistent*, so kann man unter einigen weiteren Voraussetzungen und unter Benutzung einer Integraldarstellung der Lösungen (von Thompson [8]) die auf einer Menge $\mathfrak{g} := \mathfrak{U} \cap C^\omega(B)$ gültige Fehlerabschätzung ($0 \leq t \leq T_o \leq T$)

$$\| Q_h(t)\ v_o - E_o(t)\ v_o \| \leq h^\sigma\ P_\varrho(z_1,\dots,z_\varrho) \quad \forall\, v_o \in \mathfrak{g} \tag{32}$$

gewinnen; dabei ist $\sigma > 0$, $\varrho \in \mathbb{N}$ mit $\varrho \leq \omega$ (häufig ist $\varrho = (\sigma+1)m$), und P_ϱ ein Polynom mit nichtnegativen Koeffizienten vom Höchstgrad ϱ der Variablen

$$z_\mu := \sup_{|\nu|=\mu} \| D^\nu\ v_o \|^{1/\mu} \quad (\mu = 1,\dots,\varrho),$$

wobei die z_μ nur in Potenzen von z_μ^μ auftreten [8].

Sei schließlich $\mathfrak{M} := \{ u_o \in \mathfrak{U} : D^\nu u_o \in \mathrm{Lip}\,\alpha \cap C(B)$ für $|\nu| = n \}$ ($n \in \mathbb{N}_o$, $0 < \alpha \leq 1$). Nun liefert der Satz des vorigen Abschnitts zunächst die Existenz verallgemeinerter Lösungen, d.h. die Existenz von Fortsetzungen $E(t)$ von $E_o(t)$ auf $\mathfrak{M}$. Kann man weiterhin, wie dies für wichtige Räume $\mathfrak{N} \subset C(B)$ der Fall ist, eine Kette $\{\mathfrak{g}_r^*\}_{r \in \mathcal{R}}$ linearer Teilräume der linearen Hülle $\mathfrak{g}^*$ von $\mathfrak{g}$ angeben, so daß gilt

---

8) Von dem durch die Anfangsfeldbestimmung eingeschleppten Fehler ist hier abgesehen worden, da er an den folgenden Betrachtungen nichts Wesentliches ändert.

$$\alpha^{(\mu)}_{\mathcal{G}^*}(v_o) := \sup_{|\nu|=\mu} \|D^\nu v_o\| \leq c_\mu r^\mu \|v_o\| \quad \forall v_o \in \mathcal{G}^*_r, \quad \mu = 1,\dots,\varrho$$

und

$$\sup_{r\in\mathcal{R}} \left\{ r^{n+\alpha} \inf_{v_o^* \in \mathcal{G}^*_r} \|u_o - v_o^*\| \right\} < \infty \quad \forall u_o \in \mathfrak{M} \quad ,$$

so ist (9) (mit $k_\mu = \mu$) und wegen der Bemerkung Seite 27 f. auch (11) (mit $\theta = n + \alpha$) erfüllt. Aufgrund der Bauart von P_ϱ ist damit (vgl. (14))

$$\hat{\varphi}(f_1(r,u_o), \dots, f_\varrho(r,u_o)) = P_\varrho((f_1(r,u_o))^{1/1}, \dots, (f_\varrho(r,u_o))^{1/\varrho})$$
$$= O(r^{\varrho-n-\alpha}),$$

so daß (21) mit $\beta = \varrho - n - \alpha$ erfüllt ist. In den durch diese Vorgehensweise erfaßten Fälle erhält man also als Ergebnis:
Das Differenzenverfahren konvergiert auch für schwächer strukturierte Anfangsfunktionen u_o gegen die Lösungen (bzw. sogar gegen die verallgemeinerten Lösungen), und es gilt eine Fehlerabschätzung der Art

$$\|Q_h(t)\, u_o - E(t)\, u_o\| \leq c(u_o)\, h^{\frac{n+\alpha}{\varrho}\sigma} \quad \forall u_o \in \mathfrak{M} .$$

Im Spezialfall linearer Anfangswertaufgaben gelangt man so zu den eingangs erwähnten Ergebnissen von Peetre und Thomée [5].

L i t e r a t u r

[1] MEINARDUS, G. und H.-D. STRAUER: Über Tschebyscheffsche Approximationen der Lösungen linearer Differential- und Integralgleichungen. Arch. Rat. Mech. Anal. 14 (1963), 184 - 195.

[2] ANSORGE, R.: Tschebyscheff-Approximation der Lösungen von Differentialgleichungen bei Benutzung von Differenzenverfahren. ZAMM 46 (1966), 397 - 399.

[3] COLLATZ, L.: Nichtlineare Approximation bei Randwertaufgaben. Wiss. Z. Hochschule f. Architektur u. Bauwesen, Weimar 1969.

[4] – Zur Tschebyscheff-Approximation bei Funktionen mehrerer unabhängiger Veränderlicher. Proc. Conference on Constructive Theory of Functions, Budapest 1970.

[5] PEETRE, J. and V. THOMÉE: On the rate of convergence for discrete initial-value problems. Math. Scand. 21 (1967), 159 - 176.

[6] ANSORGE, R. und C. GEIGER: Approximationstheoretische Abschätzung des Diskretisationsfehlers bei verallgemeinerten Lösungen gewisser Anfangswertaufgaben. Abh. Math. Sem. Univ. Hamburg 36 (1971), 99 - 110.

[7] HEDSTROM, G.W.: The rate of convergence of parabolic difference schemes with constant coefficients. BIT 9 (1969), 1 - 17.

[8] THOMPSON, R.J.: Difference approximations for inhomogenous and quasi-linear equations. J. Soc. Indust. Appl. Math. 12 (1964), 189 - 199.

[9] ANSORGE, R., C. GEIGER und R. HASS: Existenz und numerische Erfaßbarkeit verallgemeinerter Lösungen halblinearer Anfangswertaufgaben. ZAMM 52 (1972), 597 - 605.

[10] BUTZER, P.L. und K. SCHERER: Approximationsprozesse und Interpolationsmethoden. Mannheim/Zürich: Bibliographisches Institut 1968.

[11] ZAMANSKY, M.: Classes de saturation de certains procédés d'approximation des séries de Fourier des fonctions continues et application à quelques problèmes d'approximation. Ann. sci. École Norm. Sup. 66 (1949), 19 - 93.

[12] ANSORGE, R. und R. HASS: Konvergenz von Differenzenverfahren für lineare und nichtlineare Anfangswertaufgaben. Berlin - Heidelberg - New York: Springer 1970.

[13] FORSYTHE, G.E. and W.R. WASOW: Finite-difference methods for partial differential equations (4th printing). New York - London - Sidney: John Wiley & Sons 1967.

[14] MEINARDUS, G.: Approximation von Funktionen und ihre numerische Behandlung. Berlin - Heidelberg - New York: Springer 1964.

[15] WASOW, W.: On the truncation error in the solution of Laplace's equation by finite differences. J. Res. Nat. Bur. Standards 48 (1952), 345 - 348.

[16] WALSH, J.L. and D. YOUNG: On the accuracy of the numerical solution of the Dirichlet problem by finite differences. J. Res. Nat. Bur. Standards 51 (1953), 343 - 363.

[17] – On the degree of convergence of solutions of difference equations to the solution of the Dirichlet problems. J. Math. Phys. 33 (1954), 80 - 93.

[18] BRAMBLE, J.H., B.E. HUBBARD and V. THOMÉE: Convergence estimates for essentially positive type discrete Dirichlet problems. Math. Comput. 23 (1969), 695 - 710.

[19] DAVIS, P.J. and P. RABINOWITZ: Numerical integration. Walth. - Toronto - Lond.: Blaisdell 1967

COMPOUND SUCCESSIVE APPROXIMATIONS FOR NONLINEAR TWO POINT BOUNDARY VALUE PROBLEMS[1)]

By Jan van de Craats

1. Introduction

We consider boundary value problems for the differential equation

$$y''(t) + f(t,y(t),y'(t)) = 0. \tag{1}$$

We consider boundary conditions of the type

$$y(a) = A, \ y(b) = B \tag{2}$$

or

$$y(a) = A, \ y'(b) = m \tag{3}$$

or

$$y'(a) = m, \ y(b) = B. \tag{4}$$

$[a,b]$ is supposed to be a finite interval on the real axis. We suppose that the real function $f(t,y,z)$ satisfies the following conditions (5) and (6):

$$f(t,y,z) \text{ is continuous on } [a,b] \times R^2. \tag{5}$$

There exist continuous nonnegative functions $K(t)$ and $L(t)$ on $[a,b]$ such that

$$|f(t,y_1,z_1) - f(t,y_2,z_2)| \leq K(t)|y_1-y_2| + L(t)|z_1-z_2| \tag{6}$$

for all $a \leq t \leq b$ and all y_1,y_2,z_1,z_2.

We call problem (1),(2) the <u>first</u> boundary value problem, problem (1),(3) the <u>second</u> b.v.p., and problem (1),(4) the <u>third</u> b.v.p..

Let the differential operator Φ be defined by

$$\Phi|y|(t) \equiv y''(t) + f(t,y(t),y'(t)). \tag{7}$$

<u>Definition:</u> Φ is called regular on $[a,b]$ with respect to first b.v.p.'s - shortly <u>regular(I) on [a,b]</u> - if the b.v.p.

$\Phi[y](t) = 0, \quad y(c) = A, \ y(d) = B$

has <u>exactly one</u> solution for each $a \leq c < d \leq b$ and for all A and B.

In a similar way we define the concepts <u>regularity(II)</u> and <u>regularity(III)</u> on $[a,b]$.

Obvious problems are

(i) finding regularity conditions for all Φ in a certain class,

(ii) constructing iterative methods for approximating the solution if these regularity conditions are satisfied.

---

1) This paper describes a part of the author's thesis at Leiden University.

As far as iterative methods are concerned, we restrict our attention mainly to Picard's method of successive approximations, defined by

$$y_{n+1}''(t) + f(t,y_n(t),y_n'(t)) = 0 \qquad (8)$$

with the corresponding boundary conditions for $y_{n+1}(t)$.

For second and third b.v.p.'s the problems (i) and (ii) are simultaneously solved by the following theorems 1 and 2:

Theorem 1: Let $K(t)$, $L(t)$ be continuous and nonnegative on $[a,b]$.

(i) All Φ of the form (7), where f satisfies (5) and (6), are regular(II) on $[a,b]$ if and only if the solution $v(t)$ of

$$\Lambda[v](t) \equiv v''(t) + L(t)v'(t) + K(t)v(t) = 0$$

$$v(a) = 0, \; v'(a) = 1$$

satisfies $v'(t) > 0$ on $[a,b]$.

(ii) In that case method (8) converges to the solution of (1),(3) for all continuously differentiable starting functions $y_o(t)$.

Theorem 2: Let $K(t)$, $L(t)$ be continuous and nonnegative on $[a,b]$.

(i) All Φ of the form (7), where f satisfies (5) and (6), are regular(III) on $[a,b]$ if and only if the solution $w(t)$ of

$$M[w](t) \equiv w''(t) - L(t)w'(t) + K(t)w(t) = 0$$

$$w(b) = 0, \; w'(b) = -1$$

satisfies $w'(t) < 0$ on $[a,b]$.

(ii) In that case method (8) converges to the solution of (1), (4) for all continuously differentiable starting functions $y_o(t)$.

The theorems 1 and 2 were first proved by Coles and Sherman [3] in 1967. The regularity conditions already follow from results by Epheser [6] (1955).

For first b.v.p.'s similar results cannot hold. Although it is possible to formulate sufficient and necessary regularity conditions comparable to parts (i) of the theorems 1 and 2, it is not true that method (8) converges whenever these conditions are fulfilled. Bailey [1] gave an example of a regular b.v.p. for which method (8) diverges, even with starting functions arbitrarily close to the solution. Many authors have given sufficient convergence conditions for method (8). The first results are by Picard [7] (1893), and the most recent results can be found in Albrecht [8] and in van de Craats [4] (independently we obtained the same results following different ways).

In this paper we shall not study method (8) for first b.v.p.'s, but we shall construct a new iterative method based upon a combination of successive approximations for second and third b.v.p.'s. This method, which we shall call a method of compound successive approximations, will be shown to converge whenever the regularity conditions are satisfied, and thus problems (i) and (ii) are completely solved.

2. General outline of the method

First we shall state the regularity conditions for first b.v.p.'s.

Theorem 3: Let $K(t)$ and $L(t)$ be continuous and nonnegative on $[a,b]$.
All Φ of the form (7), where f satisfies (5) and (6), are regular(I) on $[a,b]$ if and only if there exists a point c, $a < c < b$, such that

(i) the solution $v(t)$ of
$\Lambda[v](t) = 0, \quad v(a) = 0, \; v'(a) = 1$
satisfies $v'(t) > 0$ on $[a,c]$,

(ii) the solution $w(t)$ of
$M[w](t) = 0, \quad w(b) = 0, \; w'(b) = -1$
satisfies $w'(t) < 0$ on $[c,b]$.

(the linear operators Λ and M are defined as in the theorems 1 and 2).

These regularity conditions follow from results by Epheser [6], and they are also derived in a different way by Bailey, Shampine and Waltman [2]. We shall give a sketch of the proof of the sufficiency of the regularity conditions of theorem 3 as an introduction to the construction of the method of compound successive approximations. Suppose, therefore, that there exists a c in (a,b) such that (i) and (ii) hold. Suppose that an operator Φ of the prescribed form is given. We want to prove that every first b.v.p. on $[a,b]$ with this Φ has exactly one solution. It is no severe restriction to suppose that the boundary conditions are given at the points $t = a$ and $t = b$. Thus we have the b.v.p.

$$\Phi[y](t) = 0 \tag{9}$$

$$y(a) = A, \; y(b) = B. \tag{10}$$

We remark that in view of (i) and theorem 1, Φ is regular(II) on $[a,c]$, and in view of (ii) and theorem 2 Φ is regular(III) on $[c,b]$.

Consider an integral curve $y_1(t)$ of (9) issuing from (a,A). It is clear that another integral curve $y_2(t)$ of (9) issuing from (a,A) cannot intersect $y_1(t)$ as long as $a < t \leq c$, since otherwise there must be an intermediate point where the derivatives of $y_1(t)$ and $y_2(t)$ are equal. Then we would have two distinct solutions of a second b.v.p. on a subinterval of $[a,c]$ which is impossible by virtue of the regularity(II) of Φ. Therefore, at a fixed point t_o, $a < t_o \leq c$, even the derivatives of distinct integral curves issuing from (a,A) cannot be equal. It is not difficult to prove that this family of integral curves forms a complete and continuous covering of the line $t = t_o$ for each t_o in $(a,c]$ (see e.g. [5] p.13). So in particular, if we denote by $v(t,m)$ the solution of (9) with $v(a,m) = A$, $v'(c,m) = m$, $v(c,m)$, regarded as a function of m, is a continuous, increasing function that takes all values. In the same way one can prove that if $w(t,m)$ is the solution of (9) with $w'(c,m) = m$, $w(b,m) = B$, the function (of m) $w(c,m)$ is a continuous, decreasing function of m that takes all values. Now it is clear that there must be exactly one $\bar{m}$ for which the function $\phi(m)$, defined by

$$\phi(m) = v(c,m) - w(c,m) \tag{11}$$

vanishes. For this value $\bar{m}$ we have

$v(c,\bar{m}) = w(c,\bar{m})$ and $v'(c,\bar{m}) = w'(c,\bar{m}) = \bar{m}$.

Since Lipschitz condition (6) guarantees unicity of solutions of initial value problems, we then have $v(t,\bar{m}) \equiv w(t,\bar{m})$. Thus we have found the unique integral curve connecting (a,A) and (b,B). For more details and for the proof of the necessarity of the conditions of theorem 3 we refer to [2] or [5].

The considerations given above also suggest an iterative method for approximating the solution $\bar{y}(t)$ of a first b.v.p. (1),(2). First we verify whether the regularity conditions are satisfied by constructing the functions v(t) and w(t) from (i) and (ii). Let a_1 be the first point to the right of a for which $v'(t) = 0$, and let b_1 be the first point to the left of b for which $w'(t) = 0$. The regularity conditions are satisfied if and only if $b_1 < a_1$, and then we can choose for c any point in (b_1,a_1). The idea of the method of compound successive approximations is the following:

1. Take an estimate m_o for the derivative of the solution $\bar{y}(t)$ at $t = c$.
2. Construct by successive approximations the integral curve $v(t,m_o)$ of (9) satisfying $v(a,m_o) = A$, $v'(c,m_o) = m_o$ (by virtue of theorem 1 the successive approximations converge).
3. Construct by successive approximations the integral curve $w(t,m_o)$ of (9) satisfying $w'(c,m_o) = m_o$, $w(b,m_o) = B$.

   In general $v(c,m_o) \neq w(c,m_o)$, so the estimate m_o for $\bar{y}'(c)$ was not correct. The correct value $\bar{m} = \bar{y}'(c)$ is the only zero of the continuous, increasing function $\phi(m)$, defined by (11). Therefore
4. by some rootfinding technique for $\phi(m)$ construct a "better" estimate m_1 for $\bar{m}$. From now on the process is repeated.

It is clear that if in this way a sequence $m_o, m_1, m_2, \ldots$ is generated converging to $\bar{m}$, we also have

$v(t,m_i) \rightarrow v(t,\bar{m}) = \bar{y}(t)$ on [a,c] and

$w(t,m_i) \rightarrow w(t,\bar{m}) = \bar{y}(t)$ on [c,b].

3. Technical Details

Of course, there are still some difficulties. The first arises from the fact that the process for determining $v(t,m_n)$ and $w(t,m_n)$ by successive approximations is not a finite one. In practice, therefore, one will stop the successive approximations after a finite number of steps, and then one will compute a new m_{n+1}. It is possible to give with each successive approximation $v_k(t,m_n)$ for $v(t,m_n)$ a bound for the error $|v_k(t,m_n) - v(t,m_n)|$ (see e.g. [2], ch.3). In a similar way a bound can be given for the error $|w_l(t,m_n) - w(t.m_n)|$. In particular at $t = c$, we have with each $v_k(c,m_n)$ an interval V_k with $v_k(c,m_n)$ as its center, in which the true value $v(c,m_n)$ must lie. In a similar way we have with each $w_l(c,m_n)$ an interval W_l in which $w(c,m_n)$ must lie.

It is clear that we can compute a new m_{n+1} as soon as these two intervals are disjoint.

A second problem is the choice of an appropriate rootfinding method for the function $\phi(m)$. Before we discuss this problem we establish the following lemma:

Lemma 4: There exist positive numbers T_1, T_2, only depending on the Lipschitz functions $K(t)$ and $L(t)$ and on the choice of the point c, such that for all m_1 and m_2

$$0 < T_1 \leq \frac{\phi(m_1) - \phi(m_2)}{m_1 - m_2} \leq T_2.$$

Therefore, the slope of any chord connecting two points of the graph of $\phi(m)$ remains between two fixed, positive bounds. We shall give a sketch of the proof.
First we consider the quotient

$$d(t) = \frac{v(t,m_1) - v(t,m_2)}{m_1 - m_2} \qquad \text{for some } m_1 \neq m_2.$$

We shall use the notations $v_1(t) = v(t,m_1)$, $v_2(t) = v(t,m_2)$.
We substitute v_1 and v_2 in the differential equation (1), subtract, and divide by $m_1 - m_2$ to get

$$d'' + \frac{f(t,v_1,v_1') - f(t,v_1,v_2')}{v_1' - v_2'} d' + \frac{f(t,v_1,v_2') - f(t,v_2,v_2')}{v_1 - v_2} d = 0 \qquad (12)$$

with boundary conditions

$$d(a) = 0, \; d'(c) = 1. \qquad (13)$$

Denoting the coefficients of d' and d in (12) by $p(t)$ and $q(t)$ respectively, (12), (13) reads

$$d''(t) + p(t)d'(t) + q(t)d(t) = 0 \qquad (14)$$

$$d(a) = 0, \; d'(c) = 1. \qquad (15)$$

In view of Lipschitz condition (6) we have $|p(t)| \leq L(t)$, $|q(t)| \leq K(t)$.
Let $d_1(t)$ be the solution of

$d_1''(t) - L(t)d_1'(t) - K(t)d_1(t) = 0$, $d_1(a) = 0$, $d_1'(c) = 1$,

and let $d_2(t)$ be the solution of

$d_2''(t) + L(t)d_2'(t) + K(t)d_2(t) = 0$, $d_2(a) = 0$, $d_2'(c) = 1$.

By virtue of condition (i) of theorem 3 and of theorem 1 the functions $d(t)$, $d_1(t)$ and $d_2(t)$ are increasing on $[a,c]$. Moreover it can be proved that on $[a,c]$
$0 \leq d_1(t) \leq d(t) \leq d_2(t)$ (see e.g. [5] p.68).

In a similar way, define

$$e(t) = \frac{w(t,m_1) - w(t,m_2)}{m_1 - m_2}.$$

Let $e_1(t)$ be the solution of

$e_1''(t) + L(t)e_1'(t) - K(t)e_1(t) = 0,\quad e_1'(c) = 1,\ e_1(b) = 0,$

and let $e_2(t)$ be the solution of

$e_2''(t) - L(t)e_2'(t) + K(t)e_2(t) = 0,\quad e_2'(c) = 1,\ e_2(b) = 0.$

Then $e(t)$, $e_1(t)$ and $e_2(t)$ are increasing,negative on $[c,b)$ and it can be proved that $e_2(t) \le e(t) \le e_1(t) \le 0$ on $[c,b]$.

Since $\dfrac{\phi(m_1) - \phi(m_2)}{m_1 - m_2} = d(c) - e(c)$ it follows that if

$T_1 = d_1(c) - e_1(c),\quad T_2 = d_2(c) - e_2(c)$ then

$$0 < T_1 \le \frac{\phi(m_1) - \phi(m_2)}{m_1 - m_2} \le T_2.$$

We can use the bounds T_1 and T_2 in establishing a convergent rootfinding method for $\phi(m)$. The values $\phi(m_n)$ are not computed exactly. Since we use a finite number of successive approximations, we find an approximate value $\Phi(m_n)$. Using the error bounds for the successive approximations, we get bounds $\underline{\phi}(m_n)$ and $\bar{\phi}(m_n)$ between which $\phi(m_n)$ must lie. $\underline{\phi}(m_n)$ and $\bar{\phi}(m_n)$ are of the same sign. Suppose that $|\underline{\phi}(m_n)| \le |\bar{\phi}(m_n)|$, then it is clear that the zero $\bar{m}$ of $\phi(m)$ must lie in between

$$m_n - \frac{\underline{\phi}(m_n)}{T_2} \quad \text{and} \quad m_n - \frac{\bar{\phi}(m_n)}{T_1}.$$

In this way one can compute with each evaluation of a value $\Phi(m_n)$, bounds between which the zero $\bar{m}$ lies. A converging rootfinding method is given by

$$m_{n+1} = m_n - \frac{\underline{\phi}(m_n)}{T_2},$$

but the convergence of this method will not be very fast. For this reason we use as a rule the (faster converging) secant method:

$$m_{n+1} = m_n - \frac{m_n - m_{n-1}}{\Phi(m_n) - \Phi(m_{n-1})}\,\Phi(m_n).$$

We also keep up lower bounds and upper bounds for $\bar{m}$ as described above, and if the new m_{n+1} computed by the secant method, would not lie in between them, we reject this value and take for m_{n+1} the mean of the upper bound and the lower bound.

4. Algorithm

Using the devices described in the preceding section and some minor technical tools, such as a starting procedure and a stop criterion, we constructed an algorithm for the method of compound successive approximations for solving a first b.v.p. (1),(2). The input of the algorithm consists of

1. $f(t,y,z)$,
2. a, b, A, B,
3. Lipschitz functions $K(t)$, $L(t)$,
4. a starting value m_o (arbitrary, e.g. $m_o = \frac{B-A}{b-a}$),
5. starting functions $v_o(t,m_o)$, $w_o(t,m_o)$ (arbitrary),
6. an a priori error bound $\varepsilon > 0$ for the final approximations,
7. an a priori error bound $\varepsilon' > 0$ for the derivatives of the final approximations.

The algorithm first verifies whether for this interval $[a,b]$ and these Lipschitz functions $K(t)$ and $L(t)$ the regularity conditions are fulfilled. If this is the case it computes an intermediate point c and it yields the output:

1. the final approximation $v_k(t,m_n)$ on $[a,c]$,
   the final approximation $w_l(t,m_n)$ on $[c,b]$,
2. a posteriori error bound $E_1(t)$ for $|v_k(t,m_n) - \bar{y}(t)|$ on $[a,c]$,
   a posteriori error bound $E_2(t)$ for $|w_l(t,m_n) - \bar{y}(t)|$ on $[c,b]$,
3. a posteriori error bound $|E_1'(t)|$ for $|v_k'(t,m_n) - \bar{y}'(t)|$ on $[a,c]$,
   a posteriori error bound $|E_2'(t)|$ for $|w_l'(t,m_n) - \bar{y}'(t)|$ on $[c,b]$.

For the algorithm itself and an elaboration of all technical details we refer to [5], ch.VII.

5. Numerical Illustration

We have written a computer program based on the algorithm just described. For simplicity reasons we used Lipschitz constants K and L. In the examples we shall give we chose them such that the ordinary method of successive approximations (8) could fail to converge (see [1]).
In the program we use a discretization with equidistant mesh points. We divide the interval $[a,b]$ into N subintervals of length
$h = \frac{b-a}{N}$, and we define $t^j = a + jh$ $(j = 0,\dots,N)$.
Instead of the continuous problem (1), (2) we intend to solve the corresponding discrete problem

$$\frac{y^{j+1} - 2y^j + y^{j-1}}{h^2} + f(t, y^j, \frac{y^{j+1} - y^{j-1}}{2h}) = 0 \quad (j = 1,\dots,N-1) \tag{16}$$

$$y^0 = A, \quad y^N = B. \tag{17}$$

We "translated" the continuous algorithm into a discrete "algorithm" by replacing all derivatives by corresponding divided differences. All linear b.v.p.'s thus are replaced by finite, tridiagonal systems of equations. These systems are easily and efficiently solved by Gaussian elimination. Hence the performance of the discrete "algorithm" presents no practical problems. We did not yet try to prove finiteness of the discrete version of the algorithm or to prove convergence of the output to the solution of the discrete problem (16),(17) as the a priori accuracy ε tends to zero. We shall only give two examples which illustrate that the discrete "algorithm" can work satisfactorily.

Example 1:

$y''(t) + 8y'(t) + 0.005y(t) - 0.005 = 0$

$y(0) = y(1) = 1.$

The exact solution is $y(t) \equiv 1$. This is also the exact solution of the discrete problem. We have chosen $K = 0.01$, $L = 10$, $\varepsilon = \varepsilon' = 0.1$, $N = 10$, with starting value $m_o = 2$ and starting vectors

$v_o = \{v_o^j\}$, where $v_o^j = 1 + 2jh$,

$w_o = \{w_o^j\}$, where $w_o^j = 2jh - 1$.

The algorithm yielded $c = 0.5$, $T_1 = 0.29875$, $T_2 = 36.9408$.

In fig.1 we give an illustration of the results. We don't give all approximations. We only sketch the last approximations $v_k^j(m_n)$ and $w_l^j(m_n)$ just before the computation of a new m_{n+1}. Table 1 shows the values of k, l, m_n and $\Phi(m_n)$ at these stages.

In table 2 we give the final approximations Y^j, the "a posteriori error bounds" E^j for Y^j, and the "a posteriori error bounds" D^j for the derivatives (these are the discrete versions of the continuous a posteriori error bounds. It is, however, not sure whether these discrete versions in general are real error bounds).

n	m_n	k	l	$\Phi(m_n)$
0	2	12	12	10.58
1	1.43	11	11	7.60
2	-1.611.'-2	19	19	-8.5218.'-2
3	-5.755.'-5	21	20	-3.0565.'-4
4	2.301.'-7	14	14	8.6757.'-7

table 1

j	Y^j	E^j.'3	D^j.'3
0	1	0	
1	1.00000044	4.49	29.9
2	1.00000071	5.99	9.97
3	1.00000083	6.48	3.32
4	1.00000088	6.65	1.10
5	1.00000091	6.70	0.36
5	1.000000040	6.70	0.36
6	1.000000063	6.65	1.10
7	1.000000073	6.48	3.32
8	1.000000077	5.99	9.97
9	1.000000071	4.49	29.9
10	1	0	

table 2

Example 2:

$y''(t) + 10 \cos y'(t) + 0.01 \sin y(t) + g(t) = 0$

$y(0) = 0$, $y(1) = \sin 25$,

where $g(t)$ is chosen in such a way that $y(t) = \sin 25t$ is the exact solution, i.e.

$g(t) = 625 \sin 25t - 10 \cos (25 \cos 25t) - 0.01 \sin (\sin 25t)$.

Now discretization errors will be present. The discrete problem has an (unknown) solution $\{y^j\}$ which is not exactly the same as $\{y(t^j)\} \equiv \{\sin 25t^j\}$.

We chose N = 200, K = 0.01, L = 10, $\varepsilon = \varepsilon' = 0.1$. To start the algorithm we used $m_o = 0$, $v_o^j = 0$ for all j, $w_o^j = \sin 25$ for all j. The program yielded c = 0.5, $T_1 = 0.20360$ and $T_2 = 29.912$. Table 3 shows the values of m_n, k, ℓ, and $\Phi(m_n)$ just before the computation of a new m_{n+1}.

n	m_n	k	ℓ	$\Phi(m_n)$
0	2	4	4	-2.2855.'+1
1	3.5178	4	4	-2.1363.'+1
2	2.5249.'+1	6	5	3.3160.'-1
3	2.4917.'+1	6	6	4.1156.'-3
4	2.4913.'+1	6	6	2.9677.'-5

table 3

Table 4 gives for j = 0, 10, 20, ... , 200

- the final approximations $v_k^j(m_n)$, $w_\ell^j(m_n)$ (here called Y^j)
- the a posteriori bounds E^j
- the differences $\bar{E}^j = Y^j - \sin 25t^j$
- the a posteriori error bounds D^j for the derivatives
- the differences $\bar{D}^j = \dfrac{Y^{j+1} - Y^{j-1}}{2h} - 25 \cos 25t^j$.

In fig.2 we sketch the final approximation Y^j.

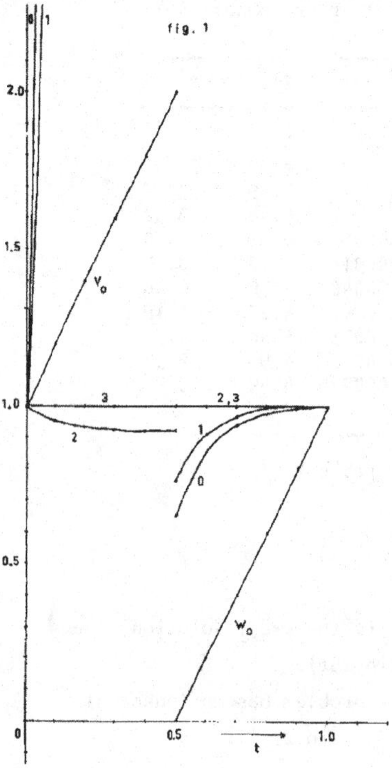

fig. 1

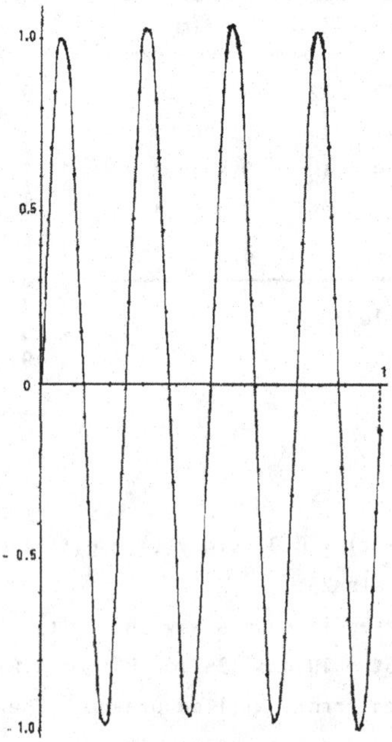

fig. 2

j	Y^j	E^j.'+3	$\bar{E}^j$.'+3	D^j.'+3	$\bar{D}^j$.'+3
0	0	0	0		
10	0.95410	2.7	5.1	41	67
20	0.60699	4.3	8.5	25	104
30	-0.56024	5.2	11.3	15	124
40	-0.94328	5.8	15.6	9.1	87
50	-0.01152	6.2	21.7	5.5	59
60	0.96533	6.4	27.3	3.3	76
70	0.65603	6.5	31.3	2.0	114
80	-0.50937	6.6	34.6	1.2	138
90	-0.92827	6.7	39.5	0.7	101
100	-0.02018	6.7	46.1	0.4	70
100	-0.02021	6.7	46.1	0.4	-138
110	0.96865	6.7	42.7	0.7	-104
120	0.68798	6.6	37.7	1.2	-69
130	-0.48555	6.5	31.3	2.0	-76
140	-0.95003	6.4	25.6	3.3	-110
150	-0.07775	6.2	21.6	5.5	-126
160	0.93169	5.8	18.7	9.1	-95
170	0.68948	5.2	14.3	15	-58
180	-0.47862	4.3	8.6	25	-64
190	-0.97899	2.7	3.4	41	-97
200	-0.13235	0	0		

table 4

The discrepancy between E^j and $\bar{E}^j$ and between D^j and $\bar{D}^j$ are probably mainly due to discretization errors.

The computations have been performed on the IBM 360/65 computer of the Centraal Reken-Instituut of Leiden University.

REFERENCES:

1. P.B.BAILEY, On the interval of convergence of Picard's Iteration, ZAMM 48, 127 - 128 (1968).
2. P.B.BAILEY, L.F.SHAMPINE, P.E.WALTMAN, Nonlinear Two Point Boundary Value Problems, New York 1968, Academic Press.
3. W.J.COLES, T.L.SHERMAN, Convergence of successive approximations for nonlinear two-point boundary value problems, SIAM J.Appl.Math.15, 426 - 433 (1967).
4. J.van de CRAATS, On the region of convergence of Picard's iteration, ZAMM 52, 487 - 491 (1972).
5. J.van de CRAATS, Successive approximations for nonlinear two point boundary value problems, thesis, Leiden University 1972.
6. H.EPHESER, Über die Existenz der Lösungen von Randwertaufgaben mit gewöhnlichen, nichtlinearen Differentialgleichungen zweiter Ordnung, Math.Z. 61, 435 - 454 (1955).
7. E.PICARD, Sur l'application des méthodes d'approximations successives à l'étude de certaines équations différentielles ordinaires, J.Math. 9, 217 - 271 (1893).
8. J.ALBRECHT, Zur Wahl der Norm beim Iterationsverfahren für Randwertaufgaben, ZAMM 52, 626 - 628 (1972).

SPEZIELLE INTERPOLATIONSQUADRATUREN VOM GAUSS'SCHEN TYP

Von Hermann Engels

Abstract

GAUSS' quadrature procedure, based upon the zeros of LEGENDRE-polynomials as nodes, may be also developed by integration of an HERMITE-interpolation-polynomial involving first derivatives of the integrand at the same nodes. All the weights of these derivatives in the resulting quadrature formula vanish, while the weights of the function values are positive. After building up a more general HERMITE-interpolation-operator (cf. |6|, |8|, |10|) general quadratures are obtained by integration. They are defined as general GAUSS-type-quadratures if and only if all weights of the derivatives of the quadrature vanish. As a consequence the weights of the function values of such a quadrature are shown to be positive. Besides the ordinary GAUSS-quadrature and variants with weight-functions within the integrand there ist not only the possibility of constructing arbitrarily sets of new GAUSS-type-quadratures but also some known quadrature-procedures are shown to be of this type, e. g. WILF-quadrature (cf. |18|, |4|) and a procedure due to MARTENSEN (cf. |15|). For the WILF-quadrature new aspects for the numerical determination of nodes and weights follow from this new point of view. It is well known that this problem is very ill conditioned. The usual GAUSS-quadrature surely is contained as a special case for the various weight-functions. It should be mentioned, that the general interpolating operator yields also interesting results in numerical differentiation (cf. |9|).

1. Einleitung und Übersicht

Die GAUSS'schen Quadraturformeln kann man bekanntlich durch Integration eines HERMITE'schen Interpolationspolynoms erhalten. Es zeigt sich, daß die Gewichte der Ableitungen der Quadraturformeln sämtlich verschwinden, und daß die Gewichte der Funktionswerte stets positiv sind. Letzteres zieht unmittelbar die Konvergenz des GAUSS'schen Quadraturverfahrens nach sich. Aber nicht allein diese wünschenswerte Eigenschaft begründet das Interesse an Quadraturverfahren, sondern auch ihre Anwendbarkeit bei der Lösung von Integral- und Integrodifferentialgleichungen sowie bei der Konstruktion von RUNGE-KUTTA-Verfahren (vgl. |7|). Ersetzt man nun das HERMITE'sche Interpolationspolynom durch einen allgemeinen HERMITE'schen Interpolationsoperator, so erhält man daraus durch Integration entsprechende andere Quadraturformeln. Diese werden genau dann als "vom GAUSS'schen Typ" bezeichnet, wenn sämtliche Gewichte der Ableitungen verschwinden. Analog zu den gewöhnlichen GAUSS'schen Quadraturformeln erweisen sich dann die Gewichte der Funktionswerte als sämtlich positiv. Die Konstruktion des HERMITE-Operators (vgl. |6|, |8|, |10|) enthält zahlreiche Freiheiten und gibt damit nicht nur die Möglichkeit zur Konstruktion neuer GAUSS-Typ-Quadraturen in fast beliebiger Vielfalt, sondern es erweist sich darüber hinaus, daß einige bekannte Quadraturverfahren in diesem Sinne GAUSS-Typ-Verfahren sind, wie überraschenderweise das WILF'sche Quadraturverfahren (vgl. |18|, |4|) und ein von MARTENSEN (vgl. |15|) angegebenes Verfahren. Darüber hinaus ergeben sich neue Aspekte für die numerische Bestimmung der Stützstellen und Gewichte der WILF-Quadraturen, ein Problem, das bekanntlich äußerst schlecht konditioniert ist. Selbstverständlich ist die gewöhnliche GAUSS-Quadratur als Spezialfall auch für die verschiedensten Gewichtsfunktionen enthalten. Es sei noch erwähnt, daß der allgemeine Interpolationsoperator auch bei der numerischen Differentiation zu interessanten Ergebnissen führt (vgl. |9|).

2. Das allgemeine Quadraturverfahren

Sei $\{x_i\}$, $i = 1(1)n$ eine Menge paarweise verschiedener reeller Zahlen oder Stützstellen und $[a,b]$ ein reelles Intervall, das im Weiteren als Stützintervall bezeichnet wird, und das gewöhnlich die Stützstellen enthält.

Dann gilt der folgende Satz:

<u>**Satz 1 :**</u> *Sei* $h_i(x)$ *eine reelle Funktionenmenge mit den Eigenschaften*

1) $h_i(x) \in C^{2n}[a,b]$, $i = 1(1)n$

2) $h_i(x_i) = 0$, $h_i'(x_i) \neq 0$, $i = 1(1)n$

3) $h_i(x_k) \neq 0$ für $i \neq k$, $i,k = 1(1)n$.

Sei $\omega_n(x)$ *eine reelle Funktion mit den Eigenschaften*

1) $\omega_n(x) \in C^{2n}[a,b]$

2) $\omega_n(x_i) = 0$, $\omega_n'(x_i) \neq 0$, $i = 1(1)n$.

Dann ist

$$H_n(f,x) = \sum_{i=1}^{n} \left\{ \left[1-2\lambda_i'(x_i) \frac{h_i(x)}{h_i'(x_i)}\right] f(x_i) + \frac{h_i(x)}{h_i'(x_i)} f'(x_i) \right\} \lambda_i^2(x)$$

mit $\lambda_i(x) = \dfrac{\omega_n(x) h_i'(x_i)}{h_i(x)\omega_n'(x_i)}$, $i = 1(1)n$

ein HERMITE'*scher Interpolationsoperator, d. h. es gilt*

$H_n(f,x_i) = f(x_i)$ *und* $H_n'(f,x_i) = f'(x_i)$.

Der Beweis erfolgt durch Einsetzen und Berücksichtigen der Eigenschaften der Funktionen $h_i(x)$, $i = 1(1)n$, und $\omega_n(x)$.

Bemerkung: **Man sieht unmittelbar, daß u. a.**

$$\omega_n(x) = \prod_{i=1}^{n} h_i(x)$$

gewählt werden kann.

Sei nun $W(x)$ eine in $[a,b]$ integrierbare Gewichtsfunktion mit einerlei Vorzeichen in $[a,b]$. Ohne Beschränkung der Allgemeinheit kann $W(x) \geq 0$ angenommen werden. Sei ferner $W(x) \not\equiv 0$. Sei außerdem der HERMITE-Operator $H_n(f,x)$ in $[a,b]$ integrierbar. Dann gilt für integrierbares $f(x)$

$$\int_a^b W(x)f(x)dx = \int_a^b W(x)H_n(f,x)dx + R_n(f) ,$$

wobei $R_n(f)$ der Quadraturfehler ist. Man erhält

$$\int_a^b W(x)H_n(f,x)dx = \sum_{i=1}^{n} (A_i f(x_i) + B_i f'(x_i)) =: Q_n(f) .$$

Bei den gewöhnlichen GAUSS-Formeln gilt nun $B_i = 0$ für $i = 1(1)n$ gleichmäßig in n. In Analogie dazu wird definiert

<u>Definition:</u> $Q_n(f)$ *heißt allgemeine* GAUSS-*Quadratur genau dann, wenn* $B_i = 0$ *für* $i = 1(1)n$ *und jedes* $n \in \mathbb{N}$ *gilt.*

Daraus folgt: Es muß

$$B_i = \int_a^b W(x) \frac{h_i(x)}{h'_i(x_i)} \lambda_i^2(x)dx = 0 \qquad , i = 1(1)n$$

gelten. Dies gilt unter Berücksichtigung der Definition von $\lambda_i(x)$ genau dann, wenn eine der Beziehungen

$$\int_a^b W(x)\lambda_i(x)\omega_n(x)dx = 0 \qquad , i = 1(1)n$$

$$\int_a^b W(x) \frac{\omega_n^2(x)}{h_i(x)} dx = 0 \qquad , i = 1(1)n$$

erfüllt ist.

Die Gleichungen $B_i = 0$, $i = 1(1)n$, bilden ein nicht lineares Gleichungssystem zur Bestimmung der Stützstellen x_i, $i = 1(1)n$. Über die Lösbarkeit dieses Systems kann natürlich in dieser Allgemeinheit keine Aussage gemacht werden.

Für die Gewichte A_i, $i = 1(1)n$, gilt

<u>Satz 2 :</u> $Q_n(f)$ *ist ein positives Quadraturverfahren, d. h. es gilt* $A_i > 0$, $i = 1(1)n$, *für jedes* $n \in \mathbb{N}$.

Beweis : Zunächst gilt

$$A_i = \int_a^b W(x) \left(1-2\lambda_i'(x_i) \frac{h_i(x)}{h_i'(x_i)}\right) \lambda_i^2(x)dx \qquad , i = 1(1)n$$

$$A_i = \int_a^b W(x)\lambda_i^2(x)dx - 2\lambda_i'(x_i)B_i \qquad , i = 1(1)n$$

$$= \int_a^b W(x)\lambda_i^2(x)dx \quad \text{wegen} \quad B_i = 0 \qquad , i = 1(1)n \ .$$

Da $W(x) \not\equiv 0$ vorausgesetzt ist und $\lambda_i(x) \not\equiv 0$ ist, folgt

$$A_i > 0 \quad \text{für} \quad i = 1(1)n \quad \text{und} \quad n \in \mathbb{N} \ .$$

Darüber hinaus läßt sich mit den klassischen Methoden eine Restdarstellung geben:

Satz 3 : *Sei $\omega_n(x)$ quadratisch integrierbar in $[a,b]$. Unter den Voraussetzungen von Satz 1 existiert dann eine Zahl $\xi \in [a,b]$ so, daß gilt*

$$R_n(f) = \frac{f^{(2n)}(\xi) - H_n^{(2n)}(f,x)(x=\xi)}{[\omega_n^2(x)]^{(2n)}(x=\xi)} \int_a^b W(x)\omega_n^2(x)dx \ ,$$

wobei weiter vorauszusetzen ist, daß $[\omega_n^2(x)]^{(2n)}(x=\xi) \neq 0$ ist.

Durch spezielle Wahl der Funktionen $\omega_n(x)$ und $h_i(x)$, $i = 1(1)n$ lassen sich nun positive Interpolationsquadraturen ableiten. Es wird im Folgenden gezeigt, daß einige bekannte Quadraturverfahren als Spezialfälle in diesem allgemeinen Konzept positiver Interpolationsquadraturen enthalten sind.

3. Spezielle positive Interpolationsquadraturen

3.1 GAUSS-Quadraturen zur JACOBI-Belegung

Sei zunächst folgende Wahl getroffen:

$$W(x) \equiv 1$$

$$h_i(x) = x - x_i \qquad , i = 1(1)n$$

$$\omega_n(x) = \prod_{i=1}^{n} (x-x_i) \ .$$

Man erhält die *klassische* GAUSS-Quadratur mit

$$A_i = \frac{1}{[\omega_n'(x_i)]^2} \int_a^b W(x) \frac{\omega_n^2(x)}{(x-x_i)^2} dx \qquad , i = 1(1)n$$

$$B_i = 0$$

genau dann, wenn

$$\int_a^b W(x)\lambda_i(x)\omega_n(x)dx = 0 \ .$$

Da $\omega_n(x) \in \mathbb{P}_n$ und $\lambda_i(x) \in \mathbb{P}_{n-1}$, so folgt, daß $\omega_n(x)$ bis auf einen konstanten Faktor das n-te LEGENDRE-Polynom ist und x_i, $i = 1(1)n$, dessen n Nullstellen sind.

Da hier $H_n(f,x) \in \mathbb{P}_{2n-1}$ und $\omega_n^2(x) \in \mathbb{P}_{2n}$ mit dem führenden Koeffizienten 1 ist, so folgt die bekannte Restdarstellung

$$R_n(f) = \frac{f^{(2n)}(\xi)}{(2n)!} \int_a^b W(x)\omega_n^2(x)dx \ .$$

Sei nun $P_n^{(\alpha,\beta)}(x)$ das durch

$$(1-x)^\alpha(1+x)^\beta P_n^{(\alpha,\beta)}(x) = \frac{(-1)^n}{2^n n!} \frac{d^n}{dx^n} \left[(1-x)^{n+\alpha}(1+x)^{n+\beta}\right] ,$$

mit $\alpha,\beta > -1$, definierte n-te JACOBI-Polynom über dem Intervall $[-1,1]$, und x_k seien seine n Nullstellen. Verwendet man diese Nullstellen als Stützstellen eines Quadraturverfahrens

$$\int_{-1}^{1} f(x)dx = \sum_{i=1}^{n} A_i f(x_i) + E_n(f) ,$$

so erhebt sich die Frage nach der Positivität der Gewichte A_i, $i = 1(1)n$. In einer jüngst erschienenen Arbeit erweitert R. ASKEY (vgl. |1|) eigene Ergebnisse sowie bekannte Resultate von FEJÉR und SZEGÖ (vgl. |2|, |11|, |13|, |14|) und zeigt, daß für eine Reihe weiterer Werte von α und β die Positivität dieses Quadraturverfahrens gesichert ist. Andererseits zeigen Resultate von SZEGÖ (|17|) und LOCHER (|13|, |14|), daß für max $(\alpha,\beta) > 3/2$ das Quadraturverfahren divergiert. Vorausgesetzt ist dabei immer, daß gilt

$$E_n(x^i) = 0 \quad \text{für} \quad i = 0(1)n-1 \ .$$

Es stellt sich dann die Frage: Gibt es eine andere Möglichkeit zur Gewinnung von Gewichten a_i statt A_i, so daß stets $a_i > 0$ gilt? Diese Frage wird im Folgenden positiv beantwortet.

Es gilt der

Satz 4 : *Seien* x_i, $i = 1(1)n$ *die Nullstellen des* JACOBI-*Polynoms* $P_n^{\alpha,\beta}(x)$ *mit* $\alpha,\beta > -1$. *Dann gibt es Gewichte* a_i, *so daß für das Quadraturverfahren*

$$\int_{-1}^{1} f(x)dx = \sum_{i=1}^{n} a_i f(x_i) + e_n(f)$$

für jedes n:

$$a_i > 0 \ , \quad i = 1(1)n$$

gilt.

Beweis : Sei

$$G(x) = \sqrt{(1-x)^\alpha(1+x)^\beta} \ , \quad h_i(x) = x - x_i \ , \quad i = 1(1)n \ ,$$

$$\omega_n(x) = G(x) \prod_{i=1}^{n} (x-x_i) \quad \text{und}$$

$$\lambda_i(x) = \frac{\omega_n(x)}{\omega_n'(x_i)(x-x_i)} \quad , \ i = 1(1)n \ .$$

Dann gilt für die Gewichte b_i der Ableitungswerte in der Quadraturformel

$$b_i = \int_{-1}^{1} (x-x_i)\ \lambda_i^2(x)dx$$

$$= \int_{-1}^{1} \left[\frac{G(x)}{G(x_i)}\right]^2 \prod_{\substack{j=1\\ \neq i}}^{n} \left[\frac{x-x_j}{x_i-x_j}\right]^2 \cdot (x-x_i)dx$$

$$= \int_{-1}^{1} (1-x)^\alpha(1+x)^\beta P_n^{\alpha,\beta}(x) g_{n-1}(x)dx$$

$$= 0 \ .$$

$g_{n-1}(x)$ ist ein Polynom von (n-1)tem Grade. Die Stützstellen sind also die Nullstellen der o. a. JACOBI-Polynome. Die Positivität der Gewichte a_i, $i = 1(1)n$ folgt aus Satz 2.

Bemerkung 1):

Aus (vgl. |5|, |6|)

$$H_n(\lambda_i^2(x),x) \equiv \lambda_i^2(x) \qquad , i = 1(1)n$$

und

$$H_n(\lambda_i^2(x)(x-x_i),x) \equiv \lambda_i^2(x)(x-x_i) \qquad , i = 1(1)n$$

folgt, daß stets Linearkombinationen von $\lambda_k^2(x)$ und $\lambda_k^2(x)(x-x_k)$ mit

$$(1+x)^\alpha(1-x)^\beta x^i \equiv \sum_{k=1}^{n} (\gamma_k+\delta_k(x-x_k))\lambda_k^2(x) \qquad , i = 0(1)2n-1$$

existieren, d. h. es gilt

$$e_n((1+x)^\alpha(1-x)^\beta x^i) = 0 \quad \text{für} \quad i = 0(1)2n-1 .$$

Bemerkung 2):

Mit derselben Beweistechnik gelingt natürlich der Nachweis dafür, daß auch für die mit den LAGUERRE- und HERMITE-Polynomen verknüpften Quadratur-Verfahren, deren Stützstellen die Nullstellen eben dieser Polynome sind (entsprechende Änderung des Integrationsbereiches und der Gewichtsfunktion sei vorausgesetzt) positive Quadraturverfahren existieren.

Bemerkung 3):

Satz 3 folgt auch aus klassischen Sätzen über GAUSS-JACOBI-Quadraturen, z. B. aus dem in |3| S. 343 angegebenen Satz.

3.2 Das WILF'sche Quadraturverfahren

In |18| leitet WILF ein Quadraturverfahren

$$W_n(f) = \sum_{i=1}^{n} A_i f(x_i) + E_n(f)$$

aus der Forderung

$$T = \sum_{\nu=0}^{\infty} E_n^2(x^\nu) = \sum_{\nu=0}^{\infty} \left\{ \int_0^1 x^\nu dx - \sum_{i=1}^{n} A_i x_i^\nu \right\}^2 \overset{!}{=} \text{Min}$$

ab. T ist eine Funktion der Stützstellen x_i, $i = 1(1)n$ und der Gewichte A_i, $i = 1(1)n$. Für die Minimalität ist notwendig

$$\frac{\partial T}{\partial A_k} = 0 \ , \quad \frac{\partial T}{\partial x_k} = 0 \quad , \ k = 0(1)n \ .$$

Dies ist ein *gekoppeltes* nichtlineares Gleichungssystem für Gewichte *und* Stützstellen. Es hat sich gezeigt, daß dieses System numerisch äußerst schlecht konditioniert ist und daher WILF'sche Formeln mit nicht vorgegebenen Stützstellen nur bis etwa $n = 5$ zuverlässig bekannt sind.

Sei nun $f(z)$ holomorph für $|z| < 1$ und für $|z| = 1$ quadratisch integrierbar. Dann existieren für die WILF'schen Quadraturformeln ableitungsfreie Fehlerschranken

$$\|E_n(f)\| \leq \sigma_n \|f\| \ ,$$

wobei σ_n nicht von f abhängt und folgende Norm gewählt ist

$$\|f\|^2 = \frac{1}{2\pi} \int_0^{2\pi} f(z)\overline{f(z)} \, ds \ ,$$

wobei s die Bogenlänge des Einheitskreises ist. WILF gab mit Majorisierungsmethoden unter Annahme der Existenz eines Minimums von T einen Konvergenznachweis.

ECKHARDT bewies in |4| u. a. folgende Resultate mit Hilfe von HILBERT-Raum-Methoden:

1) $\lim_{n\to\infty} \sigma_n = 0$

2) $a_i > 0 \ , \ i = 1(1)n \ , \ n \in \mathbb{N}$

3) $0 < x_i < 1 \ , \ i = 1(1)n \ , \ x_i$ paarweise verschieden.

Dieses Quadraturverfahren, das völlig unabhängig von Interpolationstechniken entstand, erweist sich überraschend als Interpolationsquadratur vom GAUSS'schen Typ.

Es gilt

<u>Satz 5 :</u> $W_n(f)$ *ist eine positive Interpolationsquadratur vom* GAUSS'*schen Typ.*

Beweis : Mit $a = 0$, $b = 1$, $W(x) \equiv 1$ sowie

$$h_i(x) = x - x_i \ , \quad i = 1(1)n$$

$$\omega_n(x) = \prod_{i=1}^{n} \frac{x-x_i}{1-xx_i}$$

läßt sich nach Satz 2 eine positive Interpolationsquadratur vom GAUSS'schen Typ erzeugen. Es wird nun gezeigt, daß diese Quadratur mit der WILF-Quadratur identisch ist. Zunächst sind die 2n Funktionen

$$\frac{1}{1-xx_k} \ , \quad \frac{x-x_k}{(1-xx_k)^2} \ , \quad k = 1(1)n$$

Fixelemente des zugehörigen HERMITE-Operators (vgl. |5|, |9|), d. h. sie werden sämtlich vom Fehlerfunktional der Quadratur annulliert.

Es gilt somit

$$\int_0^1 \frac{dx}{1-xx_k} = \sum_{i=1}^{n} \frac{A_i}{1-x_i x_k} \qquad , \ i,k = 1(1)n$$

$$\int_0^1 \frac{x-x_k}{(1-xx_k)^2}\,dx = \sum_{i=1}^{n} \frac{A_i(x_i-x_k)}{(1-x_i x_k)^2} \ .$$

Wegen $x \in [0,1]$ lassen sich für $x_i \in (0,1)$, $i = 1(1)n$, alle Brüche in konvergente geometrische Reihen entwickeln zu

$$\sum_{\nu=0}^{\infty} x_k^{\nu} \left\{ \int_0^1 x^{\nu} dx - \sum_{i=1}^{n} A_i x_i^{\nu} \right\} = 0$$

$$, \ i = 1(1)n$$

$$\sum_{\nu=1}^{\infty} \nu x_k^{\nu-1} \left\{ \int_0^1 x^{\nu} dx - \sum_{i=1}^{n} A_i x_i^{\nu} \right\} = 0 \ .$$

Dies sind die notwendigen Bedingungen dafür, daß gilt

$$\sum_{\nu=0}^{\infty} \left\{ \int_0^1 x^{\nu} dx - \sum_{i=1}^{n} A_i x_i^{\nu} \right\}^2 = \text{Min} \ ,$$

womit der Ausgangspunkt der WILF'schen Quadraturformeln erreicht ist.

Darüber hinaus hat man auch für die *numerische* Bestimmung der Stützstellen und Gewichte von $W_n(f)$ Wesentliches gewonnen:

Zunächst ist das nichtlineare gekoppelte Gleichungssystem für Gewichte und Stützstellen entkoppelt, denn in $B_i = 0$, $i = 1(1)n$ hat man ein nichtlineares *Gleichungssystem für die Stützstellen allein.* Zur Lösung eines solchen Systems bieten sich Iterationsverfahren an. Dazu benötigt man aber Ausgangsnäherungen und - wie Proberechnungen zeigen - sehr gute Ausgangsnäherungen.

Solche Startwerte können aber ebenfalls gewonnen werden. Setzt man etwa

$$\omega_n(x,\varepsilon) = \prod_{i=1}^{n} \frac{x-x_i}{1-\varepsilon x x_i} ,$$

so erhält man für

$\varepsilon = 1$ die WILF-Quadratur

$\varepsilon = 0$ die gewöhnliche GAUSS-Quadratur .

Damit bietet sich ein Einbettungsverfahren an, indem man bei $\varepsilon = 0$ mit den wohlbekannten GAUSS'schen Stützstellen startet und ε von 0 nach 1 wachsen läßt. Man erhält somit eine Folge von Quadraturverfahren, die für $\varepsilon = 1$ gute Näherungen für die Schlußiteration im Gleichungssystem $B_i = 0$, $i = 1(1)n$, bieten. Die bisher durchgeführten numerischen Experimente sind sehr zufriedenstellend ausgefallen.

3.3 Die MARTENSEN'sche Quadraturformel

In |15| gibt MARTENSEN die Quadraturformel

$$\int_{-\infty}^{\infty} f(x)dx = h \sum_{j=-\infty}^{\infty} f(jh) + e_h(f)$$

an. Dabei ist vorausgesetzt, daß das Integral existiert und daß die Summe konvergiert. h ist hier die Schrittweite. Sei abkürzend

$$M_h(f) = h \sum_{j=-\infty}^{\infty} f(jh)$$

gesetzt.

Es seien nun folgende Voraussetzungen erfüllt:

$f(z)$ sei reell für $z \in \mathbb{R}$.

$f(z)$ sei holomorph im Streifen $-\infty < \mathrm{Re}(z) < \infty$, $0 \leq \mathrm{Im}(z) \leq \sigma$, $\sigma > 0$.

$\lim\limits_{\mathrm{Re}(z)\to\pm\infty} |f(z)| = 0$ gleichmäßig im ganzen Streifen.

$f(z)$ sei auf $\mathrm{Im}(z) = \sigma$ absolut integrabel.

Dann existiert die ableitungsfreie Fehlerabschätzung

$$|e_h(f)| \leq (\coth \frac{\pi\sigma}{h} - 1) \int_{-\infty+i\sigma}^{\infty+i\sigma} |f(z)|ds$$

(vgl. |15|) .

Auch dieses Quadraturverfahren paßt in den Rahmen der allgemeinen Quadraturen vom GAUSS-Typ. Es gilt

<u>Satz 6 :</u> $M_n(f)$ *ist eine Interpolationsquadratur vom* GAUSS'*schen Typ.*

Beweis : Die Wahl

$a = -\infty$, $b = \infty$, $W(x) \equiv 1$

$h_i(x) = x - x_i$, $i = -\infty\ (1)\ \infty$

$\omega_h(x) = \sin \frac{\pi x}{h}$

ermöglicht die Konstruktion einer positiven Interpolationsquadratur durch Integration des hier bemerkenswert einfachen HERMITE-Operators

$$H_h(f,x) = \frac{h^2}{\pi^2} \sum_{j=-\infty}^{\infty} \left\{ f(jh) + (x-jh)f'(jh) \right\} \frac{\sin^2 \frac{\pi x}{h}}{(x-jh)^2} .$$

Das Gleichungssystem $B_i = 0$, $i = -\infty(1)\infty$ lautet ausführlich

$$\int_{-\infty}^{\infty} \frac{\sin^2 \frac{\pi x}{h}}{(x-x_j)} dx = 0 , \quad j = -\infty(1)\infty .$$

Der Zähler des Integranden ist h-periodisch; damit folgt, daß das System für $x_j = jh$ erfüllt ist.

Die Gewichte berechnen sich zu

$$A_j = \frac{h^2}{\pi^2} \int_{-\infty}^{\infty} \frac{\sin^2 \frac{\pi x}{h}}{(x-jh)^2} dx$$

$$= \frac{h^2}{\pi^2} \int_{-\infty}^{\infty} \frac{\sin^2 \frac{\pi x}{h}}{x^2} dx$$

$$= \frac{h}{\pi} \int_{-\infty}^{\infty} \frac{\sin t}{t} dt \qquad \text{mit} \qquad t = \frac{\pi x}{h}$$

$$= h \; ,$$

womit die Quadratur $M_h(f)$ verifiziert ist.

4. Abschließende Bemerkungen

Mit dem hier dargelegten Konzept positiver Interpolationsquadraturen aus HERMITE'schen Interpolationsoperatoren hat man nicht nur die Möglichkeit, neue positive Quadraturen in praktisch beliebiger Vielfalt zu erzeugen, sondern man hat auch ein sehr effizientes Klassifikationsprinzip, das einige bekannte, scheinbar völlig heterogene Quadraturverfahren auf einen gemeinsamen Kern zurückführt. Hinzu kommt die numerisch nicht zu unterschätzende Vereinfachung durch Entkopplung des nicht linearen Gleichungssystems zur Bestimmung der Stützstellen und Gewichte sowie die Möglichkeit zur Anwendung von Einbettungsverfahren in besonderen Fällen.

Schließlich ist zu erwähnen, daß sich das Konzept dieser allgemeinen Quadraturformeln auch auf Formeltypen von RADAU und LOBATTO ausdehnen läßt, Formeln bei denen ein bzw. zwei Stützstellen fest vorgeschrieben sind. Die wesentliche Schwierigkeit besteht darin, diese zusätzlichen Informationen in den HERMITE-Operator einzubauen. Dies ist aber möglich, und man erhält zusätzliche hinreichende Positivitätsbedingungen, die in den bisher untersuchten Fällen besagen, daß die festen Stützstellen auf dem Rande des Integrationsintervalles zu liegen haben. Zur numerischen Bestimmung der Gewichte und Stützstellen kann man wiederum Einbettungsmethoden heranziehen, die als Ausgangswerte die Parameter der GAUSS-Typ-Formel mit derselben Stützstellenzahl verwenden.

Literatur

|1| ASKEY, R.: Positivity of COTES numbers for some JACOBI abscissas. NUMERISCHE MATHEMATIK 19 (1972) S. 46 - 48.

|2| ASKEY, R.; FITCH, J.: Positivity of the COTES numbers for some ultraspherical abscissas. SIAM J. NUMER. ANAL. 5 (1968) S. 199 - 201.

|3| DAVIS, P. J.: Interpolation and Approximation. BLAISDELL Publ. Comp. 1965.

|4| ECKHARDT, U.: Einige Eigenschaften WILF'scher Quadraturformeln. NUMERISCHE MATHEMATIK 12 (1968) S. 1 - 7.

|5| ENGELS, H.: Über allgemeine GAUSS'sche Quadraturen. COMPUTING (im Druck).

|6| ENGELS, H.: Über einige allgemeine lineare Interpolationsoperatoren und ihre Anwendung auf Quadratur und RICHARDSON-Extrapolation. Berichte der KFA Jülich: Jül-831-MA (1972).

|7| ENGELS, H.: RUNGE-KUTTA-Verfahren auf der Basis von Quadraturformeln. Sammelband über die Tagung 'Numerische Methoden bei Differentialgleichungen', 4. 6. - 10. 6. 1972 in OBERWOLFACH, Ltg.: J. ALBRECHT und L. COLLATZ. Verlag BIRKHÄUSER (im Druck).

|8| ENGELS, H.: Eine Verallgemeinerung von NEWTON-Interpolation und NEVILLE-AITKEN-Algorithmus und deren Anwendung auf die RICHARDSON-Extrapolation. COMPUTING (im Druck).

|9| ENGELS, H.: Über ein numerisches Differentiationsverfahren mit ableitungsfreien Fehlerschranken. ZAMM (im Druck).

|10| ENGELS, H.: Allgemeine interpolierende Splines vom Grade 3. COMPUTING (im Druck).

|11| FEJER, L.: Mechanische Quadraturen mit positiven COTES'schen Zahlen. MATH. ZEITSCHRIFT 37 (1933) S. 289 - 309.

|12| HÄMMERLIN, G.: Zur numerischen Integration periodischer Funktionen. ZAMM 39 (1959) S. 80 - 82.

|13| LOCHER, F.: Norm bounds of quadrature processes. (Manuskript) erscheint demnächst.

|14| LOCHER, F.: Positivität bei Quadraturformeln. Habilitationsschrift, TÜBINGEN 1971.

|15| MARTENSEN, E.: Zur numerischen Auswertung uneigentlicher Integrale. ZAMM 48 (1968) T83 - T85.

|16| SZEGÖ, G.: Asymptotische Entwicklungen der JACOBI'schen Polynome. Schriften Königsb. Gelehrten Ges. nat. wiss. Kl. 10 (1933) S. 35 - 112.

|17| SZEGÖ, G.: Orthogonal polynomials. Amer. Math. Soc. Colloq. Pub. 23, Providence, R. I. (1967).

|18| WILF, H. S.: Exactness conditions in numerical quadrature. NUMERISCHE MATHEMATIK 6 (1964) S. 315 - 319.

ZUR DISKRETISIERUNG VON EXTREMALPROBLEMEN

Von H. Esser

1. Bezeichnungen und Problemstellung

Sei $\{E_k\}_{k=1}^{\infty}$ eine Folge linearer normierter Räume mit Normen $\|\cdot\|_k$, und E ein linearer normierter Raum mit Norm $\|\cdot\|$. Ferner sei $\{R_k\}_{k=1}^{\infty}$ eine Folge linearer und beschränkter Operatoren, $R_k : E \to E_k$, mit $\lim_{k\to\infty} \|R_k x\|_k = \|x\|$, $x \in E$. Definiert man dann eine Abbildung R, die jedem $x \in E$ eine Klasse von (äquivalenten) Folgen in $\prod_{k=1}^{\infty} E_k$ zuordnet, durch $Rx = \{\{y_k\}_{k=1}^{\infty}, y_k \in E_k, \lim_{k\to\infty} \|R_k x - y_k\|_k = 0\}$, so ist durch das Tripel $(E, \prod_{k=1}^{\infty} E_k, R)$ eine diskrete Approximation $\mathcal{A}(E, \prod_{k=1}^{\infty} E_k; R)$ von E im Sinne von F. Stummel ([22],[23]) gegeben. Sei I eine unendliche Teilfolge von $\{1,2,\dots\}$. Die Folge $\{x_k\}_{k\in I}$, $x_k \in E_k$ $(k \in I)$, konvergiert diskret (s. [22]) gegen $x \in E$, falls $\lim_{k\to\infty, k\in I} \|R_k x - x_k\|_k = 0$, in Zeichen $x_k \to x$ $(k \in I)$.

Es seien X, X_k (k=1,2,...) nichtleere Teilmengen von E bzw. E_k (k=1,2,...), und $f : E \to \mathbb{R}$, $f_k : E_k \to \mathbb{R}$ (k=1,2,...) Funktionale. Wir betrachten die Aufgaben

$$(1.1) \qquad \inf_{x\in X} f(x) \qquad \text{und}$$

$$(1.1)_k \qquad \inf_{x\in X_k} f_k(x) \qquad (k=1,2,\dots),$$

und untersuchen das Problem: Wann gilt

$$(1.2) \qquad \lim_{k\to\infty} \inf_{x\in X_k} f_k(x) = \inf_{x\in X} f(x) \qquad ?$$

D.h. wir ersetzen das kontinuierliche Problem (1.1) durch eine Folge "diskreter" Probleme $(1.1)_k$ und untersuchen die Konvergenz der Extremalwerte.

Fragestellungen dieser Art sind mit Hilfe von mengenwertigen Abbildungen naturgemäß schon untersucht worden. Etwa in dem Buch von C. Berge ([1], 1959) wird folgendes Problem behandelt: Z,E topologische Räume und X eine Abbildung von Z in $\mathcal{P}(Y)$, f ein reellwertiges Funktional auf $Z \times Y$. Wann ist das Funktional $M : Z \to \mathbb{R}$ definiert durch

$$M(z) = \sup \{f(z,y) \ , \ y \in X(z)\} \quad \text{stetig?}$$

In diesem Zusammenhang sei auf Arbeiten von W.W. Hogan ([13], 1971), A.V. Fiacco ([10], 1971), J.P. Evans und F.J. Gould ([9], 1970), H.J. Greenberg und W.P. Pierskalla ([11], 1971) und J. Pirzl ([20],[21] 1971) verwiesen, wo mit ähnlichen Begriffsbildungen, wie sie Berge ([1]) verwendete, gearbeitet wird.

Anwendungen von Stetigkeitssätzen auf Approximations- und Kontrollprobleme sind in den Arbeiten von W. Krabs ([16],[17],[18],[19] [1]) und H. Esser ([8]) gegeben.

Die aufgeführten Arbeiten haben gemeinsam, daß das Problem (1.1) und entsprechende Probleme $(1.1)_k$ in demselben Raum definiert sind, was ja in vielen Anwendungen zunächst nicht vorliegt, und durch eine Einbettung erst erreicht wird. Daher schien es angebracht, die von F. Stummel ([22],[23]) entwickelte Theorie zu verwenden. Hierbei ergeben sich Zusammenhänge mit dem von W. Daniel ([7]) angegebenen Diskretisierungssatz für Extremalprobleme. Dort werden jedoch andere Voraussetzungen getroffen, insbesondere die Existenz von Fortsetzungs- und Restriktionsoperatoren. Der hier einfach zu beweisende Diskretisierungssatz ist als Arbeitsgrundlage zu verstehen; ein schwierigeres Problem ist es, in konkreten Beispielen die Voraussetzungen zu verifizieren. Dies soll in dieser Arbeit am Beispiel eines Kontrollproblems geschehen, dessen Kontrollfunktionen stetige Funktionen sind, die nicht in einer in $C[a,b]$ kompakten Menge variieren (hierzu s. [8]).

2. Ein Diskretisierungssatz

Im folgenden seien die infima in (1.1), $(1.1)_k$ $(k=1,2,\ldots)$ endlich, und wir setzen

(2.1) $$v = \inf_{x \in X} f(x)$$

$(2.1)_k$ $$v_k = \inf_{x \in X_k} f_k(x) \qquad (k=1,2,\ldots)$$

Satz 2.1: Gegeben seien die Probleme (1.1) und $(1.1)_k$ $(k=1,2,\ldots)$; ferner seien folgende Voraussetzungen erfüllt:

A : Zu jedem $x \in X$ existiert eine Folge $\{x_k\}_{k=1}^{\infty}$, $x_k \in X_k$ $(k=1,2,\ldots)$, so daß

$$x_k \to x \quad .$$

[1] Hier findet man einen ausführlichen Literaturüberblick.

B : Zu jeder Folge $\{x_k\}_{k=1}^{\infty}$, $x_k \in X_k$ $(k=1,2,\dots)$, existiert eine Folge $\{y_k\}_{k=1}^{\infty}$, $y_k \in X$ $(k=1,2,\dots)$, so daß

$$\lim_{k\to\infty} \|R_k\, y_k - x_k\|_k = 0 \quad .$$

C : Aus $x_k \in X_k$ $(k=1,2,\dots)$, $x \in X$ und $x_k \to x$ folgt $\lim\limits_{k\to\infty} f_k(x_k) = f(x)$ [2].
(Diskrete Konvergenz der Funktionale)

D : Für die Folgen in B gilt

$$\lim_{k\to\infty} |f_k(x_k) - f(y_k)| = 0 \quad . \text{[3]}$$

Dann folgt

$$\lim_{k\to\infty} v_k = v \quad . \tag{2.2}$$

<u>Beweis:</u> Sei $\varepsilon > 0$ gegeben; dann existiert ein $x^\varepsilon \in X$, so daß $v_k - v \leq f_k(x) - f(x^\varepsilon) + \varepsilon$ für $x \in X_k$ $(k=1,2,\dots)$. Wegen Voraussetzung A existiert zu x^ε eine Folge $\{x_k\}_{k=1}^{\infty}$, $x_k \in X_k$ $(k=1,2,\dots)$, mit $x_k \to x^\varepsilon$. Daher folgt wegen Voraussetzung C

$$\overline{\lim_{k\to\infty}}\, (v_k - v) \leq 0 \quad . \tag{2.3}$$

Andererseits existiert zu beliebigem $\varepsilon > 0$ eine Folge $\{x_k^\varepsilon\}_{k=1}^{\infty}$, $x_k^\varepsilon \in X_k$ $(k=1,2,\dots)$, so daß

$$v - v_k \leq f(x) - f_k(x_k^\varepsilon) + \varepsilon \qquad \text{für alle } x \in X,\ k=1,2,\dots \quad .$$

Wegen der Voraussetzungen B und D ergibt sich daraus $\overline{\lim\limits_{k\to\infty}}\, (v-v_k) \leq 0$, was zusammen mit (2.3) (2.2) beweist.

Das Problem (1.1) heißt lösbar, falls ein $\hat{x} \in X$ existiert mit

$$\inf_{x \in X} f(x) = f(\hat{x}) \quad . \tag{2.4}$$

Jedes $\hat{x} \in X$ mit (2.4) heißt optimal oder Lösung.

Die Menge $\prod\limits_{k=1}^{\infty} X_k$ heißt diskret kompakt (s. [23]), falls jede Folge $\{x_k\}_{k=1}^{\infty}$, $x_k \in X_k$ $(k=1,2,\dots)$, eine Teilfolge enthält, die diskret gegen ein $x \in E$ konvergiert.

[2] Es reicht aus $\overline{\overline{\lim\limits_{k\to\infty}}}\, f_k(x_k) \leq f(x)$.

[3] $\overline{\lim\limits_{k\to\infty}}\, (f(y_k) - f_k(x_k)) \leq 0$ reicht aus.

Satz 2.2: Unter den Voraussetzungen von Satz 2.1 gilt:

(i) Sind die Probleme $(1.1)_k$ für k=1,2,... lösbar, und $\hat{x}_k$ (k=1,2,...) optimal mit $\hat{x}_k \to \hat{x}$, wobei $\hat{x} \in X$, dann ist $\hat{x}$ optimal (für (1.1)).

(ii) Ist E ein Banachraum, X kompakt und die Probleme $(1.1)_k$ für jedes k=1,2,... lösbar, dann existiert eine Teilfolge von Lösungen der Probleme $(1.1)_k$, $\{\hat{x}_{k_j}\}_{j=1}^{\infty}$, mit $\hat{x}_{k_j} \to \hat{x} \in X$, wobei $\hat{x}$ eine Lösung von (1.1) ist. Ist (1.1) eindeutig lösbar, dann konvergiert jede Folge von Lösungen der Probleme $(1.1)_k$ diskret gegen die Lösung von (1.1).

(iii) Existiert eine Folge linearer Operatoren $\{L_k\}_{k=1}^{\infty}$, $L_k : E_k \to E$ (k=1,2,...), die diskret gegen die Identität konvergieren [4], und gleichmäßig konsistent auf X sind, d.h. $\lim_{k\to\infty} \|L_k R_k x - x\| = 0$ glm. in X, und ist die Menge $\prod_{k=1}^{\infty} X_k$ diskret kompakt, sowie X abgeschlossen, dann enthält jede Folge von Lösungen der Probleme $(1.1)_k$ eine Teilfolge, die diskret gegen eine Lösung von (1.1) konvergiert.

(iv) Ist E ein Banachraum, und existiert ein $\hat{x} \in X$ mit

$$\delta(\varepsilon) = \inf_{\substack{x \in X \\ \|x-\hat{x}\| \geq \varepsilon}} \{f(x) - f(\hat{x})\} > 0 \text{ für jedes } \varepsilon > 0,$$

dann konvergiert jede Folge von Lösungen der Probleme $(1.1)_k$ diskret gegen die eindeutige Lösung $\hat{x}$ von (1.1).

Beweis:

(i): Sei $x \in X$, dann existiert nach Voraussetzung A von Satz 2.1 eine Folge $\{x_k\}_{k=1}^{\infty}$ mit $x_k \in X_k$ (k=1,2,...) und $x_k \to x$. Wegen

$$f_k(x_k) \geq f_k(\hat{x}_k) \quad (k=1,2,\ldots)$$

folgt hieraus für $k\to\infty$ nach Voraussetzung C $f(x) \geq f(\hat{x})$; d.h. $\hat{x}$ ist optimal.

(ii): Sei $\{\hat{x}_k\}_{k=1}^{\infty}$ eine Folge von Lösungen der Probleme $(1.1)_k$ (k=1,2,...). Dann existiert nach Voraussetzung B eine Folge $\{y_k\}_{k=1}^{\infty}$, $y_k \in X$ (k=1,2,...), mit

$$\lim_{k\to\infty} \|R_k y_k - \hat{x}_k\|_k = 0 \quad . \tag{2.5}$$

---

[4] d.h. (s. [22]) aus $x_k \to x \Rightarrow \|L_k x_k - x\| \to 0$.

Da X kompakt, existiert eine Teilfolge $\{y_{k_j}\}_{j=1}^{\infty}$ und ein $\hat{x} \in X$ mit $\lim_{j\to\infty} y_{k_j} = \hat{x}$. Für jedes $x \in E$ gilt nach Voraussetzung $\lim_{k\to\infty} \|R_k x\|_k = \|x\|$, daher folgt nach einem Lemma von Gelfand (s. z.B. [15], S. 206) die Existenz einer Konstanten $c > 0$, so daß

(2.6) $$\|R_k x\|_k \leq c\|x\| \qquad (k=1,2,\dots) \quad , \qquad x \in E \; .$$

Nun ist

$$\|\hat{x}_{k_j} - R_{k_j}\hat{x}\|_{k_j} \leq \|\hat{x}_{k_j} - R_{k_j}y_{k_j}\|_{k_j} + \|R_{k_j}y_{k_j} - R_{k_j}\hat{x}\|_{k_j} \quad ,$$

woraus sich mit (2.5) und (2.6) $\hat{x}_{k_j} \to \hat{x}$ ergibt.

Wir zeigen, daß $\hat{x}$ optimal ist. Nach Voraussetzung A existiert zu $\hat{x}$ eine Folge $\{x_k\}_{k=1}^{\infty}$, $x_k \in X_k$ $(k=1,2,\dots)$, mit $\lim_{k\to\infty} \|x_k - R_k\hat{x}\|_k = 0$. Setzt man

$$\check{x}_l = \begin{Bmatrix} \hat{x}_{k_j} & l=k_j \\ x_k & \text{sonst} \end{Bmatrix} \qquad l=1,2\dots \quad ,$$

so ist $\check{x}_l \in X_l$ $\quad l=1,2,\dots$ und $\lim_{l\to\infty} \|\check{x}_l - R_l\hat{x}\|_l = 0$.

Dann ergibt sich wegen Voraussetzung C

$$\lim_{l\to\infty} f_l(\check{x}_l) = f(\hat{x}) \; ; \quad \text{aber } f_{k_j}(\check{x}_{k_j}) = f_{k_j}(\hat{x}_{k_j}) = v_{k_j} \quad ,$$

woraus wegen (2.2) $f(\hat{x}) = v$ folgt; d.h. $\hat{x}$ ist optimal. Da jeder Häufungspunkt der Folge $\{y_k\}_{k=1}^{\infty}$ in (2.5) für (1.1) optimal ist, und (1.1) eindeutig lösbar, konvergiert $\{y_k\}_{k=1}^{\infty}$ gegen die Lösung $\hat{x}$ von (1.1), und damit $\{\hat{x}_k\}_{k=1}^{\infty}$ diskret gegen $\hat{x}$.

(iii): Sei $\{\hat{x}_k\}_{k=1}^{\infty}$ eine Folge von Lösungen der Probleme $(1.1)_k$ $(k=1,2,\dots)$. Wegen der diskreten Kompaktheit von $\prod_{k=1}^{\infty} X_k$ existiert eine Teilfolge $\{\hat{x}_{k_j}\}_{j=1}^{\infty}$ und ein $\hat{x} \in E$ mit $\hat{x}_{k_j} \to \hat{x}$. Wir zeigen, daß $\hat{x} \in X$; dann folgt wie bei dem Beweis von (ii) die Optimalität von $\hat{x}$. Nach Voraussetzung B existiert eine Folge $\{y_{k_j}\}_{j=1}^{\infty}$,

$y_{k_j} \in X$ $(j=1,2,\ldots)$, mit $\|R_{k_j} y_{k_j} - \hat{x}_{k_j}\|_{k_j} \to 0$ $(j \to \infty)$. Also gilt

$$\lim_{j\to\infty} \|R_{k_j} y_{k_j} - R_{k_j} \hat{x}\|_{k_j} \to 0 . \tag{2.7}$$

Es ist

$$\|y_{k_j} - \hat{x}\| \leq \|L_{k_j} R_{k_j} y_{k_j} - y_{k_j}\| + \|L_{k_j} R_{k_j} y_{k_j} - L_{k_j} R_{k_j} \hat{x}\| +$$
$$+ \|L_{k_j} R_{k_j} \hat{x} - \hat{x}\| = I_1^j + I_2^j + I_3^j .$$

Da L_k diskret gegen I konvergiert, gilt offensichtlich $\lim_{j\to\infty} I_3^j = 0$; und wegen der gleichmäßigen Beschränktheit der Normen $\|L_k\|$ (vergl. [22]) folgt nach (2.7) $\lim_{j\to\infty} I_2^j = 0$. Schließlich ergibt sich aus der gleichmäßigen Konsistenz auf X $\lim_{j\to\infty} I_1^j = 0$. Damit haben wir $\lim_{j\to\infty} y_{k_j} = \hat{x} \in X$, da X abgeschlossen.

(iv) Es ist klar, daß $\hat{x}$ optimal und die eindeutige Lösung von (1.1) ist. Sei $\{\hat{x}_k\}_{k=1}^{\infty}$ eine Folge von Lösungen der Probleme $(1.1)_k$ $(k=1,2,\ldots)$, dann existiert eine Folge $\{y_k\}_{k=1}^{\infty}$, $y_k \in X$ $(k=1,2,\ldots)$, mit $\lim_{k\to\infty} \|R_k y_k - \hat{x}_k\|_k = 0$. Es gilt

$$0 \leq f(y_k) - f(\hat{x}) \leq |f(y_k) - f_k(\hat{x}_k)| + |f_k(\hat{x}_k) - f(\hat{x})| =$$
$$= I_1^k + I_2^k \qquad (k=1,2,\ldots) .$$

Wegen Voraussetzung D ist $\lim_{k\to\infty} I_1^k = 0$, und wegen (2.2) gilt $\lim_{k\to\infty} I_2^k = 0$. Daher folgt aus der letzten Ungleichung aufgrund der Voraussetzung $\lim_{k\to\infty} y_k = \hat{x}$, woraus man wie bei dem Beweis von (iii) $\hat{x}_k \to \hat{x}$ schließt.

3. Ein Beispiel

Es sei $C[0,1]$ der Raum der auf $[0,1]$ stetigen Funktion versehen mit der Maximumnorm $\|\cdot\|_C$.
Für das weitere benötigen wir zwei Lemmata

<u>Lemma 3.1</u> Sei $f, g \in C[0,1]$ mit $\Delta(t) = \frac{1}{2}\{g(t) - f(t)\} \geq \frac{1}{c_o}, (c_o > 0)$, $t \in [0,1]$. Gegeben sei eine Funktion $u(t) \in C[0,1]$ mit $f(t) \leq u(t) \leq g(t)$, $t \in [0,1]$. Dann existiert eine Folge von Polynomen n-ten Grades, $\{q_n(t)\}_{n=1}^{\infty}$, so daß für genügend großes n

$$(3.1) \qquad f(t) \leq q_n(t) \leq g(t) \qquad (t \in [0,1]) \quad \text{und}$$

$$(3.2) \qquad \|u-q_n\|_C \leq \rho_n(u)\{1 + c_o(\|\tfrac{1}{2}(f+g)\|_C + \|u\|) + $$

$$+ \rho_n(\tfrac{1}{2}(f+g))c_o\{\|\tfrac{1}{2}(f+g)\|_C + \|u\|\}$$

gilt, wobei $\rho_n(\cdot)$ die beste Approximation durch algebraische Polynome n-ten Grades bzgl. $C[0,1]$ bezeichnet.

<u>Beweis:</u> Sei $\lambda_n = \min_{0 \leq t \leq 1} \{(\Delta(t) + \rho_n(u))^{-1}(\Delta(t) - \rho_n(h))\}$ mit $h = \frac{1}{2}(f+g)$.
Dann ist nach dem Satz von Weierstraß für genügend großes n $0 < \lambda_n < 1$, und es gilt $\lim_{n\to\infty} \lambda_n = 1$. Bezeichnet man mit $t_n^*(u)$, $t_n^*(h)$ die Polynome bester Approximation von u bzw. h, dann folgt

$$(1 - \lambda_n)t_n^*(h) + \lambda_n t_n^*(u) \leq (1 - \lambda_n)(h + \rho_n(h)) + \lambda_n(u + \rho_n(u))$$

$$= \lambda_n(u + \rho_n(u) - h) + h + (1 - \lambda_n)\rho_n(h)$$

$$\leq \lambda_n(g + \rho_n(u) - \tfrac{1}{2}(f+g)) + \tfrac{1}{2}(f+g) + (1 - \lambda_n)\rho_n(h)$$

$$= \lambda_n(\Delta + \rho_n(u)) + \tfrac{1}{2}(f+g) + (1 - \lambda_n)\rho_n(h)$$

$$\leq \Delta - \rho_n(h) + \tfrac{1}{2}(f+g) + (1 - \lambda_n)\rho_n(h)$$

$= g - \lambda_n\rho_n(h) \leq g$. Also ist für genügend großes n

$$(3.3) \qquad (1 - \lambda_n)t_n^*(h) + \lambda_n t_n^*(u) \leq g \qquad (t \in [0,1]).$$

Andererseits gilt

$$(1 - \lambda_n)t_n^*(h) + \lambda_n t_n^*(u) \geq (1 - \lambda_n)h + \lambda_n(u - \rho_n(u)) - (1 - \lambda_n)\rho_n(h)$$

$$\geq (1 - \lambda_n)h + \lambda_n(f - \rho_n(u)) - (1 - \lambda_n)\rho_n(h)$$

$$= h -(1 - \lambda_n)\rho_n(h) + \lambda_n(- \Delta - \rho_n(u))$$

$\geq h - (1 - \lambda_n)\rho_n(h) - (\Delta - \rho_n(h))$

$= f + \lambda_n\rho_n(h) \geq f$.

Setzt man daher $q_n(t) = (1 - \lambda_n)t_n^*(h) + \lambda_n t_n^*(u)$, dann folgt aus der letzten Ungleichung zusammen mit (3.3) (3.1), und

$$(3.4) \quad \|u-q_n\| \leq \lambda_n\|t_n^*(u)-u\|_C + (1-\lambda_n)\|t_n^*(h)-u\|_C \quad .$$

Man sieht leicht, daß für genügend großes n $0 \leq (1-\lambda_n) \leq c_o\{\rho_n(u) + \rho_n(h)\}$ gilt, woraus mit (3.4) die Behauptung (3.2) folgt.

<u>Lemma 3.2</u> Sei $T_k = \{t_{ik};\ t_{ik} = i\cdot k^{-1},\ i=0,1,\ldots k\}$ und $f,g \in C[0,1]$ mit $g(t) - f(t) > 0,\ \ t \in [0,1]$. Ferner sei $\{p_n(t)\}_{n=0}^{\infty}$ eine Folge von Polynomen n-ten Grades derart, daß für alle $n \geq N_o$

$$f(t) < p_n(t) < g(t) \qquad (t \in [0,1]) \quad \text{und}$$

$$g(t) - p_n(t) \geq c_o > 0,\quad p_n(t) - f(t) \geq c_1 > 0 \quad (t \in [0,1])$$ [5] gilt.

Ist dann $\{p_n^k(t)\}_{k=1}^{\infty}$ eine Folge von Polynomen n-ten Grades mit $f(t) \leq p_n^k(t) \leq g(t)$, $t \in T_k$, $k=1,2,\ldots$, dann existiert eine Folge von Polynomen n-ten Grades $\{q_n^k(t)\}_{k=1}^{\infty}$ und eine Konstante $A_o > 0$, so daß für $n \geq N_o$ und $k > n^2$ die Beziehungen

$$(3.5) \qquad f(t) \leq q_n^k(t) \leq g(t), \quad t \in [0,1], \qquad \text{und}$$

$$(3.6) \quad \max_{t \in T_k} |p_n^k(t) - q_n^k(t)| \leq A_o\ \{(n^2k^{-1})^2 + \max(\omega_2(g;k^{-1}),\ \omega_2(f;k^{-1}))\}$$

gültig sind. (ω_2 bezeichnet den zweiten Stetigkeitsmodul).

<u>Beweis:</u> Sei $p_n^k(t)$ mit $f(t) \leq p^k(t) \leq g(t)$, $t \in T_k$, und $L_k(p_n^k;\ t)$ die lineare Interpolation von p_n^k bzgl. T_k. Dann gilt bekanntlich $\|p_n^k - L_k(p_n^k)\|_C \leq \frac{1}{8}\ k^{-2}\|p_n^{k''}\|_C \leq \frac{1}{2}\ (n^2k^{-1})^2\|p_n^k\|_C$, wobei wir die Markov-Ungleichung benutzt haben. Daher folgt

$$\{1 - \tfrac{1}{2}\ (n^2k^{-1})^2\}\|p_n^k\|_C \leq \|L_k(p_n^k)\|_C = \max_{t \in T_k} |p_n^k(t)| \leq M$$

mit $M = \max\ \{\|f\|_C,\ \|g\|_C\}$. Mithin ist für $k > n^2$

[5] Dies ist für genügend große n stets erfüllbar.

(3.7) $$\|p_n^k\|_C \leq \{1-\tfrac{1}{2}(n^2k^{-1})^2\}^{-1}\cdot M \leq 2\cdot M \quad .$$

Für jedes $h \in C[0,1]$ ist $\|L_k h - h\|_C \leq a_o\omega_2(h;k^{-1})$ (a_o eine Konstante).
Wir setzen $d_{k,n} = (n^2k^{-1})^2 \cdot M$ und
$\varepsilon_{k,n} = d_{k,n} + a_o \max\{\omega_2(f;k^{-1}), \omega_2(g;k^{-1})\}$, dann gilt wegen (3.7)

(3.8) $$p_n^k \leq L_k p_n^k + d_{k,n} \leq L_k g + d_{k,n} \leq g + \varepsilon_{k,n} \quad , \quad \text{und}$$

(3.8)' $$p_n^k \geq L_k p_n^k - d_{k,n} \geq L_k f - d_{k,n} \geq f - \varepsilon_{k,n} \quad .$$

Definiert man

(3.9) $$\lambda_{k,n} = \min\{\min_{0\leq t\leq 1} (g(t) - p_n(t) + \varepsilon_{k,n})^{-1}(g(t) - p_n(t)) \ ,$$
$$\min_{0\leq t\leq 1} (p_n(t) - f(t) + \varepsilon_{k,n})^{-1}(p_n(t) - f(t))\},$$

und setzt

$$q_n^k(t) = (1-\lambda_{k,n})p_n(t) + \lambda_{k,n}p_n^k(t) \quad ,$$

dann folgt mit (3.8)', (3.8) und (3.9)

$$(1-\lambda_{k,n})p_n(t) + \lambda_{k,n}p_n^k(t) \geq (1-\lambda_{k,n})p_n(t) + \lambda_{k,n}f(t) - \lambda_{k,n}\varepsilon_{k,n}$$

$$= -\lambda_{k,n}(p_n(t) - f(t) + \varepsilon_{k,n}) + p_n(t)$$

$$\geq f(t) - p_n(t) + p_n(t) = f(t) \quad ,$$ und ähnlich $q_n^k(t) \leq g(t)$.

Damit ist (3.5) bewiesen. Wegen

$$\max_{t\in T_k} |p_n^k(t) - q_n^k(t)| \leq (1-\lambda_{k,n})\{\|p_n\|_C + \max_{t\in T_k} |p_n^k(t)|\}$$

folgt aus dem oben bewiesenen leicht (3.6).

Wir behandeln nun eine Diskretisierung des folgenden Kontrollproblems:
Sei $f_1, f_2 \in C[0,1]$ mit $f_2(t) - f_1(t) > 0$, $t \in [0,1]$, und

(3.10) $$X = \{x(t);\ x(t) \in C[0,1],\ f_1(t) \leq x(t) \leq f_2(t),\ t \in [0,1]\}.$$

Sei $M' = \max\{\|f_1\|_C, \|f_2\|\}$, die Funktion $F(t,y,x)$ stetig auf $[0,1]\times\mathbb{R}\times\{x;\ |x| \leq M'\}$ und genüge dort einer Lipschitzbedingung der

Form

$$(3.11) \qquad |F(t,y,x) - F(t,y',x')| \le L_1|y-y'| + L_2|x-x'|$$

für alle Paare (t,y,x), $(t,y',x') \in [0,1] \times \mathbb{R} \times \{x; |x| \le M'\}$. Dann besitzt bekanntlich das Anfangswertproblem

$$(3.12) \qquad \begin{cases} \dot{y}_x(t) = F(t,y_x(t), x(t)) , & 0 \le t \le 1 \\ y_x(0) = y_o \end{cases}$$

für jedes $x \in X$ genau eine Lösung $y_x(t)$. Sei ferner die Funktion $g(t,y,x)$ stetig auf $[0,1] \times \mathbb{R}_2$. Das Kontrollproblem lautet:

(3.13) Unter den Nebenbedingungen $x \in X$ und (3.12) ist das Funktional $f(x) = \int_0^1 g(t,y_x(t), x(t))d\alpha(t)$, wobei $\alpha(t)$ eine Funktion von beschränkter Variation ist, zu minimieren.

Die Diskretisierung solcher Probleme (mit Konvergenzbeweis für die infima) wird auch in den Arbeiten von J. Cullum ([5],[6], 1969, 1971) und Budak et al. ([2],[3],[4], 1969, 1970) behandelt. In [5],[6] ist die Differentialgleichung linear (mindestens in der Steuerung), und die Menge der Steuerungsfunktionen X besteht aus meßbaren Funktionen. Budak et al. ([2],[3],[4]) untersuchen den allgemeineren Fall. daß die Differentialgleichung nichtlinear ist, und die Menge der Steuerungsfunktionen aus meßbaren Funktionen besteht. Den Konvergenzsatz in [2] kann man auch mit Satz 2.1 beweisen. Der Fall stetiger Steuerungsfunktionen X wird von W. Krabs ([16],1972) und H. Esser ([8],1972) untersucht. In [16] ist X endlichdimensional, die Differentialgleichung linear und das Funktional konvex. In [8] wird im wesentlichen das Problem (3.13) für kompaktes X diskretisiert.

Der Übersicht wegen diskretisieren wir das Problem (3.13) in zwei Schritten:

1. Diskretisierung:
X in (3.13) wird durch

$$(3.14) \qquad X_n = \{p_n \in \mathcal{P}_n; \; f_1(t) \le p_n(t) \le f_2(t) \; , \quad t \in [0,1]\}$$

($\mathcal{P}_n$ = Menge aller algebraischen Polynome vom Grad $\le$ n) ersetzt. Dann erhalten wir das Problem

$$(3.15)_n \qquad \inf_{x \in X_n} f(x) \; ; \qquad v_n = \inf_{x \in X_n} f(x) \quad .$$

2. Diskretisierung:
Sei $T_k = \{t_{ik} ; \ t_{ik} = i \cdot k^{-1}, \quad i=0,1,\dots k\}$, $(k=1,2,\dots)$. Die Differentialgleichung in $(3.15)_n$ ersetzen wir durch ein konsistentes Runge-Kutta-Verfahren mit Verfahrensfunktion $F_k(t,y,x)$, und X_n in $(3.15)_n$ durch

$$(3.16) \qquad X_{n,k} = \{p_n^k \in \mathcal{P}_n ; \ f_1(t) \leq p_n^k(t) \leq f_2(t), \quad t \in T_k\} .$$

Das Integral in $(3.15)_n$ diskretisieren wir mit einem für jede auf $[0,1]$ stetige Funktion konvergenten Quadraturverfahren $\sum_{i=0}^{k} A_{i,k} h(t_{ik})$, $(h \in C[0,1])$. Dann erhalten wir das Problem

$$(3.17)_{n,k} \qquad \inf_{x \in X_{n,k}} f_{n,k}(x) \quad , \quad v_{n,k} = \inf_{x \in X_{n,k}} f_{n,k}(x),$$

mit $f_{n,k}(x) = \sum_{i=0}^{k} A_{i,k}\, g(t_{ik}, y_x^k(t_{ik}), x(t_{ik}))$, $(x \in X_{n,k})$, wobei $y_x^k(t)$, $(t \in T_k)$, die durch das Runge-Kutta-Verfahren definierte diskrete Lösung von (3.12) ist.
$(3.17)_{n,k}$ ist ein endlichdimensionales nichtlineares Optimierungsproblem.
<u>Satz 3.1:</u> Gegeben seien die Probleme (3.13) und $(3.15)_n$ $(n=1,2,\dots)$ $v = \inf_{x \in X} f(x)$ sei endlich. Dann gilt

$$(3.18) \qquad \lim_{n\to\infty} v_n = v \quad .$$

<u>Beweis:</u> Wir wenden Satz 2.1 an. Dazu setzen wir dort $E = E_n = C[0,1]$ versehen mit der Maximumnorm, $R_n = I$, $f = f_n$, $X = \{x;\ x \in C[0,1],\ f_1 \leq x \leq f_2\}$ und $X_n = \{p_n;\ p_n \in \mathcal{P}_n,\ f_1 \leq p_n \leq f_2\}$.
Wegen Lemma 3.1 ist die Bedingung A erfüllt. B ist trivialerweise gegeben, ebenso D. Zu zeigen bleibt daher C. Sei $x \in X$, $\|x_n - x\|_C \to 0$ $(x_n \in X_n \quad n=1,2,\dots)$, dann ist

$$|f(x)-f(x_n)| = \left|\int_0^1 \{g(t,y_x(t),x(t)) - g(t,y_{x_n}(t),x_n(t))\}d\alpha(t)\right| \leq$$

$$(3.19) \qquad \leq V_0^1\alpha^{6)} \, \|g(\cdot, y_x, x) - g(\cdot, y_{x_n}, x_n)\|_C \; .$$

---
6) $V_0^1\alpha$ bezeichnet die totale Variation von α.

Nach dem Lemma von Bellman und Gronwall existiert eine Konstante $K_o > 0$, so daß

$$(3.20) \qquad \| y_x - y_{x_n} \| \leq K_o \| x - x_n \| ,$$

woraus sich $\| y_{x_n} \| \leq \| y_{x^o} \| + K_o(\| x_n \| + \| x^o \|) \leq K_1$

ergibt, wobei x^o fest aus X ist. Daher folgt mit (3.19), (3.20) und der glm. Stetigkeit von g in einem genügend großen Kompaktum $\lim\limits_{n\to\infty} |f(x) - f(x_n)| = 0$.

Folgerung 3.1: Genügt die Funktion g(t,y,x) der Lipschitzbedingung

$$(3.21) \quad |g(t,y,x) - g(t,y',x')| \leq M_1(G)|y-y'| + M_2(G)|x-x'|$$

für alle Paare (t,y,x), (t,y',x') $\in [0,1] \times G$, wobei G kompakt in $\mathbb{R}_2$, und besitzt das Problem (3.13) eine Lösung $\hat{x} \in X$, dann gilt für genügend großes n

$$(3.22) \qquad 0 \leq v_n - v \leq A_1\{\rho_n(\hat{x}) + \rho_n(\tfrac{1}{2}(f_1 + f_2))\} .$$

Besitzt (3.13) keine Lösung in X, dann existiert zu jedem $\varepsilon > 0$ ein $\hat{x}_\varepsilon \in X$, so daß für genügend großes n

$$(3.23) \qquad 0 \leq v_n - v \leq A_1\{\rho_n(\hat{x}_\varepsilon) + \rho_n(\tfrac{1}{2}(f_1 + f_2)) + \varepsilon\}$$

gültig ist. (A_1 eine Konstante)

Beweis: Mit Lemma 3.1 folgen wie im Beweis von Satz 3.1 leicht (3.22) und (3.23).

Es ist klar, daß die Probleme $(3.15)_n$ und $(3.17)_{n,k}$ für genügend großes n lösbar sind (stetiges Funktional auf einer nichtleeren kompakten Menge).

Satz 3.2: Gegeben seien die Probleme $(3.15)_n$ und $(3.17)_{n,k}$. Dann gilt für genügend großes (aber festes) n:

(i) $\quad \lim\limits_{k\to\infty} v_{n,k} = v_n$

(ii) Es existiert eine Teilfolge $\{\hat{p}_n^{k_j}\}_{j=1}^{\infty}$ von Lösungen der Probleme $(3.17)_{n,k}$, die diskret gegen eine Lösung $\hat{p}_n$ des Problems $(3.15)_n$ konvergiert, d.h.

$$\lim_{k\to\infty} \max_{t\in T_{k_j}} |\hat{p}_n^{k_j}(t) - \hat{p}_n(t)| = 0 \quad , \quad \text{und es gilt}$$

$$\lim_{k\to\infty} \max_{t\in T_{k_j}} |y_{\hat{p}_n^{k_j}}^{k_j}(t) - y_{\hat{p}_n}(t)| = 0 \; .$$

Beweis: der Beweis verläuft mit Hilfe von Lemma 3.1, Lemma 3.2, Satz 2.1 und Satz 2.2 ähnlich wie in [8], Satz 3.1.

Für $h \in C(T_k)$ sei $\|h\|_k = \max_{t\in T_k} |h(t)|$, und die Abbildung $R_k : C[0,1] \to C(T_k)$ sei definiert durch $R_k f = f|_{T_k}$, $(f \in C[0,1])$. Schließlich sei mit Q_k das Fehlerfunktional des Quadraturverfahrens in $(3.17)_{n,k}$ bezeichnet.

Wir werden unter Differenzierbarkeitsvoraussetzungen für das Problem (3.13) eine Folge von diskreten Problemen (nämlich $(3.17)_{n,n^3}$) angeben, deren infima gegen das von (3.13) konvergieren. Besitzt (3.13) eine hinreichend glatte Lösung $\hat{x}$ (etwa $\hat{x} \in C^2[0,1]$), dann ist die Konvergenz von der Ordnung $O(n^{-2})$.

Lemma 3.3: Gegeben seien die Probleme $(3.15)_n$ und $(3.17)_{n,k}$. Die Funktion g genüge (3.21).
Dann gilt für genügend großes n (aber fest) und $k > n^2$

$$|v_{n,k} - v_n| \le A_2\{ \sup_{x\in X_n} |Q_k(g(\cdot, y_x(\cdot), x(\cdot))| + \sup_{x\in X_n} \|y_x - y_x^k\|_k + H_{n,k}\} \; , \tag{3.24}$$

wobei $H_{n,k} = \sup_{\bar{x}\in X_{n,k}} \inf_{x\in X_n} \|R_k x - \bar{x}\|_k$.

Beweis: Sei $\hat{x}_n$ eine Lösung von $(3.15)_n$, dann ist $v_{n,k} - v_n \le f_{n,k}(x) - f(\hat{x}_n)$ für jedes $x \in X_{n,k}$. Also gilt, da $R_k x \in X_{n,k}$, für $x \in X_n$

$$v_{n,k} - v_n \le \sup_{x\in X_n} |f_{n,k}(R_k x) - f(x)| \; . \tag{3.25}$$

Ebenso einfach folgt

$$v_n - v_{n,k} \leq \sup_{x \in X_n} |f_{n,k}(R_k x) - f(x)| + |f_{n,k}(R_k x) - f_{n,k}(\hat{x}_{n,k})|$$

für jedes $x \in X_n$, wobei $\hat{x}_{n,k}$ eine Lösung von $(3.17)_{n,k}$ ist. Daher gilt für jedes $x \in X_n$

$$(3.26) \qquad |v_{n,k} - v_n| \leq \sup_{x \in X_n} |f_{n,k}(R_k x) - f(x)| +$$

$$+ |f_{n,k}(R_k x) - f_{n,k}(\hat{x}_{n,k})|$$

$$= I_1 + I_2(x) \quad .$$

Es ist $|f(x) - f_{n,k}(R_k x)| \leq$

$$\leq \Big|\int_0^1 g(t,y_x(t),x(t))d\alpha(t) - \sum_{i=0}^{k} A_{ik}\, g(t_{ik}, y_x(t_{ik}), x(t_{ik}))\Big|$$

$$+ \sum_{i=0}^{k} |A_{ik}| \cdot |g(t_{ik}, y_x(t_{ik}), x(t_{ik})) - g(t_{ik}, y_x^k(t_{ik}), x(t_{ik}))|$$

$$= I_1^1 + I_1^2 \; .$$

$$I_1^1 \leq \sup_{x \in X_n} |Q_k(g(., y_x(.), x(.)))| \; .$$

Da die rechte Seite der Differentialgleichung in (3.13) einer Lipschitzbedingung genügt, genügt die ein Runge-Kutta-Verfahren definierende Funktion F_k in $(3.17)_{n,k}$ ebenfalls einer Lipschitzbedingung (s. [12]). Daraus erhält man mit der üblichen Schlußweise ein diskretes Lemma von Bellman und Gronwall:

$$(3.27) \qquad \| y_x^k - y_{x'}^k \|_k \leq a_1 \|x - x'\|_C \quad (k=1,2,\ldots,\ a_1 \text{ eine Konstante})$$

$x, x' \in C[0,1]$. Sei x^o fest aus $C[0,1]$, dann folgt mit (3.27) für jedes $x \in X_n$

$$\|y_x^k\|_k \leq \|y_{x^o}^k\|_k + a_1(\|x^o\| + \|x\|) \quad .$$

Da aber das Runge-Kutta-Verfahren diskret konvergiert, folgt

$$\sup_{x \in X_n} \|y_x^k\|_k \leq a_2 \quad k=1,2,\ldots \qquad n \geq N_o.$$

Daraus ergibt sich mit (3.21) unter Beachtung der Konvergenz des Quadraturverfahrens ($\sum_{j=0}^{k} |A_{jk}| \leq a_3$, k=1,2,...)

(3.28) $$I_1^2 \leq a_4 \sup_{x \in X_n} \|y_x - y_x^k\|_k,$$ woraus

(3.29) $$I_1 \leq \sup_{x \in X_n} |Q_k(g)| + a_4 \sup_{x \in X_n} \|y_x - y_x^k\|_k$$ folgt.

Mit (3.7) zeigt man ähnlich

(3.30) $$I_2(x) \leq a_5 \|R_k x - \hat{x}_{n,k}\|_k \quad .$$

Hieraus ergibt sich dann mit (3.29) die Behauptung.

Die ersten beiden Summanden auf der rechten Seite der Ungleichung (3.24) kann man im Prinzip durch geeignete Wahl des Runge-Kutta-Verfahrens und des Quadraturverfahrens unter genügend starken Differenzierbarkeitsbedingungen mit beliebiger Ordnung $O(k^{-\sigma})$ (n fest) approximieren. Aber für $H_{n,k}$ haben wir nach Lemma 3.2 nur $H_{n,k} = O((n^2k^{-1})^2)$, wenn $f_1, f_2 \in C^2[0,1]$; d.h. wir können nur $|v_{n,k} - v_n| = O_n(k^{-2})$ erreichen. Daher verwenden wir als Runge-Kutta-Verfahren in $(3.17)_{n,k}$ das verbesserte Euler-Cauchy-Verfahren und als Quadraturverfahren die zusammengesetzte Trapezregel. Koppelt man jetzt in $(3.17)_{n,k}$ k mit n durch $k = n^3$, dann gilt

<u>Satz 3.3:</u> Gegeben seien die Probleme (3.13) und $(3.17)_{n,n^3}$ (d.h. in $(3.17)_{n,k}$ ist $k = n^3$ gesetzt). Das Runge-Kutta-Verfahren in $(3.17)_{n,n^3}$ sei das verbesserte Euler-Cauchy-Verfahren definiert durch

$$y_x^h(t+h) = y_x^h(t)+h\ F(t+h/2,\ y_x^h(t) + h/2\ F(t,y_x^h(t),\ x(t)),\ x(t+h/2)),$$

$t \in T_h'$, $x \in X_{n,n^3}$, wobei $h = n^{-3} = k^{-1}$ und $T_h' = T_k - \{1\}$ gesetzt ist. Das Quadraturverfahren sei die zusammengesetzte Trapezregel, die man durch Integration der linearen Interpolation einer Funktion bzgl. $\alpha(t)$ erhält. Ferner seien zusätzlich folgende Differenzierbarkeitsbedingungen erfüllt: $F(t,y,x)$, $g(t,y,x) \in C^2([0,1] \times \mathbb{R}_2)$ und $f_1(t)$, $f_2(t) \in C^2[0,1]$. Besitzt (3.13) eine Lösung $\hat{x} \in X$, dann gilt

(3.31) $$|v - v_{n,n^3}| = O(\rho_n(\hat{x}) + n^{-2}) \qquad (n \to \infty) \quad .$$

Besitzt (3.13) keine Lösung in X, dann existiert zu jedem $\varepsilon > 0$ ein $\hat{x}_\varepsilon \in X$ mit

$$(3.32) \qquad |v - v_{n,n^p}| = O(\rho_n(\hat{x}_\varepsilon) + n^{-2} + \varepsilon) \qquad (n \to \infty) \quad .$$

<u>Beweis</u>: Wegen $\rho_n(1/2\ (f_1 + f_2)) = O(n^{-2})$ (Jackson-Satz) reicht es mit Folgerung 3.1 offensichtlich aus,

$$(3.33) \qquad |v_n - v_{n,n^p}| = O(n^{-2})$$

zu zeigen. Dazu schätzen wir für $k = n^3$ jeden Summanden in der Ungleichung (3.24) ab. Wegen (3.6) ist $H_{n,n^3} = O(n^{-2})$.

Für jedes $h(t) \in C^2[0,1]$ gilt bekanntlich $|Q_k(h)| \leq a_6 k^{-2} \|\ddot{h}(t)\|_C$. Daraus folgt nach etwas Rechnung die Existenz von positiven Konstanten a_7, a_8, a_9, a_{10}, so daß

$$(3.34) \qquad \left|\frac{d^2}{dt^2}\ g(t, y_x(t), x(t))\right| \leq a_7 + a_8 \|\dot{x}\|_C + a_9 \|\dot{x}^2\|_C + a_{10} \|\ddot{x}\|_C, \qquad t \in [0,1],\ x \in X_n.$$

Aus (3.34) ergibt sich aber mit der Markov-Ungleichung

$$(3.35) \qquad \sup_{x \in X_n} |Q_k(g(.,\, y_x(.),\, x(.)))| \leq A_3 k^{-2} \cdot n^4 = O(n^{-2}) \ ,$$

wobei A_3 eine Konstante ist.

$$(3.36) \qquad \sup_{x \in X_n} \|y_x - y_x^k\|_k$$

schätzen wir nach der üblichen Methode ab (s. [12], [14]); d.h. wir schätzen den Abbruchfehler $\tau_k(t)$ $(t \in T_k')$ ab, und erhalten daraus durch Rekursion eine Abschätzung für (3.36).

Für das verbesserte Euler-Cauchy-Verfahren rechnet man nach, daß positive Konstanten a_{11}, a_{12}, a_{13}, a_{14} existieren, so daß für $x \in X_n$

$$(3.37) \qquad |\tau_k(t)| \leq k^{-2}\{a_{11} + a_{12} \|\dot{x}\|_C + a_{13} \|\dot{x}^2\|_C + a_{14} \|\ddot{x}\|_C\}$$

gilt. Daraus erhält man wieder unter Beachtung der Markov Ungleichung

$$(3.38) \qquad \sup_{x \in X_n} |\tau_k(t)| \leq A_4 k^{-2} n^4 = O(n^{-2}) \quad ,$$

woraus sich $\sup_{x \in X_n} \|y_x - y_x^k\|_k = O(n^{-2})$, ergibt. Damit ist alles bewiesen.

Literatur

[1] Berge, C. : Topological Spaces. Oliver u. Boyd, Edinburgh, 1963.

[2] Budak, B.M. Berkovich, E.M. und E.N. Solov'eva : Difference Approximations in Optimal Control Problems. SIAM J. on Control 7 (1969), 18 - 31.

[3] ——— " ——— : The Convergence of Difference Approximations for Optimal Control Problems. USSR Comput. Math. and math. Phys. 9 (1969), 30 - 65.

[4] Budak, B.M. und A.I. Ivanov : Difference Approximations of Differential Games with Phase Constraints. Ž. Vyčisl. Mat. i. Mat. Fiz 10 (1970), 868 - 884.

[5] Cullum, J. : Discrete Approximations to Continuous Optimal Control Problems. SIAM J. on Control 7 (1969), 32 - 50.

[6] ——— " ——— : An explicit Procedure for Discretizing continuous Optimal Control Problems. JOTA 8 (1971), 15 - 34.

[7] Daniel, J.W. : The approximate Minimization of Functionals. Prentice Hall, Englewood Cliffs, N.J., 1971.

[8] Esser. H. : Über die Stetigkeit des Extremalwertes nichtkonvexer Optimierungsprobleme mit einer Anwendung auf die Diskretisierung von Kontrollproblemen. ZAMM 52 (1972), 535 - 542.

[9] Evans, J.P. und F.J. Gould : Stability in nonlinear Programming. Oper. Res. 18 (1970), 107 - 118.

[10] Fiacco, A. V. : Convergence Proporties of local Solutions of Sequences of Mathematical Programming Problems in general Spaces.
The George Washington University School of Engeneering and Applied Science. Institute for Management Science and Engeneering. Serial T-254, 1971.

[11] Greenberg, H.J. und W.P. Pierskalla : Extensions of the Evans-Gould Stability Theorems for Mathematical Programs.
Oper. Res. 19 (1971), 143 - 153.

[12] Grigorieff, R.D. : Numerik gewöhnlicher Differentialgleichungen.
Teubner Studienbücher, Teubner Verlag, Stuttgart 1972.

[13] Hogan, W.W. : Point to set maps in Mathematical Programming.
Western Management Science Institute, University of California, Los Angeles, Working Paper No. 170, 1971.

[14] Henrici, P. : Discrete Variable Methods in Ordinary Differential Equations.
John Wiley, New York 1962.

[15] Kantorowitsch, L.W. u. G.P. Akilov : Funktionalanalysis in normierten Räumen.
Akademie Verlag, Berlin 1964.

[16] Krabs, W. : Zur stetigen Abhängigkeit des Extremalwertes eines konvexen Optimierungsproblems von einer stetigen Änderung des Problems.
ZAMM 52 (1972), 359 - 368.

[17] Krabs, W. : Stetigkeitsfragen bei der Diskretisierung konvexer Optimierungsprobleme. Tagungsbericht Oberwolfach, erscheint demnächst in der Reihe ISNM, Birkhäuser Verlag, 1973.

[18] ——— " ——— : On Discretization in Generalized Rational Approximation. Erscheint in: Abh. Math. Sem. der Universität, Hamburg, 1973.

[19] ——— " ——— : Stabilität und Stetigkeit bei nichtlinearer Optimierung. Erscheint demnächst in: Methoden des Operation Research, Herausgeber R. Henn, 1973.

[20] Pirzl, J. : Optimierung unter Nebenbedingungen, Struktur einer Klasse von Algorithmen I. Computing 8 (1971), 121 - 142.

[21] ——— " ——— : Optimierung unter Nebenbedingungen, Struktur einer Klasse von Algorithmen II. Computing 8 (1971), 272 - 283.

[22] Stummel, F. : Diskrete Konvergenz linearer Operatoren I. Math. Anal. 190 (1970), 45 - 92.

[23] ——— " ——— : Discrete Convergence of Mappings. Erscheint demnächst in: Proceedings of the Conference on Numerical Analysis, Dublin, August 1972.

DIFFERENZENVERFAHREN ZUR LÖSUNG QUASILINEARER DIFFUSIONSGLEICHUNGEN IN ZYLINDERSYMMETRIE+

Von Karl Graf Finck von Finckenstein

1. Wir betrachten das folgende Anfangs-Randwertproblem, das sich aus Diffusionsvorgängen in zylindersymmetrisch angeordneten magnetisierten Plasmen ergibt:

$$
\begin{aligned}
\frac{\partial u(r,t)}{\partial t} &= \frac{1}{r}\cdot\frac{\partial}{\partial r}\left(r\cdot\varphi(u(r,t))\cdot\frac{\partial u(r,t)}{\partial r}\right), \quad \varphi(u)>0 \text{ für } u>0 \\
u(r,0) &= f(r)>0, \quad 0\le r\le R \\
u(R,t) &= g(t)>0, \quad 0\le t\le T
\end{aligned}
\tag{1}
$$

wobei φ, f, g hinreichend glatt vorausgesetzt seien.
Probleme dieser Art sind bisher nur vereinzelt behandelt worden; für den linearen Fall (φ = const.) liegen einige numerische Untersuchungen von Albasiny, Eisen, Kreiss, Mitchell und Pearce vor (vgl. [1], [3], [4], [5], [10], [11]). Für die speziellen quasilinearen Fälle $\varphi = u'$ und $\varphi = \frac{\partial u}{\partial r}$ werden in [6] und [7] Konvergenzbeweise für explizite Differenzenverfahren gebracht. Allerdings erhält man hier nur Konvergenzaussagen in einem möglicherweise gegenüber $[0,T]$ eingeschränkten Zeitintervall.

Die vorliegende Arbeit befaßt sich mit der Konvergenz von impliziten Differenzenmethoden, die zu (1) von 2. Ordnung konsistent in Δr sind. Zur Lösung der Differenzengleichungen bedarf es nur der Auflösung eines linearen Gleichungssystems mit tridiagonaler Koeffizientenmatrix. Dies bedeutet für die Praxis gewisse Vorteile, da solche Gleichungssysteme nicht iterativ gelöst zu werden brauchen. Die Konvergenzaussagen gelten in der Maximum-Norm sowie in dem ganzen betrachteten Zeitintervall $[0,T]$. Ferner erweist sich die Konvergenzordnung als genauso hoch, wie die Konsistenz des Verfahrens.

Alle Überlegungen lassen sich ohne weiteres auf den ebenen Fall:

+ "Diese Arbeit wurde im Rahmen des Assoziationsvertrages zwischen dem Max-Planck-Institut für Plasmaphysik und EURATOM durchgeführt".

$$\frac{\partial u}{\partial t} = \frac{\partial}{\partial x}\left(\varphi(u)\cdot\frac{\partial u}{\partial x}\right)$$

(jetzt mit zwei Randbedingungen) übertragen, allerdings erhält man hier dieselben Aussagen aus [9], wo allgemeinere Randwertprobleme 1. Art mit der Monotoniemethode behandelt werden.

2. Wir nehmen an, daß (1) in dem abgeschlossenen Bereich $G = \{(r,t): 0 \leq r \leq R,\ 0 \leq t \leq T\}$ eine hinreichend glatte Lösung $u(r,t)$ besitzt. "Hinreichend glatt" soll heißen: $u(r,t)$ ist so oft stetig differenzierbar, wie es für die folgenden Betrachtungen erforderlich ist. Man überlegt sich leicht, daß für (1) das Minimum-Maximum-Prinzip gilt, d.h. jeder Wert der Lösung u ist zwischen den Zahlen

$$c = \mathrm{Min}\{f(r), g(t) : (r,t) \in G\}, \quad d = \mathrm{Max}\{f(r), g(t) : (r,t) \in G\} \tag{2}$$

eingeschlossen. Ferner folgt aus (1) für kleine r und $0 \leq t \leq T$:

$$\frac{\partial^\nu u}{\partial r^\nu} = \mathcal{O}(r), \quad \nu > 0\text{, ungerade}. \tag{3}$$

Wir transformieren (1) auf folgende Weise: Sei $v = \psi(u) = \int_c^u \varphi(s)\,ds$.

$\psi(u)$ ist für $u \geq c$ streng monoton wachsend und nichtnegativ. Es existiert also die Umkehrfunktion $u = \tilde{\psi}(v) \geq c$.

Mit: $w = \chi(v) = \varphi(\tilde{\psi}(v))$, $\tilde{f}(r) = \int_c^{f(r)} \varphi(s)\,ds$, $\tilde{g}(t) = \int_c^{g(t)} \varphi(s)\,ds$ geht (1) über in:

$$\begin{aligned} \frac{\partial v}{\partial t} &= \chi(v)\cdot\frac{1}{r}\cdot\frac{\partial}{\partial r}\left(r\cdot\frac{\partial v}{\partial r}\right), \quad \chi(v) > 0 \text{ für } v > 0 \\ v(r,0) &= \tilde{f}(r) \geq 0, \quad 0 \leq r \leq R \\ v(R,t) &= \tilde{g}(t) \geq 0, \quad 0 \leq t \leq T. \end{aligned} \tag{1*}$$

Mit $\tilde{d} = \int_c^d \varphi(s)\,ds$ folgt sofort:

$$0 \leq v(r,t) \leq \tilde{d} \quad \text{für} \quad (r,t) \in G. \tag{4}$$

Ferner definieren wir:

$$C = \mathrm{Max}\{\chi(v),\ 0 \leq v \leq \tilde{d}\} \tag{5}$$

Wir versehen nun G mit einem rechteckigen Gitter mit den Gitterpunkten:

$$G_{hk} = \{((i+\tfrac{1}{2})\cdot h,\ j\cdot k)\ ;\ i = 0,\cdots,N\ ;\ j = 0,\cdots,M\}. \tag{6}$$

Dabei ist $(N+\frac{1}{2})\cdot h = R$, $M\cdot k = T$. Für alle in G bzw. G_{hk} auftretenden Funktionen F(r,t) schreiben wir zur Abkürzung: $F_{ij} = F((i+\frac{1}{2})\cdot h, j\cdot k)$.
Außerdem setzen wir $\lambda = k\cdot h^{-2}$. Mit $0 \leq \vartheta \leq 1$ folgert man aus (1*) durch Taylorentwicklung:

$$v_{ij+1} = v_{ij} + \lambda w_{ij}\{\vartheta\cdot[\frac{2i}{2i+1}v_{i-1j+1} - 2v_{ij+1} + \frac{2i+2}{2i+1}v_{i+1j+1}] + \\ +(1-\vartheta)\cdot[\frac{2i}{2i+1}v_{i-1j} - 2v_{ij} + \frac{2i+2}{2i+1}v_{i+1j}]\} + a_{ij}kh^2 + b_{ij}k^2 \tag{7}$$

$$i = 0, \cdots, N-1 \; ; \; j = 0, \cdots, M-1 .$$

Anmerkung: Für i = 0 sind die Terme v_{i-1j} nicht definiert. Das stört aber nicht, da die davorstehenden Faktoren ohnehin verschwinden.

Die a_{ij}, b_{ij} sind von höheren Ableitungen der Lösung $u(r,t)$ abhängige Gitterfunktionen. Wegen der vorausgesetzten Glattheit von u existieren positive Zahlen a, b mit:

$$\sup\{|a_{ij}| \; ; \; i=0,\cdots,N-1 \; ; \; j=0,\cdots,M-1 \; ; \; h,k>0\} = a \\ \sup\{|b_{ij}| \; ; \; i=0,\cdots,N-1 \; ; \; j=0,\cdots,M-1 ; h,k>0\} = b \tag{8}$$

3. Wir wollen das zu (7) gehörige Differenzenverfahren betrachten. Mit V_{ij} bezeichnen wir die Lösung der durch (7) definierten Differenzengleichungen. Sei $W_{ij} = \chi(V_{ij})$. Ferner definieren wir:

$$B = \begin{pmatrix} 2 & -2 & & & & 0 \\ -\frac{2}{3} & 2 & -\frac{4}{3} & & & \\ & \ddots & \ddots & \ddots & & \\ & -\frac{2i}{2i+1} & 2 & -\frac{2i+2}{2i+1} & & \\ & & \ddots & \ddots & \ddots & \\ & & & & & -\frac{2N-2}{2N-3} \\ 0 & & & & -\frac{2N-2}{2N-1} & 2 \end{pmatrix}, \quad \begin{aligned} \mathfrak{M}_j &= \mathrm{Diag}(W_{0j}, \cdots, W_{N-1j}) , \\ V_j &= (V_{0j}, \cdots, V_{N-1j})^T , \\ \tilde{V}_j &= \left(0, \cdots, 0, \frac{2\lambda N}{2N-1}\cdot W_{N-1j}[\vartheta V_{Nj+1} + (1-\vartheta)V_{Nj}]\right)^T . \end{aligned}$$

Mit diesen Bezeichnungen können wir die Differenzengleichungen schreiben in der Form:

$$(I + \vartheta\lambda\mathfrak{M}_j B)\cdot V_{j+1} = (I - (1-\vartheta)\lambda\mathfrak{M}_j B)\cdot V_j + \tilde{V}_j \tag{9}$$

für j = 0, ⋯, M-1.

Satz 1: Es gelte:

$$\lambda \leq \frac{1}{2(1-\vartheta)C} \tag{10}$$

Dann folgt für die in (9) auftretenden Vektoren V_j:

$$0 \leq V_{ij} \leq \tilde{d} \; ; \; i = 0, \cdots, N \; ; \; j = 0, \cdots, M \; . \tag{11}$$

Beweis: Wir führen Induktion über j. Für j = 0 ist (11) wegen (4) richtig, da für t = 0 exakte Lösung und Näherungslösung übereinstimmen. Ferner ist (11) auch für i = N und alle j richtig wegen $V_{Nj} = \tilde{g}(jk)$. Da $I + \vartheta\lambda \mathfrak{M}_j B$ eine M-Matrix ist, folgt: $A_j = (\alpha^{(j)}_{\nu\mu}) = (I + \vartheta\lambda \mathfrak{M}_j B)^{-1} \geq 0$. Addieren wir in der Matrixgleichung $A_j \cdot (I + \vartheta\lambda \mathfrak{M}_j B) = I$ die Elemente jeder Zeile, dann folgt:

$$\sum_{\mu=0}^{N-1} \alpha^{(j)}_{\nu\mu} + \frac{2N}{2N-1} \lambda \vartheta W_{N-1j} \, \alpha^{(j)}_{\nu N-1} = 1 \; ; \; \nu = 0, \cdots, N-1 . \tag{12}$$

Berechnet man aus (9) die ν-te Komponente von V_{j+1}, dann erhält man eine Gleichung der Form:

$$V_{\nu j+1} = \sum_{\mu=0}^{N} \beta^{(j)}_{\nu\mu} V_{\mu j} + \beta^{(j)}_{\nu N+1} V_{N j+1} \; .$$

Da wegen (10) die Matrix $I - (1-\vartheta)\lambda \mathfrak{M}_j B$ nichtnegativ ist, folgt $\beta^{(j)}_{\nu\mu} \geq 0$ für $\mu = 0, \cdots, N+1$. Ferner ergibt sich aus der Rechnung sofort:

$$\sum_{\mu=0}^{N+1} \beta^{(j)}_{\nu\mu} = \sum_{\mu=0}^{N-1} \alpha^{(j)}_{\nu\mu} + \frac{2N}{2N-1} \lambda \vartheta W_{N-1j} \, \alpha^{(j)}_{\nu N-1}$$

d.h., diese Summe ist 1 wegen (12). Es ist also $V_{\nu j+1}$ in der konvexen Hülle der Zahlen $V_{0j}, \cdots, V_{Nj}, V_{Nj+1}$ enthalten; da dies für $\nu = 0, \cdots, N-1$ gilt, folgt damit wegen der Induktionsvoraussetzung die Behauptung (11) für j + 1. q.e.d.

Wir beweisen jetzt die Konvergenz des Differenzenverfahrens (9). Es sei $z_{ij} = v_{ij} - V_{ij}$ der Fehler zwischen exakter Lösung und Näherungslösung. Ferner bezeichne $\|\cdot\|$ im folgenden die Maximumsnorm, bzw. bei Matrizen die Zeilensummennorm.

Satz 2: Es gelte die Bedingung (10). Dann existiert eine Konstante $K \geq 0$ mit:

$$\|z\| \leq K^{-1} (e^{KT} - 1) \cdot (ah^2 + bk) \; , \tag{13}$$

wobei $\| z \| = \underset{i,j}{Max} \{ |z_{ij}| \}$ ist. Die a, b sind in (8) erklärt.

Beweis: Zunächst erhält man durch Entwicklung:

$$\left.\begin{aligned} w_{ij} v_{\mu j} - W_{ij} V_{\mu j} &= W_{ij} z_{\mu j} + \tilde{W}'_{ij} v_{\mu j} z_{ij} \\ \text{bezw.: } w_{ij} v_{\mu j+1} - W_{ij} V_{\mu j+1} &= W_{ij} z_{\mu j+1} + \tilde{W}'_{ij} v_{\mu j+1} z_{ij} \end{aligned}\right\} \text{für } \mu = i-1, i, i+1 , \tag{14}$$

wobei $\tilde{W}'_{ij} = \frac{d\chi}{dv}(\tilde{V}_{ij})$ und $\tilde{V}_{ij}$ ein Zwischenwert zwischen v_{ij} und V_{ij} bezeichnet. Subtrahiert man für ein festes j die Gleichungen (9) von den Gleichungen (7), dann erhält man mit Hilfe von (14):

$$\left(I + \vartheta \lambda \mathcal{M}_j B \right) \cdot z_{j+1} = \left(I - (1-\vartheta)\lambda \mathcal{M}_j B + \tilde{D}_j \right) \cdot z_j + a_j k h^2 + b_j k^2 . \tag{15}$$

Dabei ist:

$$\tilde{D}_j = \lambda \cdot \mathcal{D}iag \left(\tilde{W}'_{ij} \cdot [\vartheta \cdot \delta^2 v_{ij+1} + (1-\vartheta) \delta^2 v_{ij}] \; ; \; i = 0, \cdots, N-1 \right)$$

und:

$$\delta^2 v_{i\mu} = \frac{2i}{2i+1} v_{i-1\mu} - 2 v_{i\mu} + \frac{2i+2}{2i+1} v_{i+1\mu} \quad \text{für } \mu = j, j+1 .$$

Man folgert leicht, daß gilt:

$$\frac{1}{h^2} \cdot \delta^2 v_{ij} = \frac{1}{r} \cdot \frac{\partial}{\partial r}\left(r \cdot \frac{\partial v_{ij}}{\partial r} \right) + \mathcal{O}(h^2) .$$

Weiter folgt aus (3), daß $Max \{ |\frac{1}{r} \frac{\partial}{\partial r}(r \frac{\partial v(r,t)}{\partial r})| \; ; \; (r,t) \in G \}$ existiert.
Wir definieren jetzt:

$$K = Max \{ \frac{d\chi(v)}{dv} \; ; \; 0 \le v \le \tilde{d} \} \cdot Max \{ |\frac{1}{r} \frac{\partial}{\partial r}(r \cdot \frac{\partial v(r,t)}{\partial r})| \; ; \; (r,t) \in G \} . \tag{16}$$

Hieraus ergibt sich: $\| \tilde{D}_j \| \le K \cdot k$ für j = 0, ···, M-1, denn wegen (11) und des Minimum-Maximum-Prinzips bei (1*) sind auch die Zwischenwerte $\tilde{V}_{ij}$ zwischen 0 und $\tilde{d}$ eingeschlossen. Wegen (12) haben wir: $\| (I + \vartheta \lambda \mathcal{M}_j B)^{-1} \| \le 1$, und wegen (10) und (11) prüft man sofort nach: $\| I - (1-\vartheta)\lambda \mathcal{M}_j B \| = 1$, denn diese Matrix ist nichtnegativ, und zur Normbildung braucht man bloß die Zeilenelemente zu addieren; alle diese Summen ergeben 1. Wegen $\| z_0 \| = 0$ erhalten wir nun aus (15) das Differenzen-Anfangswertproblem:

$$\begin{aligned} \| z_{j+1} \| &\le (1 + K \cdot k) \cdot \| z_j \| + a k h^2 + b k^2 \; ; \; j = 0, \cdots, M-1 \\ \| z_0 \| &= 0 . \end{aligned} \tag{17}$$

Hieraus ergibt sich (13), wenn man $\| z_j \|$ durch die Lösung der zu (17) gehörigen gewöhnlichen Differentialgleichung abschätzt. q.e.d.

Literatur

[1] Albasiny, E.L.: On the numerical solution of a cylindrical heat conduction problem. Quart. Journ. Mech. Appl. Math. 13 (1960), 374-384

[2] Courant, R., E. Isaacson, M. Rees: On the solution of nonlinear hyperbolic differential equations by finite differences. Comm. Pure Appl. Math. 5 (1952), 243-255

[3] Eisen, D.: Stability and convergence of finite difference schemes with singular coefficients.
SIAM Journ. on Num. Analysis 3 (1966), 545-552

[4] Eisen, D.: On the numerical solution of $u_t = u_{rr} + 2r^{-1} u_r$.
Num. Math. 10 (1967), 397-409

[5] Eisen, D.: Consistency conditions for difference schemes with singular coefficients.
Math. of Comp. 22 (1968), 347-351

[6] v. Finckenstein, K., K.v. Hagenow: Konvergenz eines Differenzenverfahrens für quasilineare parabolische Anfangs-Randwertprobleme in Zylindersymmetrie.
Erscheint in Num. Math. 1973.

[7] v. Finckenstein, K., K.v. Hagenow: Über explizite Differenzenmethoden zur Lösung nichtlinearer Diffusionsgleichungen in Zylindersymmetrie.
Meth. u. Verf. der Mathem. Physik, B I Hochschulskripten 1972.

[8] Gorenflo, R.: Differenzenschemata monotoner Art für schwach gekoppelte Systeme parabolischer Differentialgleichungen.
Computing 8 (1971), 343-362

[9] Kolar, W.: Über allgemeine monotone Differenzenverfahren zur Lösung des ersten Randwertproblems bei parabolischen Differentialgleichungen.
Dissert. RWTH Aachen 1970; Bericht Jül.-672-MA der K.F.A. Jülich

[10] Kreiss, H.O.: On the numerical solution of the spherically symmetric diffusion equation.
Num. Math. 12 (1968), 223-225

[11] Mitchell, A.R., R.P. Pearce: Explicit difference methods for solving the cylindrical heat conduction equation.
Math. of Comp. 17 (1963), 426-432.

MORREY SPACE METHODS IN THE THEORY OF ELLIPTIC DIFFERENCE EQUATIONS

Von Jens Frehse

Introduction. Stability theorems and error estimates for the approximation of elliptic boundary value problems by the method of finite differences are usually obtained only by discrete maximum principles, if one desires results which hold with respect to the ***maximum norm up to the boundary of the basic domain*** (see GERSCHGORIN[13], COLLATZ [6], FORSYTHE-WASOW [8], GREENSPAN[14], BRAMBLE-HUBBARD-THOMÉE [2], CIARLET [4] and others). This method requires that the difference operators involved are "mean value operators", which excludes a lot of natural approximations of the differential operator or requires, in particular in non-linear cases, hard restrictions on the coefficients, see e.g. Mc ALLISTER[18], or STEPLEMAN[29]. Also *systems* of elliptic differential equations cannot be handled by maximum principle methods.

So, other devices have been developed. There are several methods to obtain maximum estimates for the solutions of elliptic difference equations or the error of the approximation; important examples follow:

(i) discrete analogues of SCHAUDER's a-priori-estimates for solutions of elliptic equations; this yields a-priori-estimates for the discrete Hölder norm of the solutions (resp. error functions) u_h and their second difference quotients $\nabla_h^2 u_h$ in the case of, say second order, elliptic difference equations. See THOMEE[33].

(ii) L^p-estimates for $\nabla_h^2 u_h$ using discrete CALDERON-ZYGMUND estimates of the discrete fundamental solution of the difference operator. See VOIGTMANN[35]. Then the maximum estimate for u_h follows by the discrete SOBOLEV lemma [26].

(iii) L^2-estimates for $\nabla_h^2 u_h$ by the discrete Fourier transform, see

During the preparation of the paper, the author was a guest of the Scuola Normale Superiore in Pisa, supported by the German Research Association (=Deutsche Forschungsgemeinschaft).

THOMEE-WESTERGREN [31] . This yields maximum estimates for u_h in the case of two independent variables. Estimating higher difference quotients by this method, one can treat the case of more than two variables, too.

(iv) L^2-estimates for $\nabla_h^2 u_h$ by a discrete analogue of the proof of the so-called second fundamental inequality for second-order elliptic operators. This was first done by NITSCHE-NITSCHE [24]; see also Mc ALLISTER [19] and THOMÉE [32]. Again, this yields maximum estimates in the case of two variables.
(v) L^2-estimates for the higher difference quotients $\nabla_h^m u_h$ using coercive bilinear forms; see STUMMEL [30]. The maximum estimates follow from the discrete SOBOLEV lemma.

(vi) A discrete analogue of STAMPACCHIA's truncation method [27] which yields maximum estimates, too. This was done by CIARLET and RAVIART in [5] for related problems (finite element method); it was found independently by the author [9] who used a discrete analogue of F. CONTI's refinement [7] of STAMPACCHIA's method.

(vii) Let us remark that we tried a discrete analogue of MOSER's technique [22], too. But there are the same difficulties as in (vi), see the foot note.

All these "non-maximum-principle methods" (except (vi) and (vii)) do not work up to the boundary of the basic domain, in general. (Besides, they are restricted to linear problems). Some of these methods would work in the case of cubes or half spaces as domain. (For the method (iii) rather general results in this direction can be found in GRIGORIEFF's paper [15].)

Thus, the contribution in this paper is of interest. We treat a natural difference approximation of the general non-linear elliptic differential equation in divergence form. In the case of two independent variables, we show that the maximum of the error behaves like the truncation error. This holds up to the boundary of the basic domain. Furthermore, stability theorems for the solutions of the difference equation and their first difference quotients with respect to the Hölder norm are established.

We did not publish this result, because we did not see whether this technique would give more general results than the discrete maximum principle. In fact, later it was shown by V. Thomeé that this is not the case.

The basic domain has only to satisfy a cone property, and the non-linear differential operator has to be uniformly elliptic. No assumptions concerning mean value properties of the difference operator are made. The result is also new in the linear case.

The method of the proof works in analogy to MORREY's regularity theory [20], [21] for non-linear elliptic systems, because we estimate a discrete analogue of the Morrey norm of the difference quotients of the error resp. the solution of the difference equation.

The method described here also gives an approach to obtain existence theorems for classical solutions of (variational) elliptic systems in two dimensions via the difference equations (avoiding Sobolev spaces).

Notations.

Ω = bounded open subset of the n-dimensional euclidean space R^n, n=2,3..

$L^p(\Omega)$ (shorter: L^p) = Lebesgue space on Ω with norm

$\|u\|_p = (\int|u|^p dx)^{1/p}$, $1 \leq p < \infty$, $\|u\|_\infty$ = ess sup$\{|u(x)|, x \in \Omega\}$.

$\int$ = integration over Ω in the sense of Lebesgue.

$H^{m,p}(\Omega)$ (shorter: $H^{m,p}$) = Sobolev space on Ω with norm

$\|u\|_{m,p} = \Sigma_k \|\nabla^k u\|_p$, (k=0,...,m). $H^m = H^{m,2}$.

$\nabla^k u$ = vector of generalized derivatives of u of order k.

∂_i = derivative with respect to the i-th argument. ∂_0 = identity.

$C(B)$ resp. $C(\bar{B})$, $C^\alpha(B)$ resp. $C^\alpha(\bar{B})$, $C^{m+\alpha}(B)$ = space of real functions on the open subset $B \subseteq R^n$ which are continuous in B resp. $\bar{B}$, Hölder continuous on interior domains of B resp. uniformly Hölder continuous on $\bar{B}$ with exponent $\alpha \in (0,1)$, m-times Hölder continuously differentiable in B.

$C_0^\infty(\Omega)$ = space of test functions on Ω.

$H_0^1(\Omega)$ = closure of $C_0^\infty(\Omega)$ in $H^1(\Omega)$.

The boundary value problem. Let F_i, i=0,...,n, be real valued functions on $\bar{\Omega} X R^{r(n+1)}$ for which some smoothness assumptions will be made. We consider the generalized DIRICHLET problem: *Find* $u \in H_0^1$ *such that*

(1) $\langle Tu,v\rangle := \Sigma_i \int F_i(x,u,\nabla u)\cdot\partial_i v \, dx = 0$ (i=0,...,n) *for all* $v \in H_0^1$.

If the functions F_i are the partial derivatives of a function F on $\bar{\Omega}XR^{n+1}$, then equation (1) is the Euler equation to the *variational problem : Find* $u \in H_o^1$ *which minimizes*

(2) $\quad I(u) := \int F(x,u,\nabla u)dx$ *on* H_o^1 .

The discrete boundary value problem. Notations.

Let $h>0$, $h\in R^1$. $R_h^n = \{x\in R^n | x=(m_1,\dots,m_n)h;\ m_i \text{ integers}\}$

Ω_h = set of all those grid points $x\in R_h^n$ such that the cube $[x-\vec{h},x+\vec{h}) \in \Omega$

$\vec{h} = (h,\dots,h)\in R^n$; e_i = unit vector of R^n, $i=1,\dots,n$.

$\bar{\Omega}_h = \Omega_h \cup \{x\pm he_i | x\in\Omega, i=1,\dots,n\} \cup \{x-he_i+he_k | x\in\Omega, i,k=1,\dots,n\}$

$\Omega_h' = \Omega_h \cup \{x | x+he_i \in \Omega_h \text{ for some } i=1,\dots,n\}$, $\partial\Omega_h = \bar{\Omega}_h - \Omega_h$.

Φ_h = set of real grid functions on R_h^n with finite support.

V_h = set of real grid functions on R_h^n with support in Ω_h .

$D_i^{\pm h}w(x) = \pm(w(x\pm he_i)-w(x))/h$, $\nabla_h w = (D_1^h w,\dots,D_n^h w)$. $D_o^h w = w$, $D_o^{-h} = -w$.

$\|u\|_{p,h} = (h^n \Sigma_x |u(x)|^p)^{1/p}$, $(x\in R_h^n)$, for $u\in\Phi_h$, $1\le p<\infty$.

$\|u\|_{p,h,W} = (h^n \Sigma_x |u(x)|^p)^{1/p}$, $(x\in W)$, for $W\subseteq R_h^n$ and $u\in\Phi_h$.

$\|u\|_{\infty,h}$ resp. $\|u\|_{\infty,h,W} = \max\{|u(x)| \,|\, x\in R_h^n$ resp. $x\in W\}$, $u\in\Phi_h$, $W\subseteq R_h^n$.

To each grid function $v\in\Phi_h$ we assign the stepfunction

$Jv(x) = v(y)$, $x\in[y,y-\vec{h})$, $y\in R_h^n$.

The *finite difference analogue* of (1): *Find a function* $u_h\in V_h$ *such that*

(3) $\quad h^n \Sigma_x \Sigma_i F_i(x,u_h(x),\nabla_h u_h(x))\cdot D_i^h v_h(x) = 0$ *for all* $v_h \in V_h$,
$(x\in\Omega_h'$, $i=0,\dots,n)$

(Note that supp $D_i^h v_h \subseteq \Omega_h'$). This is equivalent to the difference equation *Find* $u_h\in V_h$ *such that*

(4) $\quad -\Sigma_i D_i^{-h} F_i(\cdot,u_h,\nabla u_h)(x) = 0$ *for* $x\in\Omega_h$, $(i=0,\dots,n)$.

If the functions F_i are the partial derivatives of a function F on $\bar{\Omega}XR^{n+1}$, then equ. (3) or (4) is the discrete Euler equation of the *discrete variational problem : Find* $u_h\in V_h$ *which minimizes*

(5) $\quad I_h(u_h) = h^n \Sigma_x F(\cdot,u_h,\nabla u_h)(x)$, $(x\in\Omega_h')$, *on* V_h .

Near the boundary $\partial\Omega$, the approximations (4) and (5) are rather crude. Thus we shall allow that the functions F and F_i depend on h in order to obtain better approximations. For the following theorems, it is only important that there hold uniform coerciveness and growth conditions for F and F_i . One natural approximation of (5) would be: *Minimize*

(5') $\qquad I_h'(u_h) = h^n \Sigma_x F(\cdot, u_h, A_h \nabla_h u_h)$, $(x \in \Omega_h')$, *on* V_h .

where A_h is an nXn matrix depending on x ; the i-th component of $A_h \nabla_h u_h(x)$ is equal to $D_i^h u_h(x)$ if x and $x+he_i \in \Omega_h$; otherwise, for $x \in \partial\Omega_h$ or $x+he_i \in \partial\Omega_h$, A_h is defined in such a way that

$$|A_h \nabla_h u(x) - \nabla u(x)| = O(h) \quad \text{for } u \in C^2(\bar{\Omega}),\ x \in \Omega_h'$$

Then the discrete boundary value problem has the form: *Find* $u_h \in V_h$ *such that*

(3') $\qquad h^n \Sigma_x \Sigma_i F_i(\cdot, u_h, A_h \nabla_h u_h) \cdot (A_h \nabla_h v_h)_i(x) = 0$, $(x \in \Omega_h'$, $i = 0, \dots, n)$

for all $v_h \in V_h$.

It is not hard to put natural conditions on the coefficients F_i which assure the solvability of the difference equations (4), see e.g. [10]. Furthermore, it is possible to prove *qualitative convergence* results under approximately the same conditions which are used in order to prove the existence of weak solutions of (1). The main tools for proving existence theorems for (1) are the *direct methods* (I) and the theory of *monotone operators* (II). In the following we state the discrete analogues of these two methods because we shall refine them later with the aid of the MORREY space method. We restrict ourselves here to to the case that the non-linearity in the F_i is of linear growth and give simplified versions.

I. *Direct methods*. Let the function F in (2) satisfy the following conditions:

(i) *Continuity*. $\quad F \in C(\bar{\Omega} X R^{n+1})$

(ii) *Coerciveness and growth condition*. There exist constants K,K', and c>0 such that

$$c|\eta|^2 - K'(1+|u|) \le F(x,u,\eta) \le K(1+|u|^2+|\eta|^2) \ ,x \in \bar{\Omega}, u \in R^1, \eta \in R^n \ .$$

(iii) *Convexity*. $F(x,u,\eta)$ is convex in $\eta \in R^n$.

THEOREM 1 . *Let Λ be a null sequence of positive numbers* h . *Under the assumptions I,(i)-(iii), the discrete problems (5) have a solution* u_h , *and there is a subsequence $\Lambda' \subset \Lambda$ and a solution $u \in H_0^1$ of (2) such that* $Ju_h \to u$ *in* L^2 , $J\nabla_h u_h \to \nabla u$ *weakly in* L^2 , $(h \to 0,\ h \in \Lambda')$.

This theorem is derived with the aid of our generalized lower semi-continuity theorem of [11] and the technique described there. The essential difficulty is the proof of the relation $I(u) \le \underline{\lim}\, I_h(u_h)$, $(h \to 0)$. That is why we used in [11] a certain truncation technique, combined with the BANACH-SAKS theorem. This simplifies the proof compared to the usual techniques of proving lower semicontinuity, see e.g. [21], chap.4.

II. *Existence and convergence via monotone operators.* Let the functions F_i in (1) satisfy the following conditions:

(i) *Continuity.* $F_i \in C(\bar{\Omega} X R^{n+1})$, $i=0,\dots,n$.

(ii) *Growth condition.* There exists a constant K such that

$$|F_i(x,\zeta)| \le K(1+|\zeta|) ,\quad x \in \bar{\Omega} ,\quad \zeta \in R^{n+1} ,\quad i=0,\dots,n .$$

(iii) *Coerciveness condition.* There exist constants K' and c>0 such that

$$\Sigma_{i=0}^{n} F_i(x,u,\eta)\eta_i \ge c|\eta|^2 - K'(1+|u|) ,\quad x \in \bar{\Omega},\ u \in R^1,\ \eta=(\eta_1,\dots,\eta_n) \in R^n .$$

(iv) *Monotonicity.* $\Sigma_{i=1}^{n} [F_i(x,u,\xi)-F_i(x,u,\eta)](\xi_i-\eta_i) > 0$,
$x \in \bar{\Omega},\ u \in R^1,\ \xi=(\xi_1,\dots,\xi_n) \in R^n,\ \eta=(\eta_1,\dots,\eta_n) \in R^n$.

THEOREM 2 . *Let Λ be null sequence of positive numbers* h . *Under the assumptions II,(i)-(iv), the discrete problems (3) have a solution* u_h , *and there is a subsequence $\Lambda' \subset \Lambda$ and a solution $u \in H_0^1$ of (1) such that* $Ju_h \to u$ *in* L^2 , $J\nabla_h u_h \to \nabla u$ *weakly in* L^2 , *and in measure* $(h \to 0,\ h \in \Lambda')$.

For the proof, see e.g. [10]. Theorem 2 is essentially a discrete version of the theorem of LERAY-LIONS [17] concerning the existence of weak solutions of elliptic boundary problems. (Note our slight refinement of the theorem of LERAY-LIONS: we do not need any additional coerciveness condition for the principal part of the differential operator; see [12]).

The MORREY space method.

Let $Q_\rho = \{x=(x_1,\dots,x_n)\in R^n \mid 0\le x_i\le\rho,\ i=1,\dots,n\}$, $\mathcal{Q}_\rho = \{x+Q_\rho \mid x\in R^n\}$

$\mathcal{Q} = \bigcup_\rho \mathcal{Q}_\rho$, $(\rho\ge 0)$;

$Q_\rho^h = Q_\rho \cap R_h^n$, $\mathcal{Q}_\rho^h = \{x+Q_\rho^h \mid x\in R_h^n\}$, $\mathcal{Q}^h = \bigcup_\rho \mathcal{Q}_\rho^h$, $(\rho\ge 0)$.

Definition of the discrete MORREY norm $|||\cdot|||_{\alpha,h}$

Let u be a real grid function on R_h^n with finite support and $\alpha\in(0,n)$ be given. Then we define

$$|||u|||_{\alpha,h} = \sup\{(\rho+h)^{-\alpha}\|u\|_{2,h,W(\rho)} \mid W(\rho)\in\mathcal{Q}_\rho^h\ ,\rho\in R^1, \rho\ge 0\}$$

(In words: the supremum is taken over all cubes consisting of grid points $\in R_h^n$ with sides parallel to unit vectors e_i and with grid points as corners).

Definition of the discrete HÖLDER norm

$$[u]_{\alpha,h} = \sup\{|x-y|^{-\alpha}|u(x)-u(y)| \ \big|\ x,y\in R_h^n\ ,x\neq y\} .$$

These norms are analogues of the corresponding "continuous" norms

$$|||u|||_\alpha = \sup\{\rho^{-\alpha}(\int_{W(\rho)}|u(x)|^2dx)^{1/2} \mid W(\rho)\in\mathcal{Q}_\rho, \rho>0\}$$

$$[u]_\alpha = \sup\{|x-y|^{-\alpha}|u(x)-u(y)| \ \big|\ x,y\in R^n, x\neq y\}$$

The importance of the MORREY norm consists in the following

LEMMA 1 . *Let* $u_h\in\Phi_h$ *and* $n=2$. *Then, for every* $\alpha\in(0,1)$, *there exists a constant* K_α *such that* $[u_h]_{\alpha,h} \le K_\alpha|||\nabla_h u_h|||_{\alpha,h}$.

PROOF: Let

$$\tilde\omega_h(\xi) = \begin{cases} 1-\xi h^{-1}, & 0\le\xi\le 1\\ 1+\xi h^{-1}, & -1\le\xi<0\\ 0 & , |\xi|>1\end{cases}$$

and $\omega_h(x) = \Pi_{i=1}^n\tilde\omega_h(x_i)$. Then $\omega_h\in H^{1,\infty}$. For $u_h\in\Phi_h$, let

$$J^1u_h = \Sigma_t\omega_h(t-x)u_h(x) = (\omega_h * u_h)(x)\ , \ (t\in R_h^n) .$$

J^1u_h is called the "multilinear extension of u_h". Since $J^1u_h\in H^{1,\infty}$, we have by the continuous version of MORREY's lemma (see [21],3.5.2.)

$$[J^1u_h]_\alpha \le K'_\alpha|||\nabla J^1u_h|||_\alpha$$

Since $J^1u_h(x) = u_h(x)$ for $x\in R_h^n$, we may estimate $[u_h]_{\alpha,h} \le [J^1u_h]_\alpha$, (in fact, there holds the equality), and the lemma is proved if we show that $|||\nabla J^1 u_h|||_\alpha \leq C_\alpha |||\nabla_h u_h|||_{\alpha,h}$. This is done in the next lemma.

LEMMA 2 . *Let* $\alpha\in(0,\frac{n}{2})$. *Then there exists a constant* C_α *such that*

$$|||\nabla J^1 u_h|||_\alpha \leq C_\alpha |||\nabla_h u_h|||_{\alpha,h} \quad .$$

PROOF: From the definition of J^1 we conclude

$\partial_i J^1 u_h(x) = \omega_h^{(i)} * D_i^h u_h(x) = \Sigma_t \omega_h^{(i)}(t-x) D_i^h u_h(t)$, $(t\in R_h^n)$, where $\omega_h^{(i)}(x) = \omega_h^0(x_i)\Pi_k \bar{\omega}_h(x_k)$, $(k=1,\dots,n;\ k\neq i)$, $\omega_h^0(\xi)=1$ for $0\leq\xi\leq h$ and $=0$ else.- We have

$$\int_{Q_\rho}|\partial_i J^1 u_h|dx = \int_{Q_\rho}|\Sigma_t \omega_h^{(i)}(t-x)D_i^h u_h(t)|dx \leq \Sigma_t |\int_{Q_\rho}\omega_h^{(i)}(t-x)dx|\,|D_i^h u_h(t)|, (t\in R_h^n)$$

The summation Σ_t must be extended only over those $t\in R_h^n$ which have the property $\omega_h^{(i)}(t-x)\neq 0$ for some $x\in Q_\rho$. These t are contained in a cube $W\in\mathcal{Q}^h$ with side-length smaller than $\rho+2h$. Thus

$$\int_{Q_\rho}|\partial_i J^1 u_h|dx \leq h^n \Sigma_t |D_i^h u_h(x)|dx \quad ,(t\in W)$$

and similarly

$$\text{ess sup}\{|\partial_i J^1 u_h|(x)|x\in Q_\rho\} \leq \max\{|D_i^h u_h(x)|, x\in W\} \quad .$$

By the Riesz convexity theorem, we derive from the last two inequalities that
$$\int_{Q_\rho}|\partial_i J^1 u_h|^2 dx \leq h^n \Sigma_y |D_i^h u_h(y)|^2 \quad ,(y\in W) \quad .$$

If $\rho\geq h$, then $\rho+2h\leq 3\rho$ with equality in the worst case $\rho=h$. Thus

$$(6)\quad \rho^{-\alpha}(\int_{Q_\rho}|\nabla J^1 u_h|^2 dx)^{1/2} \leq 3^\alpha(\rho+2h)^{-\alpha}(h^n \Sigma_y |\nabla_h u_h(y)|^2),\ (y\in W)$$
$$\leq 3^\alpha |||\nabla_h u_h|||_{\alpha,h}$$

since the side-length of W is smaller than $\rho+2h$.

If $0<\rho\leq h$, then

$$\int_{Q_\rho}|\partial_i J^1 u_h|^2 dx = \int_{Q_\rho}|\Sigma_t \omega_h^{(i)}(t-x)D_i^h u_h(x)|^2 dx \leq \max\{|D_i^h u_h(y)|^2\big|\, y\in W\}\cdot$$
$$\cdot\int|\Sigma_t \omega_h^{(i)}(t-x)|^2 dx = \rho^n \max\{|D_i^h u_h(y)|^2|y\in W\} \quad ,\ (t\in W) \quad .$$

The maximum is attained in some $y_0\in W$. Thus

$$\rho^{-\alpha}(\int_{Q_\rho}|\nabla J^1 u_h|^2 dx)^{1/2} \leq \sqrt{n}(\rho/h)^{n/2-\alpha} h^{-\alpha}(h^n|\nabla_h u_h(y_0)|^2)^{1/2} \leq \sqrt{n}|||\nabla_h u_h|||_{\alpha,h}$$

By translation we obtain (6) and the last inequality for all $U\in\mathcal{Q}_\rho$, and we arrive at the desired inequality

$$|||\nabla J^1 u_h|||_\alpha \leq C_\alpha |||\nabla_h u_h|||_{\alpha,h}$$

with $C_\alpha=\max\{\sqrt{n},3^\alpha\}$. (This constant is not the best one).

Thus, in order to achieve maximum and Hölder estimates in the case of two dimensions, one can try to estimate the MORREY norm of the error and the solutions of (3). In fact, this is possible as it will be shown in the following.- For the boundary $\partial\Omega$ of Ω we shall sometimes assume the following *regularity condition:*

(7) There exists a finite covering O_i of $\partial\Omega$ by open sets O_i and corresponding open cones C_i such that, for $x\in O_i\cap\partial\Omega$, $(x+C_i)\cap O_i\cap\Omega=\emptyset$. The cones C_i have an angle which is larger then 90^0 .

For the functions F_i in (3) we shall assume the coerciveness condition

(8) $\Sigma_{i=0}^n F_i(x,\zeta)\zeta_i \geq c\Sigma_{i=1}^n \zeta_i^2 - K'(1+\zeta_0^2)$, $x\in\bar{\Omega}_h$, $\zeta=(\zeta_0,\ldots,\zeta_n)\in R^{n+1}$

for some constants $K',c>0$.

THEOREM 3 . *Let the coefficients* F_i *in* (3) *satisfy the growth condition* II,(*ii*), *of theorem* 2 *and the coerciveness condition* (8). *Assume the regularity condition* (7) *for* $\partial\Omega$ *and let* $p>2$ *be given . Then there exist constants* C *and* $\alpha\in(0,1)$ *not depending on* h *such that for every solution* u_h *of* (3) *the following inequality holds*

$$|||\nabla_h u_h|||_{\alpha,h} \leq C(\|u_h\|_{p,h} + 1) .$$

PROOF: Let $W\in\mathcal{Q}_\rho^h$, $W = x_0+Q_\rho^h$, $x_0\in R_h^n$, $W' = x_0 + Q_{\rho+h}^h$, $W''=x_0-(\overrightarrow{\rho+h})+Q_{3\rho+3h}^h$

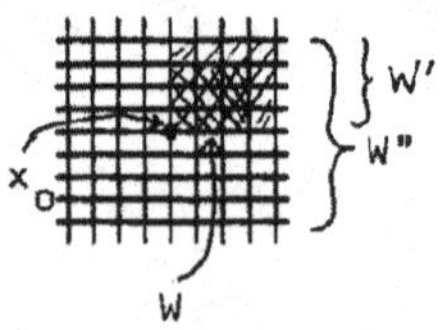

where $(\overrightarrow{\rho+h})$ is the n-dimensional vector with equal components $\rho+h$.- Let $\tau\in\Phi_h$ such that $\tau=1$ on W' , $\tau=0$ on R_h^n-W'' and $\partial W''$, (here $\partial W''$ = $= R_h^n\cap\partial(\text{convex hull of } W''))$, and $|D_i^h\tau| \leq(\rho+h)^{-1}$.

Such a function τ can easily be constructed. Note that

(9) $$D_i^h\tau = 0 \text{ on } W .$$

We set $\phi=\tau(u_h-b)$ where the constant b will be defined later, b=0 if $W''\not\subset\Omega_h\cup\partial\Omega_h$. Then $\phi\in V_h$, and we may insert $v_h=\phi$ into the difference equation. Applying the discrete Leibniz' rule

$$D_i^h[\tau(u_h-b)] = \tau D_i^h u_h + (E_i^h u_h-b)D_i^h\tau \quad , E_i^h u_h(x)=u_h(x+he_i),$$

we obtain

$$h^n\Sigma_x\Sigma_i[\tau F_i D_i^h u_h](x) \leq h^n\Sigma_x\Sigma_i|F_i||E_i^h u_h-b||D_i^h\tau|(x) \leq h^n\Sigma_x\Sigma_i(\tfrac{1}{\varepsilon}|F_i|^2 +$$

$+ \frac{\varepsilon}{4}|D_i^h\tau|^2|E_i^h u_h - b|^2)(x)$, ($x\in\Omega_h'$, $i=0,\dots,n$, $i=1,\dots,n$); E_0^h= identity, with the abbreviation $F_i = F_i(x,u_h(x),\nabla_h u_h(x))$.- Now the left hand side of the last inequality may be estimated by the coerciveness condition (8), the right hand side by the growth condition II,(ii) . Using the facts that $D_i^h\tau = 0$ on $W\cup(R_h^n - W")$, $\tau=1$ on W, and $|D_i^h\tau| \leq (\rho+h)^{-1}$, we obtain

$$(10)\quad c\,\|\nabla_h u_h\|^2_{2,h,W} - K'\|u_h\|^2_{2,h,W"} - K'(3\rho+3h)^2 \leq$$
$$\leq \varepsilon^{-1}K"(\|\nabla_h u_h\|^2_{2,h,W"-W} + \|u_h\|^2_{2,h,W"} + (3\rho+3h)^2) +$$
$$+ \frac{\varepsilon}{4}(\rho+h)^{-2}\Sigma_i\|E_i^h u_h - b\|^2_{2,h,W"-W} \quad ,(i=0,\dots,n).$$

Let W" be the cube of grid points which is concentric to W and has side length $3(\rho+h)+2h$,i.e., $W"' = x_0 - (\overrightarrow{\rho+2h}) + Q^h_{3\rho+5h}$. Then

$$(11)\qquad \Sigma_i\|E_i^h u_h - b\|^2_{2,h,W"-W} \leq (n+1)\|u_h - b\|^2_{2,h,W"'} \ , (i=0,\dots,n).$$

Setting $b=\Sigma_x u_h/\Sigma_x 1$,$(x\in W"')$, if $W"\subset \Omega_h\cup\partial\Omega_h$, we have the inhomogeneous POINCARE inequality (lemma 3)

$$\|u_h - b\|^2_{2,h,W"'} \leq \frac{1}{4}(3\rho+5h)^2\|\nabla_h u_h\|^2_{2,h,W"'}$$

Applying this to (10) and (11) we obtain

$$(12)\ \|\nabla_h u_h\|^2_{2,h,W} \leq \text{GOOD TERMS} + \varepsilon^{-1}K_1\|\nabla_h u_h\|^2_{2,h,W"-W} + \varepsilon K_0\|\nabla_h u_h\|^2_{2,h,W"'}$$

where $\text{GOOD TERMS} = K_2\left[(1+\varepsilon^{-1})\|u_h\|^2_{2,h,W"} + (3\rho+3h)^2\right]$.

If $W"\not\subset\Omega_h\cup\partial\Omega_h$, we had to set b=0 . In this case, by the cone property (7) of $\partial\Omega$ there is an $i\in\{1,\dots,n\}$ and a constant $\gamma\geq 1$ depending only on Ω such that the discrete cuboid $P = \{W"'\pm\lambda e_i\ \ 0\leq\lambda\leq\gamma(\rho+h)\}\cap R_h^n$ has one discrete hyper-surface H lying entirely outside of $\bar{\Omega}_h$. Here we

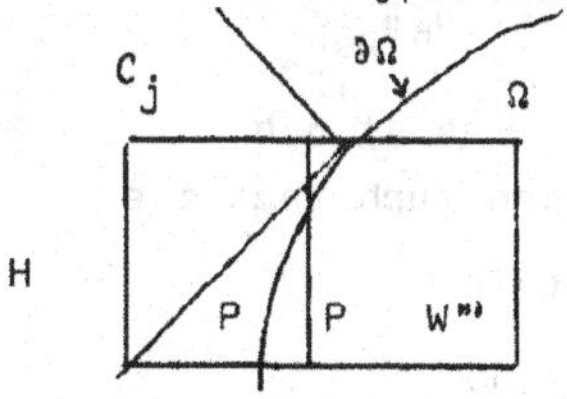

used the hypothesis that C_j has an angle which is larger then 90^0 .

For $x\in W"'$, there is an $x_0\in H$ such that $u_h(x) = h\Sigma_m D_i^h u_h(x_0+mhe_i)$, $(m=0,\dots,M)$, $x_0+Mhe_i=x$ .This is true because $u_h=0$ on H,

By HÖLDER's inequality and by summation over W"' we obtain

$$\|u_h\|^2_{2,h,W"'} \leq (\gamma\rho)(3\rho+5h)\|D_i^h u_h\|^2_{2,h,P}$$

Thus, in the case $W"\not\subset\Omega_h\cup\partial\Omega_h$, we can use the last inequality instead of POINCARE's . Replacing the constant K_0 by a larger one if necessary we obtain

(13) $\|\nabla_h u_h\|^2_{2,h,W} \le$ GOOD TERMS $+ \varepsilon^{-1}K\|\nabla_h u_h\|^2_{2,h,W''-W} + \varepsilon K'\|\nabla_h u_h\|^2_{2,h,W^*}$

Here, W^* is the smallest cube of grid points $\in \mathcal{Q}^h$ which contains P and is concentric to W . (P is set equal to W‴ in the case $W''' \subset \bar{\Omega}_h$). Obviously, the side length of W^* is smaller than $\max\{2\gamma(\rho+h), 3\rho+5h\} = L(\rho+h)$ with $L = \max\{2\gamma, 5\}$.

Now, we add the term $\varepsilon^{-1}K\|\nabla_h u_h\|^2_{2,h,W}$ to both parts of inequality (13), that is, we fill up the "hole" in the summation $\Sigma_{x\in W''-W}$ (†). Dividing the arising inequality by $1+\varepsilon^{-1}K$, we obtain

$$|\nabla_h u_h|^2_{2,h,W} \le K(K+\varepsilon)\|\nabla_h u_h\|^2_{2,h,W^*} + \varepsilon^2 K'(K+\varepsilon)^{-1}\|\nabla_h u_h\|^2_{2,h,W^*} + \varepsilon(K+\varepsilon)^{-1}\text{GOOD TERMS}$$

Choosing $\varepsilon>0$ small enough, say $\varepsilon=(2K')^{-1}$, we have $(K+\varepsilon^2K')/(K+\varepsilon)<\theta<1$, and we arrive at the inequality

(14) $\|\nabla_h u_h\|^2_{2,h,W} \le \theta\|\nabla_h u_h\|^2_{2,h,W^*} + K''\cdot$GOOD TERMS

By Hölder's inequality $\|u_h\|^2_{2,h,W''} \le K_3(\rho+h)^{2/q}\|u_h\|^2_{p,h,W''}$, $q=p/(p-2)$

Thus we may estimate

GOOD TERMS $\le K_4(\rho+h)^{2/q}(\|u_h\|^2_{p,h,W''} + 1)$, where K_4 does not depend on W or h . Applying this to (14) we obtain

$$\|\nabla_h u_h\|^2_{2,h,W} \le \theta\|\nabla_h u_h\|^2_{2,h,W^*} + K_5(\rho+h)^{2/q}(\|u_h\|^2_{p,h} + 1)$$

Choosing $\alpha\in(0,2/q)$ such that $L^{2\alpha}\theta<\ell<1$, and dividing the above inequality by $(\rho+h)^{2\alpha}$ we obtain

(15) $(\rho+h)^{-2\alpha}\|\nabla_h u_h\|^2_{2,h,W} \le \ell(L(\rho+h))^{-2\alpha}\|\nabla_h u_h\|^2_{2,h,W^*} + K_5(\rho+h)^{2/q-2\alpha}\cdot(\|u_h\|^2_{p,h} + 1)$

The term $(L(\rho+h))^{-2\alpha}\|\nabla_h u_h\|^2_{2,h,W^*}$ can be estimated by $|||\nabla_h u_h|||^2_{\alpha,h}$ (see the definition of the MORREY norm). If W is chosen such that the right-hand side of (15) is equal to $|||\nabla_h u_h|||^2_{\alpha,h}$ we obtain

$$|||\nabla_h u_h|||^2_{\alpha,h} \le (1-\ell)^{-1}K_5(\rho+h)^{2/q-2\alpha}(\|u_h\|^2_{p,h} + 1)$$

and the theorem is proved.

(†) The technique "filling up the hole" was used by WIDMAN [36] in his proof of the Hölder continuity of solutions of higher order non-linear elliptic systems.

For the proof of theorem 3 we have used

LEMMA 3 . (*Discrete Poincaré inequality*). *Let* u *be a real grid function on* R_h^n *and* $W \in Q_\rho^h$ *, Then*

$$\| u - \bar{u}_W \|_{2,h,W} \le \tfrac{1}{2}\rho \| \nabla_h u \|_{2,h,W}$$

where $\bar{u}_W = \Sigma_x u(x)/\Sigma_x 1$, $x \in W$.

PROOF: The notation becomes simpler if we prove the lemma in the case of complex grid functions u . Let $\rho = Nh$ and

$e_\alpha(x) = \Pi_j \exp(2i\pi\alpha_j x_j/(h+Nh))$,$(j=1,\ldots,n)$, $\alpha=(\alpha_1,\ldots,\alpha_n)$,$0\le\alpha_j\le N$, $i=\sqrt{-1}$. The e_α are orthogonal with respect to the scalar product $(v,w)_\rho = h^n\Sigma_x v(x)\,\bar{w}(x)$, $(x\in Q_\rho^h)$, for $\alpha\neq\beta$. Indeed, this follows from the one-dimensional orthogonality relation

$$\Sigma_\xi \exp(2i\pi\alpha_j\xi/(h+Nh))\,\overline{\exp(2i\pi\beta_j\xi/(h+Nh))} = \Sigma_k(\exp(2i\pi(\alpha_j-\beta_j)/(1+N)))^k =$$
$$[(\exp(2i\pi(\alpha_j-\beta_j)/(1+N)))^{N+1}-1]\cdot[\exp(2i\pi(\alpha_j-\beta_j)/(1+N))-1]^{-1} = 0$$
$$(\xi=kh,\ k=0,\ldots,N),\ \alpha_j\neq\beta_j .$$

Furthermore, $h^n\Sigma_x|e_\alpha(x)|^2 = |Q_\rho| = \rho^n$,$(x\in Q_\rho^h)$. Thus, the functions e_α may be taken as basis for the $(N+1)^n$-dimensional space of complex grid functions on Q_ρ^h , and we may expand any complex grid function u on Q_ρ^h by $u(x) = \Sigma_\alpha c_\alpha e_\alpha(x)$ $(0\le\alpha_j\le N)$. Note that

(16) $$c_{(0,\ldots,0)} = \bar{u}_\rho = \Sigma_x u(x)/\Sigma_x 1 \ ,(x\in Q_\rho^h).$$

This follows from the relation $(1,u)_\rho = (1,c_{(0,\ldots,0)})_\rho$ since $(1,e_\alpha)_\rho=0$, $\alpha\neq(0,\ldots,0)$.- Now, there hold the following identities

$$D_j^h e_\alpha(x) = h^{-1}(\exp(2i\pi\alpha_j/(N+1)) - 1)\, e_\alpha(x) \ ,\ \alpha_j\neq 0 \ ,\ D_j^h e_\alpha(x)=0 \ ,\ \alpha_j=0 .$$

Thus

$$\| D_j^h u\|^2_{2,h,Q_\rho} = h^n\Sigma_x\Sigma_\alpha|c_\alpha|^2 h^{-2}|\exp(2i\pi\alpha_j/(1+N)) - 1|^2 \ ,(x\in Q_\rho^h,0\le\alpha_k\le N,k=1,..n)$$

The last sum is equal to

$$\rho^n\Sigma_\alpha|c_\alpha|^2 h^{-2}|\exp(2i\pi\alpha_j/(N+1)) - 1|^2 \ ,(x\in Q_\rho^h \ ,\ 0\le\alpha_k\le N \ ,k=1,\ldots,n,\alpha_j\neq 0).$$

We have to estimate the factor $\beta_\alpha=|\exp(2i\pi\alpha_j/(N+1)) - 1|^2$.

Obviously, $\beta_\alpha=2-2\cos(2\pi\alpha_j/(N+1))\ge(2-2\cos(2\pi/(N+1))$,$\alpha_j\neq 0$). Furthermore, $1-\cos\xi\ge 2\xi^2/\pi^2$. (This follows from the following elementary consideration: $1-\cos\xi-2\xi^2\pi^{-2}=0$ for $\xi=0$ and $\xi=\pi$. Furthermore, $(1-\cos\xi-2\xi^2\pi^{-2})' =$

$= \sin\xi - 4\xi/\pi^2 < 0$ for $\xi=\pi$; thus $1-\cos\xi-2\xi^2/\pi^2>0$ in a left neighbourhood of $\xi=\pi$. If $1-\cos\xi-2\xi^2/\pi^2$ were <0 for some $\xi\in(0,\pi)$, then there would be 2 zeros of $\sin\xi-4\xi/\pi^2$ in $(0,\pi)$, and therefore 3 zeros of $\sin\xi-4\xi/\pi^2$ in $[0,\pi]$. But this is impossible since $\sin\xi$ is concave in $[0,\pi]$ on account of $\sin''\xi=-\sin\xi\leq 0$).

Therefore, $h^{-2}\beta_\alpha \geq h^{-2}(2-2\cos(2\pi/(N+1))) \geq h^{-2}2\cdot 2(2\pi/(N+1))^2\pi^{-2} = 16\rho^{-2}(N/(N+1))^2 \geq 4\rho^{-2}$, $N\geq 1$.

By this, we arrive at the inequality

$\|D_j^h u\|^2_{2,h;Q_\rho} \geq 4\rho^{-2}\Sigma_\alpha \rho^n |c_\alpha|^2$, $(0\leq\alpha_k\leq N,\ k=1,\dots,n,\ \alpha_j\neq 0)$ and

$\|\nabla_h u\|^2_{2,h;Q_\rho} \geq 4\rho^{-2}\Sigma_\alpha \rho^n |c_\alpha|^2$, $(0\leq\alpha_k\leq N,\ k=1,\dots,n,\ \alpha\neq(0,\dots,0)$)

The last term is equal to $4\rho^{-2}\|u-c_{(0,\dots,0)}\|^2_{2,h;Q_\rho}$ with $c_{(0,\dots,0)}$ defined by (16). This yields the statement of the lemma for $W=Q_\rho^h$. The general case $W\in\mathcal{Q}_\rho^h$ follows by translation.

An immediate consequence of theorem 3 is

THEOREM 4 . *Let* n=2 , *assume the regularity condition* (7) *for* $\partial\Omega$, *and the hypotheses of theorem 2 or theorem 1 and 3 . Then the convergence* $Ju_h \to u$ $(h\to 0)$ *in theorem 1 or 2 is uniform and the solution u is Hölder continuous up to the boundary* $\partial\Omega$.

PROOF: By the hypotheses, the discrete L^2-norms of $\nabla_h u_h$ are uniformly bounded for $h\to 0$. By the discrete SOBOLEV theorem [26] the discrete L^p-norms of the grid functions u_h are uniformly bounded for $h\to 0$; this holds for any fixed p* since n=2 .(SOBOLEV proved this theorem reducing the discrete case to the continuous one via the multilinear extensions of grid functions. Another way consists in a discrete analogue of GAGLIARDO's proof (see e.g. [23]) of the continuous SOBOLEV lemma). Thus, by theorem 3 , there follows the uniform boundedness of $|||\nabla_h u_h|||_{\alpha,h}$ for some $\alpha\in(0,1)$ and hence, by lemma 1, a uniform bound for $[u_h]_{\alpha,h}$. From this, there follows the uniform convergence $Ju_h\to u$ and the Hölder continuity of u .

* with p>2

REMARKS:

(i) The continuity of the solution u of (1) was obtained by the uniform estimate of the discrete Hölder norms of the solutions of the difference equations. This worked up to the boundary of the basic domain. Another interesting method by which one can obtain the continuity of u up to the boundary via the difference equations was presented in HAINER[16] He used a discrete analogue of PERRON's method[25]. This approach is restricted to linear problems and mean value operators (=discrete maximum principle), but it is not restricted to the case $n=2$.

(ii) In the case $n\geq 3$, the uniform estimate of the discrete MORREYnorm of $\nabla_h u_h$ in theorem 3 yields only the uniform boundedness of the discrete L^q-norms of u_h for $h\to 0$ where the exponent q is slightly better than the SOBOLEV exponent $2n/(n-2)$. This follows from discrete analogues of theorems of CAMPANATO [3] and STAMPACCHIA[28] .

(iii) It is also possible to prove interior estimates similar to the one in theorem 3 . Mostly, this will give a better Hölder exponent.

In the following, we shall assume that the differential equation (1) is uniformly elliptic , i.e. there hold the following conditions

III.(i) *Differentiability.* $F_i \in C^1(\bar{\Omega} X R^{n+1})$, $i=0,\dots,n$.

(ii) *Growth conditions.* For every $C>0$ there exists a constant K_C such that

$$|\nabla_x F_i(x,u,\eta)| \leq K_C(1+|\eta|) \;,\quad x\in\bar{\Omega},\ |u|\leq C \;,\ u\in R^1 \;,\quad \eta\in R^n$$

$$\left|\frac{\partial}{\partial u}F_i(x,u,\eta)\right|\leq K_C \;,\qquad x\in\bar{\Omega},\ |u|\leq C \;,\ u\in R^1 \;,\quad \eta\in R^n$$

$$|\nabla_\eta F_i(x,u,\eta)|\leq K_C \;,\qquad x\in\bar{\Omega},\ |u|\leq C \;,\ u\in R^1 \;,\quad \eta\in R^n \;,$$

$i=0,\dots n$.

(iii) *Ellipticity.* There exists a constant $\gamma>0$ such that

$$\Sigma_{i,k=1}^n \frac{\partial}{\partial\eta}F_i(x,u,\eta)\xi_i\xi_k \geq \gamma|\xi|^2 \;,\ x\in\bar{\Omega}, |u|\leq C, u\in R^1,\ \eta\in R^n$$

$$\xi=(\xi_1,\dots,\xi_n)\in R^n.$$

THEOREM 5 . *Let the coefficients* F_i *in (3) satisfy the growth and coerciveness condition II,(i)-(iii) of theorem 2, assume the ellipticity condition III above , and let n=2 . Then, for every pair* Ω_1 , $\Omega_0 \subset \Omega$ *of open subsets with* $\Omega_1 \subset\subset \Omega_0 \subset\subset \Omega$ *, there is a constant K such that*

$$|||\nabla_h D_j^h u_h|||_{\alpha,h,\Omega_1} \le K(\|\nabla_h D_j^h u_h\|_{2,h,\Omega_0} + 1) \text{ for some } \alpha\in(0,1), j=1,\dots,n.$$

Here, $|||w|||_{\alpha,h,G}$ denotes the MORREY norm taken over G :

$$|||w|||_{\alpha,h,G} = \sup\{(\rho+h)^{-\alpha}\|w\|_{2,h,U} \mid U\in Q^h , U\subset G\}$$

PROOF: By remark (iii) of theorem 4, we may assume that u_h is locally uniformly bounded such that we may assume locally the existence of the constant K_C of condition III . Then it is well known how to establish uniform bounds for the discrete L^2-norm of the second difference quotients of the solutions of the difference equations over interior domains of Ω .(In fact, one has to insert the grid - function $v_h = D_j^{-h}(\tau D_j^h u_h)$ into the difference equation (3)). The function τ is defined as in the proof of theorem 3 .
Now, the above estimate of theorem 5 can be derived similarly to the corresponding one in theorem 3 . We use the notations of the proof of th. 3 . Insert the function $v_h = D_j^{-h}(\tau(D_j^h u - b))$ into equation (3), where b is the mean value of $D_j^h u_h$, taken over W"' (for the definition of W"' see the proof of th. 3). The function τ must have its support in Ω_1 . By the technique "filling up the hole" and SOBOLEV's inequality we obtain the estimate

$$\|D_j^h \nabla_h u_h\|^2_{2,h,W} \le \theta\|D_j^h\nabla_h u_h\|^2_{2,h,W''} + K'(\rho+h)^{\beta},$$

with $\theta,\beta\in(0,1)$. We may assume that $\theta 5^{\beta} = \sigma < 1$. Since the side-length of W"' is smaller than five times the side length of W , we obtain by iteration

$$\|D_j^h \nabla_h u_h\|^2_{2,h,W} \le \theta^N \|D_j^h \nabla_h u_h\|^2_{2,h,W_N} + K'(\rho+h)^{\beta}\Sigma_{i=1}^N \theta^i 5^{i\beta}$$

Here W_N is a cube with side-length smaller than $5^N(\rho+h)$, and $W_N \subset \Omega_1$. From the last inequality, we conclude

$$(\rho+h)^{-\beta}\|D_j^h \nabla_h u_h\|^2_{2,h,W} \le \sigma^N R^{-\beta}\|D_j^h \nabla_h u_h\|^2_{2,h,W_R} + K''$$

where $R = 5^N(\rho+h)$ and W_R the cube with side-length R which is concentric to W. The theorem follows from the last inequality.

A consequence of theorem 5 and lemma 1 is

THEOREM 6 . *Assume the conditions II,(i)-(iii) of theorem 2 and the uniform ellipticity condition III,(i)-(iii) of theorem 5, and let n=2 . Then the convergence* $J\nabla_h u_h \to \nabla u$ *is uniform on interior domains and the gradient* ∇u *of the solution of the boundary value problem (1) is Hölder continuous on interior domains of* Ω .

REMARK: By a simple linearization argument, the nonlinear difference equation can be seen as a linear elliptic difference equation with coefficients which have locally uniform Hölder constants (on account of theorem 6). By THOMÉE's discrete SCHAUDER estimates [33] , there follows that the *second* difference quotients $\nabla_h^2 u_h$ have locally uniform Hölder constants and that they converge uniformly on interior domains of Ω to $\nabla^2 u$ which must be eo ipso continuous. For this reason, u must be a classical solution.

Finally we discuss the possibility how one can obtain error estimates by the MORREY space method.

Let $u \in C^2(\bar{\Omega})$ be a solution of the boundary problem (1) and let $R_h u$ be the grid function V_h which is defined by $R_h u = u$ on Ω_h and $R_h u = 0$ else . The *truncation error* is the grid function $e_h \in V_h$ which is defined by

$$h^n \Sigma_x e_h(x) w_h(x) := h^n \Sigma_i \Sigma_y \left[F_i(\cdot, R_h u, \nabla_h R_h u) D_i^h w_h\right](y) \quad ,(x \in \Omega_h , y \in \Omega_h' , i=0,\dots,n) \quad w_h \in V_h .$$

In the case of approximation (3'), we have to set

$$h^n \Sigma_x e_h(x) w_h(x) = h^n \Sigma_i \Sigma_y \left[F_i(\cdot, R_h u, A_h R_h \nabla_h u) \cdot (A_h \nabla_h w_h)_i\right](y) \; , \quad (x \in \Omega_h \; , \; y \in \Omega_h' \; , \; i=0,\dots,n)$$

THEOREM 7 . *Assume the growth and coerciveness condition II,(i)-(iii) of theorem 2 and the uniform ellipticity condition III,(i)-(iii) of theorem 5 . Let n=2 and assume the regularity condition (7) for* $\partial\Omega$. *Then, for any* $p>0$ *there exist constants* K *and* $\alpha \in (0,1)$ *such that the solutions* u_h *of* (3) *resp.* (3') *satisfy the following inequality*

$$|||\nabla_h R_h u - \nabla_h u_h|||_{\alpha,h} \le K(\|R_h u - u_h\|_{p,h} + \|e_h\|_{1,h})$$

PROOF: Since u_h is a solution of (3) resp. (3'), we obtain together with the definition of e_h .

(17) $h^n \Sigma_x \Sigma_{ik} F^o_{ik} D^h_k (R_h u - u_h) D^h_i \phi = h^n \Sigma_x e_h \phi \quad , (x \in \Omega'_h \; , i,k = 0,\dots,n)$

where F^o_{ik} is the value of the second partial derivative

$F_{ik}(x,\eta) = \frac{\partial}{\partial \eta_k} F_i(x,\eta)$, $\eta = (\eta_o,\dots,\eta_n)$ at a convex linear combination of $(x, R_h u, \nabla_h R_h u)$ and $(x, u_h, \nabla_h u_h)$. Since we know by theorem 4 that the u_h are uniformly bounded we have the existence of the constant K_C of condition III .

Now, we set $v_h = R_h u - u_h$ and shall insert the function $\phi = \tau(v_h - b)$ *into equation (17). Here τ and the following cubes of grid points are defined as in the proof of theorem 3 . Again, using the ellipticity condition we conclude with the technique "filling up the hole" from theorem 3 the estimate

$$\|\nabla_h v_h\|^2_{2,h,W} \le \theta \|\nabla_h v_h\|^2_{2,h,W} + K'\left[(\rho+h)^\beta \|v_h\|^2_{p,h} + \|e_h\|_{1,h} \|v_h - b\|_{\infty,h,W''}\right]$$

where $\theta < 1$ and $\beta > 0$. Choosing $\alpha \in (0,1)$ small enough we obtain similar to th. 3

$$|||\nabla_h v_h|||^2_{\alpha,h} \le q |||\nabla_h v_h|||^2_{\alpha,h} + K''\left\{\|v_h\|^2_{p,h} + \|e_h\|_{1,h} \|v_h - b\|_{(o)} (\rho+h)^{-\beta}\right\}, \; q<1 ,$$

if W is the cube in which the sup in the definition of the MORREY norm is attained. Estimating $\|v_h - b\|_{(o)}$ by $K_1 |||\nabla_h v_h|||_{\alpha,h} (\rho+h)^{+\beta}$ there follows the statement of the theorem.

On account of lemma 3, the estimate of theorem 7 gives also an estimate of the Hölder norm and the maximum norm of the error function v_h. It is easy to see that $\|e_h\|_{1,h} = O(h)$ if

(18) $F_i \in C^1$ and $u \in C^3$

and if one of the following conditions is satisfied:

(19) Use approximation (3'), or (19') Use approximation (3), and $\partial\Omega_h \subset \Omega_h$

---

* b is set equal to the mean value of v_h over W" , $(o) = \infty, h, W''$

In order to obtain $\|v_h\|_{p,h}=O(h)$ it suffices to prove $\|\nabla_h v_h\|_{2,h}=O(h)$ *. Again, this follows from condition (18) and (19) or (19'), if one assumes, for example, a uniform ellipticity condition

(20) $$(F_{ik})_{i,k=0}^{n} > \gamma > 0$$

We state the last considerations as theorem:

THEOREM 8 . *Assume the conditions of theorem 7 and further condition (18)-(20) . Then the maximal error of the approximation (3) resp. (3') behaves like O(h).*

REMARKS: For simplicity, we assumed that u is a scalar function. Thus, formally, we did not treat systems of differential equations, but it is clear that the MORREY space method presented here works also in the case of vector functions u.

The MORREY space method is not very related to the special approximation (3) of the differential equation. Also other approximations are feasable, e.g., those which use (for n=2) nine points instead of five in the definition of the difference operator. It is only of importance that the non-linear forms which occur in the proof of the theorems are coercive, and that the lemmata 1-3 hold for the special approximation of the derivatives.-This possibility is of importance, if on wants to extend our results to approximations which have a higher accuracy or have other desired properties; e.g. satisfy a discrete maximum principle as in [34].

In the case of the approximation (3'), one can prove theorem 8 under the weaker condition $u \in C^2$.

---

* For this problem, see also [1]

REFERENCES

[1] BRAMBLE, J.H.; KELLOGG, R.B.; THOMEE, V.: On the rate of convergence of some difference schemes for second order elliptic equations. BIT 8 (1968), 154-173.

[2] BRAMBLE, J.H.; HUBBARD, B.E.; THOMEE, V.: Convergence estimates for essentially positive type discrete ptoblems. Math. Comp. 23 (1969), 695-710.

[3] CAMPANATO, S.: Proprietà di inclusione per spazi di Morrey. Ricerche di Mat.; 12 (1963), 67-86.

[4] CIARLET, P.G.: Discrete maximum principle for finite-difference operators. Aequationes Math. 4 (1970), 338-352.

[5] CIARLET, P.G.; RAVIART, P.-A.: Maximum principle and uniform convergence for the finite element method. To appear.

[6] COLLATZ, L.: Bemerkungen zur Fehlerabschätzung für das Differenzenverfahren bei partiellen Differentialgleichungen. Zeitschr. Angew.Math.Mech. 13 (1933), 56-57.

[7] CONTI, F.: Su alcuni spazi funzionale e loro applicazioni ad equazioni differenziale di tipo ellittico. Boll. Un. Mat. Ital. 4 (1969), 554-569.

[8] FORSYTHE, G.E.; WASOW, W.: Finite difference methods for partial differential equations. New York: Wiley 1960.

[9] FREHSE, J.: Seminar in Jülich, 1970.

[10] FREHSE, J.: Existenz und Konvergenz von Lösungen nichtlinearer elliptischer Differenzengleichungen unter Dirichletrandbedingungen, Math. Z. 109 (1969), 311-343.

[11] FREHSE, J.: Über die Konvergenz von Differenzen- und anderen Näherungsverfahren bei nichtlinearen Variationsproblemen. Intern. Schriftenreihe z. Numerischen Math.15(1970), 29-44.

[12] FREHSE, J.: Über pseudomonotone nichtlineare Differentialoperatoren. Anhang zur Habilitationsschrift 1970.

[13] GERSCHGORIN, S.: Fehlerabschätzung für das Differenzenverfahren zur Lösung partieller Differentialgleichungen. Zeitschr. Angew. Math. Mech. 10 (1930), 373-382.

[14] GREENSPAN, D.: Introductory numerical analysis of elliptic boundary value problems. New York: Harper and Row, 1965.

[15] GRIGORIEFF, R.D.: Über die Koerzivität linearer elliptischer Differenzenoperatoren unter allgemeinen Randbedingungen. Frankfurt a.M.: Dissertation 1967.

[16] HAINER, K.: Differenzenapproximation elliptischer Differentialgleichungen zweiter Ordnung mit Hilfe des Maximumprinzips und eine allgemeine Konvergenztheorie. Frankfurt a.M.: Dissertation 1968.

[17] LERAY, J., LIONS, J.L.: Quelques résultats de Visik sur les problèmes elliptiques non linéaires par la méthode de Minty-Browder, Bull. Soc. Math. France 93 (1965) 97-107.

[18] McALLISTER, G.T.: Quasilinear uniformly elliptic partial differential equations and difference equations, J. SIAM Numer. Anal. 3 (1966) 13-33.

[19] McALLISTER, G.T.: Dirichlet problems for mildly nonlinear elliptic difference equations. Journ. Math. Anal. Appl. 27 (1969), 338-366.

[20] MORREY, C. B.: Multiple integral problems in the calculus of variations and related topics, Univ. California, Publ.Math., New.Ser., Vol.1(1943), 1-30.

[21] MORREY, C.B.: Multiple integrals in the calculus of variations, Springer, Berlin-Heidelberg-New York, 1966.

[22] MOSER, J.: A new proof of de Giorgi's theorem concerning the regularity problem for elliptic differential equations,

Comm. Pure and Appl. Math. 13 (1960), 457-468.

[23] NIRENBERG, L.: On elliptic partial differential equations, Ann. Scuola Norm. Sup. Pisa, Ser, 3, 13 (1959), 115-162.

[24] NITSCHE, J., NITSCHE, J.C.C.: Error estimates for the numerical solution of elliptic differential equations. Arch. Rational Mech. Anal. 5 (1960), 293-306.

[25] PERRON, O.: Eine neue Behandlung der ersten Randwertaufgabe für $\Delta u = o$. Math. Zeitschr. 18 (1923), 42-54.

[26] SOBOLEV, S.L.: On estimates for certain sums for functions defined on a grid, Izv. Akad. Nauk. SSSR, Ser.Math. 4 (1940), 5-16. (Russian).

[27] STAMPACCHIA, G.: Le problème de Dirichlet pour les équations elliptiques du second ordre à coefficients discontinus. Ann. Inst. Fourier (Grenoble) 15 (1965), 189-258.

[28] STAMPACCHIA, G.: The spaces $L^{p,\lambda}$, $N^{p,\lambda}$ and interpolation. Ann. Scuola Norm. Sup. Pisa 19 (1965), 443-462.

[29] STEPLEMAN, R.: Finite dimensional analogues of variational and quasilinear elliptic Dirichlet problems. Ph. D. Diss. Univ. Maryland 1969.

[30] STUMMEL, F.: Elliptische Differenzenoperatoren unter Dirichletrandbedingungen, Math. Zeitschr. 97 (1967), 169-211.

[31] THOMEE, V., WESTERGREN, B.: Elliptic difference equations and interior regularity, Num. Math. 11 (1968), 196-210.

[32] THOMEE, V.: Some convergence results in elliptic difference equations. Proc. Roy. Soc. Lond. A. 323 (1971), 191-199.

[33] THOMEE, V.: Discrete interior Schauder estimates for elliptic difference operators, SIAM J. Numer. Anal. 5(1968), 626-645.

[34] TÖRNIG, W.; MEIS, Th.: Diskretisierung des Dirichletproblems nichtlinearer elliptischer Differentialgleichungen. Erscheint demnächst.

[35] VOIGTMANN, K.: Dissertation, Frankfurt a.M. 1971.

[36] WIDMAN, K. O.: Hölder continuity of solutions of elliptic systems. To appear in Manuscripta Math

ZUR BERECHNUNG PERIODISCHER LÖSUNGEN BEI PARABOLISCHEN RANDWERTPROBLEMEN

Von Eckart Gekeler

1. Problemstellung

Es sei $\mathbb{R}$ die Menge der reellen Zahlen, $G = \{x \in \mathbb{R},\ 0 < x < 1\}$, $Z = G \times \{-\infty,\infty\}$ und $\overline{Z}$ die abgeschlossene Hülle von Z. Die Ableitung einer Funktion $f: \mathbb{R} \ni x \longrightarrow f(x) \in \mathbb{R}$ wird mit f_x bezeichnet. Wir gehen aus von dem Randwertproblem (RWP)

$$(1)\qquad \begin{aligned} u_t &= [a(x,t)u_x]_x + f(x,t,u,u_x), && (x,t) \in Z,\\ u(0,t) &= s^-(t),\ u(1,t) = s^+(t), && -\infty < t < \infty, \end{aligned}$$

dessen Koeffizienten in der Zeit t periodisch mit der Periode $\tau > 0$ sein sollen. Probleme dieser Form ergeben sich zum Beispiel, wenn ein Draht an den Enden periodisch erwärmt und abgekühlt wird und die dabei auftretende Wärmeverteilung berechnet werden soll *).

Existenz und Eindeutigkeit periodischer Lösungen bei parabolischen Problemen wurden von Šmulev [8], Kružkov [4] und Kolesov [3] untersucht.

Zur numerischen Lösung überdecken wir das Rechteck $G \times \{0,\tau\}$ mit einem Gitter der Maschenweite $\Delta x = 1/(M+1)$ und $\Delta t = \tau/N$, $M,N \in \mathbb{N} = \{1,2,\ldots\}$. Wegen der Periodizität der Koeffizienten in (1) können wir uns dann auf die Berechnung der Lösung an den Gitterpunkten im Innern des Rechtecks $G \times \{0,\tau\}$ und auf der Linie $t = \tau$ beschränken. Es sei

$$(2)\qquad \begin{aligned} v_k^n &= v(k\Delta x,\ n\Delta t),\ f_k^n(p,q) = f(k\Delta x, n\Delta t, p, q),\ k=1,\ldots,M,\ n=1,\ldots,N,\\ V^n &= (v_1^n,\ \ldots\ ,v_M^n)^T, \qquad n=1,\ldots,N,\\ V &= (V^1,\ \ldots\ ,V^N)^T. \end{aligned}$$

Wir ersetzen die Differentialgleichung in (1) durch die Differenzennäherung

$$(3)\qquad \frac{v_k^n - v_k^{n-1}}{\Delta t} - \frac{a_{k+1/2}^n(v_{k+1}^n - v_k^n) - a_{k-1/2}^n(v_k^n - v_{k-1}^n)}{\Delta x^2} - f_k^n\Big(v_k^n, \frac{v_{k+1}^n - v_{k-1}^n}{2\Delta x}\Big) = 0,$$

$$k=1,\ldots,M,\ n=1,\ldots,N,$$

mit den Randwerten

*) Nach einem Hinweis von Herrn Prof. Dr. W. Törnig (Darmstadt) treten derartige Probleme ferner bei der Bewetterung von Bergwerken auf.

(4) $$v_o^n = (s^-)^n = s^-(n\Delta t),\ v_{M+1}^n = (s^+)^n = s^+(n\Delta t),\ n=1,\dots,N.$$

Die MN Gleichungen (3) müssen simultan gelöst werden, da auf dem oberen und unteren Rand von $G\times(0,\tau)$ keine Randwerte vorgeschrieben sind; ihre Zusammenfassung in der üblichen Reihenfolge ergibt unter Berücksichtigung von (4) ein Gleichungssystem der Form

(5) $$P(V) := (S + A)V + F(V) = H.$$

Hier besteht also eine Ähnlichkeit zu elliptischen Problemen, aber im Unterschied dazu ist die Funktionalmatrix von (5) stets nichtsymmetrisch.

Im nächsten Abschnitt wollen wir unter geeigneten Voraussetzungen die Existenz einer Lösung V_Δ des Systems (5) beweisen und zeigen, daß V_Δ komponentenweise gegen den Blockvektor U der Lösung des analytischen Problems an den Gitterpunkten konvergiert. Für das Problem (1) mit der Differentialgleichung (DGl) $u_t = u_{xx}$ wurde dieser Sachverhalt von Tee [9] und Osborne [7] bewiesen; die dabei verwendeten Methoden eignen sich jedoch schlecht für den vorliegenden allgemeineren Fall. Wir greifen deshalb auf Eigenschaften des Systems (5) zurück, die im linearen Fall ein diskretes Maximumprinzip ergeben.

2. Konvergenz

In dem näherungsweisen Gleichungssystem (5) ist

$$H = (H^1, \dots, H^N)^T,$$

$$H^n = (\underbrace{a_{1/2}^n (s^-)^n, 0, \dots, 0, a_{M+1/2}^n (s^+)^n}_{M \text{ Komponenten}})^T / \Delta x^2,$$

$F: \mathbb{R}^{MN} \to \mathbb{R}^{MN}$ baut sich wie der Vektor V in (2) auf, A ist die Blockdiagonalmatrix $A = (A^1, \dots, A^N)$ mit den Untermatrizen

(6) $$A^n = \frac{1}{\Delta x^2}\begin{pmatrix} (a_{3/2}^n + a_{1/2}^n) & -a_{3/2}^n & & \mathbf{0} \\ -a_{3/2}^n & \ddots & \ddots & \\ & \ddots & \ddots & -a_{M-1/2}^n \\ \mathbf{0} & & -a_{M-1/2}^n & (a_{M+1/2}^n + a_{M-1/2}^n) \end{pmatrix}$$

und

(7) $$S = \frac{1}{\Delta t}\begin{pmatrix} E & 0 & & & -E \\ -E & E & & \bigcirc & \vdots \\ \vdots & \ddots & \ddots & & \\ \vdots & \bigcirc & \ddots & \ddots & \bigcirc \\ 0 & & & -E & E \end{pmatrix}$$

eine N-zyklische Blockmatrix (vgl. Varga [10, Definition 4.1], E = Einheitsmatrix der Ordnung M).

Wir betrachten zunächst das RWP (1) mit der linearen DGl

(8) $$u_t = [a(x,t)u_x]_x + b(x,t)u_x - c(x,t)u + d(x,t), \qquad (x,t) \in Z,$$

unter der folgenden für das semilineare Problem formulierten

Voraussetzung:

(i) Es seien a_{xxx} sowie alle niedrigeren Ableitungen von a nach x stetig in Z und beschränkt in $\overline{Z}$; es seien in Z

$$0 < \alpha_1 \leq a(x,t) \leq \alpha_2, \quad |a_x(x,t)| \leq \eta.$$

(ii) Es seien $f: Z \times \mathbb{R} \times \mathbb{R} \ni (x,t,p,q) \longmapsto f(x,t,p,q) \in \mathbb{R}$ sowie f_p, f_q stetig in Z für alle $p,q \in \mathbb{R}$ und beschränkt in $\overline{Z}$ für beschränkte

(9) $p,q \in \mathbb{R}$; es sei überdies in Z für alle $p,q \in \mathbb{R}$

$$|f_q(x,t,p,q)| \leq \mu, \quad f_p(x,t,p,q) \leq 0.$$

(iii) Es seien a, f, s^-, s^+ τ-periodisch in t (d.h. $s^-(t+\tau) = s^-(t)$ usw.).

(iv) Es existiere eine eindeutige klassische in t τ-periodische Lösung u von (1), und es seien u_{tt}, u_{xxxx} sowie alle niedrigeren Ableitungen von u stetig in Z und beschränkt in $\overline{Z}$.

Das System (5) hat dann die Form

(10) $$PV := (S + T + C)V = D + H$$

mit den Blockvektoren $D = (D^1,\dots,D^N)^T$, $H = (H^1,\dots,H^N)^T$

$$D^n = (d_1^n, \dots, d_M^n)^T,$$

$$H^n = (\underbrace{(2a^n_{1/2}-\Delta x b^n_1)(s^-)^n, 0, \dots, 0, (2a^n_{M+1/2}+\Delta x b^n_M)(s^+)^n}_{\text{M Komponenten}})^T/2\Delta x^2,$$

der Diagonalmatrix

$$C = (C^1, \dots, C^N), \quad C^n = (c_1^n, \dots, c_M^n),$$

der Blockdiagonalmatrix $B = (B^1, \dots, B^N)$ mit

$$B^n = \frac{1}{2\Delta x}\begin{pmatrix} 0 & -b_1^n & & & \\ b_2^n & 0 & \ddots & & 0 \\ & \ddots & \ddots & \ddots & \\ & & \ddots & \ddots & -b_{M-1}^n \\ 0 & & & b_M^n & 0 \end{pmatrix}$$

und $T = A + B$. Es sei

$$Z_\Delta = \{(k\Delta x, n\Delta t),\ k = 1,\dots,M,\ n = 1,\dots,N\}$$

und

$$\partial^* Z_\Delta = \{(k\Delta x, n\Delta t),\ k = 0, M+1,\ n = 1,\dots,N\}.$$

Für eine beliebige Gitterfunktion $v\colon Z_\Delta \cup \partial^* Z_\Delta \ni (k\Delta x, n\Delta t) \longmapsto v(k\Delta x, n\Delta t) = v_k^n \in \mathbb{R}$ bezeichnen wir mit $V_{\partial^* Z_\Delta} = (V^1_{\partial^* Z_\Delta}, \dots, V^N_{\partial^* Z_\Delta})^T$ den Blockvektor mit den Elementen

$$V^n_{\partial^* Z_\Delta} = (\underbrace{v_o^n, 0, \dots, 0, v_{M+1}^n}_{M \text{ Komponenten}})^T .$$

Schließlich sei $\dot{T} = (\dot{T}^1, \dots, \dot{T}^N)$ die Blockdiagonalmatrix mit den diagonalen Untermatrizen

$$\dot{T}^n = ((2a^n_{1/2} - \Delta x b_1^n), 0, \dots, 0, (2a^n_{M+1/2} + \Delta x b_M^n))/2\Delta x^2.$$

<u>Lemma 1 (Max-Min-Prinzip). Das RWP (1) mit der linearen DGl (8) genüge der Voraussetzung (9), ferner sei $0 < \Delta x < 2\alpha_1/\mu$ und Z_Δ fest gegeben. Ist dann v eine auf $Z_\Delta \cup \partial^* Z_\Delta$ definierte Gitterfunktion, die der Gleichung</u>

$$(S + T + C)V = \dot{T} V_{\partial^* Z_\Delta}$$

<u>genügt, so nimmt v das Maximum und Minimum auf $\partial^* Z_\Delta$ an.</u>

Der Beweis ist für $C = 0$ der gleiche wie bei Greenspan [2, Theorem 2.1]. Der Fall $C \geq 0$ stellt eine (beweistechnisch) unbedeutende Verallgemeinerung dar. Unter den Voraussetzungen des Lemmas gilt für die Gitterfunktion v

$$\min_n \{0, v_o^n, v_{M+1}^n\} \leq v_k^n \leq \max_n \{0, v_o^n, v_{M+1}^n\},$$

$$k=1,\dots,M,\ n=1,\dots,N,$$

also insbesondere

$$\max_{\substack{1 \leq k \leq M \\ 1 \leq n \leq N}} \{|v_k^n|\} \leq \max_{1 \leq n \leq N} \{|v_o^n|, |v_{M+1}^n|\}.$$

Das diskrete Maximumprinzip ist hier erfüllt, weil S + T für $0 < \Delta x < 2\alpha_1/\mu$ eine irreduzibel diagonaldominante Matrix mit positiven Diagonalelementen und ansonsten nichtpositiven Elementen also eine M-Matrix ist (vgl. die Definition bei Varga [10, S. 23 und S. 85]). Dies sind die wesentlichen Eigenschaften, die beim Beweis der Stabilität für Differenzennäherungssysteme zweiter Ordnung bei elliptischen RWPen verwendet werden, wenn der zugrunde liegende elliptische Operator nicht selbstadjungiert ist (vgl. etwa Greenspan [2]). Im nachfolgenden Satz nützen wir die genannten Eigenschaften direkt zur Abschätzung von $\|(S + T + C)^{-1}\|_\infty$ aus. Die Verwandtschaft mit elliptischen Problemen erstreckt sich naturgemäß auch auf die Beweisführung. Wir beweisen zunächst eine Eigenschaft der Matrix T:

Lemma 2. Das RWP (1) mit der linearen DGl (8) genüge der Voraussetzung (9); ferner sei

(11) $$0 < \Delta x < 2\alpha_1/\mu,$$

(12) $$16(\alpha_2 + \mu\Delta x/2)\varepsilon_1(x,\Delta x)/\Delta x^2 \leq 1,$$

(13) $$16(\alpha_2 + \mu\Delta x/2)\varepsilon_2(x,\Delta x)/\Delta x^2 \leq 1$$

mit

$$\varepsilon_{\substack{1\\(2)}}(x,\Delta x) = \left(1 + \frac{\Delta x}{1+x}\right)^\sigma - 1 \underset{(+)}{-} \frac{\sigma\Delta x}{1+x} - \frac{\sigma(\sigma-1)\Delta x^2}{2(1+x)^2} = \mathcal{O}(\Delta x^3),$$

und es sei

(14) $$\sigma = (1 + \alpha_2 + 2\eta + 2\mu)/\alpha_1.$$

Dann existiert auf $\bar{G} = \{x \in \mathbb{R},\ 0 \leq x \leq 1\}$ eine Funktion ψ mit $\psi(x) \geq 0$ derart, daß

(15) $$t_\Delta(x,s;\psi(x)) := \{-[a(x+\Delta x/2,s)+a(x-\Delta x/2,s)]\psi(x) + [a(x+\Delta x/2,s)+b(x,s)\Delta x/2]\psi(x+\Delta x) + [a(x-\Delta x/2,s)-b(x,s)\Delta x/2]\psi(x-\Delta x)\}/\Delta x^2$$

für $0 < s \leq \tau$ und $x \in G_\Delta = \{x \in \mathbb{R},\ \Delta x \leq x \leq 1 - \Delta x\}$ $(0 < \Delta x < 1)$ positiv ist und die Abschätzung

(16) $$\frac{\displaystyle\operatorname*{Max}_{0\leq x\leq 1}\{\psi(x)\}}{\displaystyle\operatorname*{Min}_{\substack{\Delta x\leq x\leq 1-\Delta x\\ 0<s\leq\tau}}\{t_\Delta(x,s;\psi(x))\}} \leq \kappa = 2^{1+\sigma}/\sigma$$

gilt.

Beweis. Für $\psi(x) = (1 + x)^\sigma$ ergibt sich

$$t_\Delta(x,s;\psi(x)) = \frac{(1+x)^\sigma}{(\Delta x)^2}\Big[-[a(x+\Delta x/2,s) + a(x-\Delta x/2,s)]$$

$$+ [a(x+\Delta x/2,s) + b(x,s)\Delta x/2](1 + \frac{\Delta x}{1+x})^\sigma$$

$$+ [a(x-\Delta x/2,s) - b(x,s)\Delta x/2](1 - \frac{\Delta x}{1+x})^\sigma\Big].$$

Wir setzen

$$(1 \pm \frac{\Delta x}{1+x})^\sigma = 1 \pm \frac{\sigma\Delta x}{1+x} + \frac{\sigma(\sigma-1)\Delta x^2}{2(1+x)^2} + \varepsilon_{\substack{1\\(2)}}(x,\Delta x), \quad \varepsilon_{\substack{1\\(2)}}(x,\Delta x) = \mathcal{O}((\Delta x)^3),$$

ein und erhalten nach einer leichten Umformung die Abschätzung

$$t_\Delta(x,s;\psi(x)) \geq \sigma[-(\alpha_2 + 2\eta + 2\mu) + \alpha_1\sigma$$

$$- 4(\alpha_2 + \mu\Delta x/2)\varepsilon_1(x,\Delta x)/\Delta x^2\sigma - 4(\alpha_2 + \mu\Delta x/2)\varepsilon_2(x,\Delta x)/\Delta x^2\sigma].$$

Daraus folgt $t_\Delta(x,s;\psi(x)) \geq \sigma/2$ nach Voraussetzung.

Satz. Das RWP (1) mit der linearen DGl (8) genüge der Voraussetzung (9). Dann besitzt das Differenzennäherungssystem (10) für $0 < \Delta x < 2\alpha_1/\mu$ genau eine Lösung V_Δ. Genügt Δx außerdem den beiden Ungleichungen (12) und (13) und stellt U den Blockvektor der Lösung des analytischen Problems an den Gitterpunkten dar, so gilt

(17) $$\|V_\Delta - U\|_\infty \leq \kappa_1\Delta t + \kappa_2\Delta x^2$$

mit von Δt und Δx unabhängigen Konstanten κ_1 und κ_2.

Beweis. Für $0 < \Delta x < 2\alpha_1/\mu$ genügt $P = S + T + C$ als irreduzibel diagonaldominante Matrix dem schwachen Zeilensummenkriterium und ist daher invertierbar; außerdem gilt $P^{-1} > 0$ und $(P - C)^{-1} > 0$ elementweise (vgl. Varga [10, S. 85]). Nach Ortega-Rheinboldt [6, 2.4.11] ist $P^{-1} \leq (P - C)^{-1}$ elementweise, und wir können uns deshalb auf den Fall $C = 0$ Beschränken. Es sei nun

$$\tilde V: Z_\Delta \cup \partial^* Z_\Delta \ni (k\Delta x, n\Delta t) \longmapsto 1,$$

dann folgt aus (3) mit (8)

$$\tilde V = (S + T)^{-1}\dot T\, \tilde V_{\partial^* Z_\Delta} ,$$

also

(18) $$\|(S + T)^{-1}\,\dot{T}\|_\infty = 1,$$

da $(S + T)^{-1} > 0$, $\dot{T} \geq 0$ elementweise gilt und $\tilde{V}$ aus lauter Einsen besteht. Für die Gitterfunktion

$$\psi:\ (k\Delta x, n\Delta t) \longrightarrow (1 + k\Delta x)^\sigma$$

($\sigma = (1 + \alpha_2 + 2\eta + 2\mu)/\alpha_1$)ergibt sich aus Lemma 2 $-(S + T)\Psi + \dot{T}\Psi_{\partial^* Z_\Delta} > 0$, wenn Δx die dortigen Voraussetzungen erfüllt, und der Blockvektor Ψ hat positive Elemente ψ_k^n. Wegen

$$(S + T)^{-1}(\,-(S + T)\Psi + \dot{T}\Psi_{\partial^* Z_\Delta}) + \Psi = (S + T)^{-1}\,\dot{T}\Psi_{\partial^* Z_\Delta}$$

folgt

$$\|(S + T)^{-1}\|_\infty \leq \frac{\|(S + T)^{-1}\,\dot{T}\|_\infty \|\Psi_{\partial^* Z_\Delta}\|_\infty}{\min\limits_{\substack{1\leq k\leq M\\ 1\leq n\leq N}} \{t_\Delta(k\Delta x, n\Delta t; \psi(k\Delta x))\}},$$

da die Komponenten des Vektors $-(S + T)\Psi + \dot{T}\Psi_{\partial^* Z_\Delta}$ die Form $t_\Delta(k\Delta x, n\Delta t; \psi(k\Delta x))$ haben. Mit (16) und (18) folgt

$$\|(S + T)^{-1}\|_\infty \leq 2^{1+\sigma}/\sigma.$$

Die Behauptung des Satzes ergibt sich nun wegen

$$\|V_\Delta - U\|_\infty = \|P^{-1}PV_\Delta - P^{-1}PU\|_\infty \leq \|P^{-1}\|_\infty \|PU - (D + H)\|_\infty$$

aus der Abschätzung

$$\|PU - (D + H)\|_\infty = \mathcal{O}(\Delta t) + \mathcal{O}(\Delta x^2),$$

die man unter Beachtung von $U^o = U^N$ mit Hilfe des üblichen Taylorabgleichs erhält.

Lemma 2 und die Behauptung des Satzes bleiben richtig unabhängig davon, ob in die Funktion b (neben t) noch weitere Parameter eingehen, solange nur der Wertebereich von b beschränkt bleibt. Nach dem Satz von Hadamard [6, 5.3.10] erweist sich somit das Differenzennäherungssystem (5) als stabil, wenn $f_q(x,t,p,q)$ unabhängig von p, $q \in \mathbb{R}$ beschränkt ist:

<u>Folgerung</u>. <u>Das RWP</u> (1) <u>genüge der Voraussetzung</u> (9). <u>Dann gilt die Aussage des Satzes für das Differenzennäherungssystem</u> (5).

<u>Bemerkung 1.</u> Es sei $G \subset \mathbb{R}^2$ ein beschränktes einfach zusammenhängendes Gebiet und der Rand ∂G sei hinreichend glatt. Wählt man dann das Gitter in G wie bei Greenspan [2, Kap. I], so gilt für das RWP

$$u_t = [a_1(x,y,t)u_x]_x + [a_2(x,y,t)u_y]_y + f(x,y,t,u,u_x,u_y)$$
$$(x,y,t) \in G\times\{-\infty,\infty\} ,$$
$$u(x,y,t) = s(x,y,t) \qquad (x,y,t) \in \partial G\times\{-\infty,\infty\} ,$$

sinngemäß das im Satz beschriebene Ergebnis, wobei aber (17) durch

$$\|V_\Delta - U\|_\infty \leq \kappa_1\Delta t + \kappa_2\Delta x + \kappa_3\Delta y$$

zu ersetzen ist. (Δx und Δy bezeichnen jetzt den maximalen Abstand der Gitterpunkte in x- und in y-Richtung.) Der Beweis verläuft ähnlich wie hier (vgl. auch Greenspan [2]).

Bemerkung 2. Im Gegensatz zu Anfangs-Randwertproblemen liefert hier auch die Differenzennäherung

$$u_t(k\Delta x,n\Delta t) = \frac{u(k\Delta x,(n+1)\Delta t) - u(k\Delta x,(n-1)\Delta t)}{2\Delta t} + \mathcal{O}(\Delta t^2)$$

stabile Approximationen, falls die Dgl in (1) linear oder fastlinear ist. Wir werden darauf an anderer Stelle ausführlich eingehen.

Bemerkung 3. Unter der Voraussetzung $0 < \Delta x < 2\alpha_1/\mu$ ist die Funktionalmatrix $P'(V)$ des Systems (5) eine M-Matrix (vgl. Varga [10, S. 85]). P selbst stellt also nach Moré-Rheinboldt [5, Theorem 5.2] eine M-Funktion dar. Für derartige Funktionen geben Ortega-Rheinboldt [6, Kap. 13] Lösungsverfahren an, die mit dem SOR-Verfahren ("successive overrelaxation method") verwandt sind. Ist das System (5) linear und sind die Diagonalelemente von P zu Eins normiert, so konvergiert das SOR-Verfahren nach [10, Theorem 3.16] für Relaxationsfaktoren $0 < \omega \leq 1$. Hängen die Koeffizienten der zugrunde liegenden Dgl nicht von der Zeit ab und ist diese linear, so konvergiert das SLOR-Verfahren ("successive line overrelaxation method") nach [10, Theorem 4.4] für $0 < \omega < N/(N-1)$.

3. Zur Voraussetzung $|f_q(x,t,p,q)| \leq \mu$

Sind für die Lösung u von (1) und deren x-Ableitung a-priori-Abschätzungen

$$|u(x,t)| \leq p_o/2, \quad |u_x(x,t)| \leq q_o/2, \qquad (x,t) \in Z,$$

bekannt und gilt $\Delta t = \mathcal{O}(\Delta x)$, so läßt sich die einschränkende Voraussetzung $|f_q(x,t,p,q)| \leq \mu$ in (9) dadurch erzwingen, daß man f durch eine Funktion $\tilde{f}$ mit den in (9) genannten Eigenschaften ersetzt, die für

$(x,t) \in Z$, $|p| \leq p_o$, $|q| \leq q_o$ mit f übereinstimmt. Hat dann das RWP (1) mit $\tilde{f}$ statt f genau eine Lösung, so stimmt diese mit der Lösung u von (1) überein. Stellt $\tilde{V}_\Delta$ die Lösung des zugehörigen Systems (5) dar und bezeichnet $\tilde{V}_{\Delta,x}$ den Blockvektor mit den Elementen

$$\frac{\tilde{v}^n_{k+1} - \tilde{v}^n_{k-1}}{2\Delta x}, \quad k = 1,\dots,M, \ n = 1,\dots,N,$$

dann gilt für hinreichend kleine Δx und Δt

$$(19) \qquad \|\tilde{V}_\Delta\|_\infty \leq p_o, \ \|\tilde{V}_{\Delta,x}\|_\infty \leq q_o$$

wegen (17) und

$$\left|\frac{\tilde{v}^n_{k+1} - \tilde{v}^n_{k-1}}{2\Delta x} - u_x(k\Delta x, n\Delta t)\right|$$

$$\leq \left|\frac{\tilde{v}^n_{k+1} - \tilde{v}^n_{k-1}}{2\Delta x} - \frac{u^n_{k+1} - u^n_{k-1}}{2\Delta x}\right| + \left|\frac{u^n_{k+1} - u^n_{k-1}}{2\Delta x} - u_x(k\Delta x, n\Delta t)\right|$$

$$= \mathcal{O}(1) + \mathcal{O}(\Delta x^2) = \mathcal{O}(1).$$

Aus (19) folgt, daß $\tilde{V}_\Delta$ für hinreichend kleine Δx und Δt mit der Lösung V_Δ des zu (1) gehörenden Systems (5) übereinstimmt, da (5) unter den obigen Voraussetzungen eindeutig lösbar ist.

Beispiel 1. Die Funktion $f: (x,t,p,q) \longmapsto f(x,t,p,q) = g(x,t,p) + b(x,t,q)$ genüge der Voraussetzung (9), aber es sei $b(x,t,q)$ nur für beschränkte q beschränkt. Dann setzen wir etwa

$$(20) \qquad \tilde{g}(x,t,p) = \begin{cases} g(x,t,p_o) + 1 - \exp(\alpha(p - p_o)) & , \ p > p_o \\ g(x,t,p) & , \ |p| \leq p_o \\ g(x,t,-p_o) - 1 + \exp(-\beta(p + p_o)), & p < -p_o \end{cases}$$

mit $\alpha = g_p(x,t,p_o)$, $\beta = g_p(x,t,-p_o)$ (vgl. Greenspan-Parter [1]) und

$$\tilde{b}(x,t,q) = \begin{cases} b(x,t,q_o) + b_q(x,t,q_o)(q - q_o) & , \ q > q_o, \\ b(x,t,q) & , |q| \leq q_o, \\ b(x,t,-q_o) + b_q(x,t,-q_o)(q + q_o), & q < -q_o. \end{cases}$$

Die Funktion $\tilde{f}: (x,t,p,q) \longmapsto \tilde{f}(x,t,p,q) = \tilde{g}(x,t,p) + \tilde{b}(x,t,q)$ genügt dann der Voraussetzung (9), und es ist sogar

$$|\tilde{g}(x,t,p)| \leq \max_{|p| \leq p_o} \{|g(x,t,p)| + 1\}.$$

Beispiel 2. Die Funktion $f: (x,t,p,q) \longmapsto f(x,t,p,q) = g(x,t,p)b(x,t,q)$ genüge der Voraussetzung (9), aber es sei f_q nur für beschränkte p,q beschränkt. Ferner sei $g_p(x,t,p) \leq 0$ und $b(x,t,q) \geq 0$ für $(x,t) \in Z$ und alle $p,q \in \mathbb{R}$. Wir wählen g wie in (20) und setzen

$$\tilde{b}(x,t,q) = \begin{cases} b(x,t,q_o), & q \geq q_1 = q_o + |b(x,t,q_o)/\gamma|, \\ b(x,t,q_o) + \\ + \gamma(q-q_o)\exp\left[1-(1-[\gamma(q-q_o)/b(x,t,q_o)]^2)^{-1}\right], & q_1 > q > q_o, \\ b(x,t,q)\ , & |q| \leq q_o, \\ b(x,t,-q_o) + \\ + \delta(q+q_o)\exp\left[1-(1-[\delta(q+q_o)/b(x,t,-q_o)]^2)^{-1}\right], & q_2 < q < -q_o, \\ b(x,t,-q_o), & q \leq q_2 = -q_o - |b(x,t,-q_o)/\delta|, \end{cases}$$

mit $\gamma = b_q(x,t,q_o)$, $\delta = b_q(x,t,-q_o)$. Ist $b(x,t,q_o) = 0$, so folgt wegen $b(x,t,q) \geq 0$ und der Stetigkeit von b_q, daß $\gamma = 0$ sein muß. In diesem Fall wird $q_1 = q_o$ gesetzt, entsprechend sei $q_2 = -q_o$ im Fall $\delta = 0$. Man rechnet nach, daß $\tilde{b}(x,t,q) \geq 0$ für alle $q \in \mathbb{R}$ gilt. Die Funktion $\tilde{f}: (x,t,p,q) \longmapsto \tilde{g}(x,t,p)\tilde{b}(x,t,q)$ genügt dann der Voraussetzung (9).

Literatur

1. Greenspan, D., und Parter, S. V.: Mildly nonlinear elliptic partial differential equations and their numerical solution, II. Numer. Math. 7, 129-146 (1965).
2. Greenspan, D.: Introductory Numerical Analysis of Elliptic Boundary Value Problems. Harper-Row, New York-London 1965.
3. Kolesov, Ju. S.: Periodic solutions of quasilinear parabolic equations of second order. Transact. Moscow Math. Soc. 21, 114-146 (1970).
4. Kružkov, S. N.: Periodic solutions of nonlinear second order parabolic equations (russ.). Differencial'nye Uravnenija 6, 731-740 (1970).
5. Moré, J. J., und Rheinboldt, W. C.: On P- and S-functions and related classes of n-dimensional nonlinear mappings. Techn. Report 70-120, University of Maryland (1970.
6. Ortega, J. M., und Rheinboldt, W. C.: Iterative Solution of Nonlinear Equations in Several Variables. Academic Press, New York-London 1970.
7. Osborne,M. R.: A note on the numerical solution of a periodic parabolic problem. Numer. Math. 7, 155-158 (1965).
8. Šmulev, I. I.: Periodic solutions of the first boundary problem for parabolic equations. AMS Translations (2) 79, 215-229 (1969).
9. Tee, G. J.: An application of p-cyclic matrices. Numer. Math. 6, 143-158 (1964).
10. Varga, R. S.: Matrix Iterative Analysis. Prentice-Hall Englewood Cliffs, N. J., 1962.

ÜBER S.GERSCHGORINS METHODE DER FEHLERABSCHÄTZUNG BEI DIFFERENZENVERFAHREN

Von Rudolf Gorenflo

Herrn Professor Dr. J. Weissinger mit den besten Wünschen zur Vollendung seines 60. Lebensjahres gewidmet.

§ 0 Einleitung

Viele elliptische und parabolische Randwertaufgaben sind invers-isoton,[1] d.h. die gesuchte Lösung hängt isoton von den Daten (Inhomogenitäten, Randwerten, Anfangswerten) ab. Man kann versuchen, durch Ausnutzung der Invers-Isotonie Einschließungen der Lösung zu finden. Hierzu vgl. man beispielsweise Collatz (1952 und 1964), Protter und Weinberger (1967), Walter (1964). Verwendet man zur Ermittlung von Näherungslösungen invers-isotone Differenzenverfahren, so kann man in vielen Fällen ihre Konvergenz durch Ausnützung der Invers-Isotonie beweisen.

Im Jahre 1930 bewies S. Gerschgorin für das Dirichlet-Problem bei linearen elliptischen Randwertaufgaben zweiter Ordnung ohne gemischte Ableitungen unter geeigneten Voraussetzungen, die u.a. die Invers-Isotonie garantieren, die Konvergenz des üblichen Fünf-Punkte-Differenzenschemas bei quadratischem Gitter; er untersuchte speziell die Poisson-Gleichung, über die in dieser Einleitung berichtet werden soll. Auf randbenachbarten inneren Gitterpunkten setzte er den Näherungswert gleich einem längs einer Gitterlinie benachbarten Randwert. Das Verfahren hat die Konvergenzordnung 1 in der Maschenweite, im Sonderfall achsenparallelen nur auf Gitterlinien liegenden Randes jedoch die Konvergenzordnung 2. Collatz verbesserte 1933 das Verfahren durch seine Methode der linearen Interpolation für randbenachbarte Gitterpunkte und erzielte dadurch die Konvergenzordnung 2. Shortley und Weller führten 1938 auch für randbenachbarte Gitterpunkte eine 5-Punkt-Approximation ein, indem sie den verschiedenen Maschenweiten entsprechend eine diskrete Mittelungsformel mit i.a. verschiedenen Gewichten aufstellten. Obwohl nun diese Approximation in der Nachbarschaft des Randes i.a. nur in der Ordnung 1 zum Differentialoperator konsistent ist, ist 1962 von Bramble und Hubbard bewiesen worden, daß das Verfahren in der Ordnung 2 konvergiert.

Aus der umfangreichen Literatur kann hier nur eine Auswahl zitiert werden. Eine neuere Darstellung von Gerschgorins Methode findet man bei

Forsythe und Wasow (1960), Verallgemeinerungen und Verfeinerungen (mit speziellen Abschätzungen von Summen diskreter Greenscher Funktionen auf Teilmengen des Gitters - vor allem einerseits über innere Gitterpunkte, andrerseits über randnahe Gitterpunkte) findet man bei Bramble und Hubbard (1962), Bramble, Hubbard und Thomée (1969), Ciarlet 1970. Man vgl. auch Hainer (1972). V. Thuraisamy (1969) behandelt gemischte Randbedingungen. Zwei-Punkt-Randwertaufgaben zweiter Ordnung findet man bei Babuška, Präger und Vitásek (1966) untersucht. Zu erwähnen ist auch die wichtige Arbeit von Bramble und Hubbard (1964), in der für eine spezielle Zwei-Punkt-Randwertaufgabe eine Approximation der Konsistenzordnung 4 in den inneren Gitterpunkten mit Abstand $\geq 2h$, der Konsistenzordnung 2 in den Gitterpunkten mit Abstand h vom Rande untersucht wird, wobei sich trotzdem 4 als Konvergenzordnung ergibt.

Eine Modifikation des Verfahrens läßt sich anwenden auf invers-isotone Differenzenschemata für parabolische Anfangs-Randwert-Aufgaben zweiter Ordnung. Hierzu seien zitiert Kolar (1970) und Gorenflo (1971), bei denen man weitere Literaturhinweise findet.

Wir wollen Gerschgorins Methode formalisieren und verallgemeinern und ihre Anwendung an einfachen Beispielen elliptischer Aufgaben illustrieren. Ohne explizite Einführung diskreter Greenscher Funktionen und Abschätzung ihrer Teilsummen über Teilmengen des Differenzengitters werden wir herausarbeiten, woran es eigentlich liegt, daß manchmal schlechtere Approximation des Differentialoperators in Randnähe nicht schadet. Im Interesse einer durchsichtigen Darstellung der wesentlichen Ideen beschränken wir uns auf den linearen Fall. Alle auftretenden Zahlen und zahlenwertigen Funktionen seien reell.

§ 1 Invers-Isotonie und Einschließungsprinzip

U und S seien halbgeordnete lineare Räume über $\mathbb{R}$, in jedem der Räume sei " $\leq$ " das Verknüpfungszeichen der Halbordnung. Wie üblich - man vgl. Collatz (1964) - sei die Halbordnung jeweils reflexiv, transitiv, antisymmetrisch, und mit der linearen Struktur verträglich. Durch $x \leq y \Longleftrightarrow y \geq x$ sei das Zeichen " $\geq$ " erklärt. In $\mathbb{R}$ sollen die Zeichen "$\leq$" und "$\geq$" die übliche Bedeutung haben. Bequemlichkeitshalber bezeichnen wir unterschiedslos das Nullelement eines beliebigen linearen Raumes mit 0. Mit D(A) sei der Definitionsbereich eines Operators A bezeichnet. Die Schreibweise $|u| \leq v$ sei gleichbedeutend mit $-v \leq u \leq v$. Aus $|u| \leq v$ folgt dann $v \geq 0$.

Der Operator $A : D(A) \subset U \to S$ heißt isoton, wenn $x \leq y \Rightarrow Ax \leq Ay$ $\forall x,y \in D(A)$. Der Operator $A : D(A) \subset U \to S$ heißt invers-isoton, wenn $Ax \leq Ay \Rightarrow x \leq y$ $\forall x,y \in D(A)$.

Endlich-dimensionale Räume $\mathbb{R}^m$ seien auf natürliche Weise halbgeordnet, also $x \leq y \Leftrightarrow x_j \leq y_j$ für $j=1,2,\dots,m$. Hierbei sind x_j und y_j die Komponenten von x und y. Lineare Operatoren $A : \mathbb{R}^m \to \mathbb{R}^m$ identifizieren wird in naheliegender Weise mit Matrizen.

Bekanntlich gelten die leicht einzusehenden Sätze 1.1, 1.2 und 1.3.

Satz 1.1: Wenn $A : D(A) \subset U \to S$ invers-isoton ist, so gibt es zu gegebenem $s \in S$ höchstens ein $u \in D(A)$ derart, daß $Au = s$ ist.

Satz 1.2: Wenn $A : \mathbb{R}^m \to \mathbb{R}^m$ linear und invers-isoton ist, so existiert $A^{-1} : \mathbb{R}^m \to \mathbb{R}^m$, und alle Elemente der zu A^{-1} gehörenden Matrix sind ≥ 0.

Satz 1.3 (Einschließungsprinzip): Wenn $A : D(A) \subset U \to S$ linear und invers-isoton ist, so gilt

$$|Au| \leq Ay \Rightarrow |u| \leq y \quad .$$

Im Hinblick auf die Anwendungen, die wir im Auge haben, treffen wir folgende Vereinbarung.

Vereinbarung: U sei ein linearer Raum vektorwertiger Funktionen, deren Werte jeweils in einem endlich-dimensionalen reellen Raum liegen. Jede Komponente sei definiert auf einem Teilkompaktum eines Kompaktums $\bar{\Omega} \subset \mathbb{R}^m$, $m \in \mathbb{N}$, und dort mindestens stetig. Die Halbordnung in U sei die natürliche, " $\leq$ " bedeute also, "komponenten- und punktweise kleiner oder gleich im Sinne der reellen Analysis". Der Operator A sei für $u \in U$ auf gewissen disjunkten Teilmengen von $\bar{\Omega}$ durch verschiedene Vorschriften erklärt (etwa im Inneren von $\bar{\Omega}$, wenn $\bar{\Omega}$ der Abschluß einer offenen Menge Ω ist, durch einen Differentialoperator höherer Ordnung, auf $\partial\Omega$ durch einen Differentialoperator niedrigerer Ordnung). Dementsprechend ist auch S ein linearer Raum endlich-dimensional reell-vektorwertiger Funktionen, jede Komponente definiert auf einer Teilmenge von $\bar{\Omega}$. Einfachheitshalber seien auch hier alle Komponenten stetig und stetig fortsetzbar auf den Rand des jeweiligen Definitionsbereiches, so daß wir auch jede Komponente als auf einem Teilkompaktum von $\bar{\Omega}$ definiert annehmen können. S sei ebenfalls auf natürliche Weise halbgeordnet. Weil A linear ist, ist $D(A)$ ein linearer Raum, und es ist oft zweckmäßig, $U = D(A)$ zu nehmen. Jeden der Räume U und S normieren wir mittels der Maximumnorm. $\|.\|$ bedeute also stets die Maximum-Norm im jeweiligen Raum; es wird immer klar sein, in welchem Raum wir uns gerade befinden, so daß wir die Norm nicht mit dem jeweiligen Raum als Index verzieren müssen. Maximum-Normierung bedeutet: zuerst werden die Betragsmaxima aller Komponenten gebildet, dann wird von diesen das Maximum genommen.

Zur Illustration ein Beispiel.

Beispiel (B): $\bar{\Omega} = [0,1]$, $U = C^2[0,1] = D(A)$, $S = C[0,1] \times \mathbb{R} \times \mathbb{R}$, $p,q \in C[0,1]$, $\varphi,\psi \in [0, \frac{\pi}{2}]$, $q(x) \geq 0 \quad \forall\, x \in [0,1]$, $q(x) \not\equiv 0$ im Sonderfalle $\varphi = \psi = 0$. In der Definition von S steht $\mathbb{R} \times \mathbb{R}$ für $C\{0\} \times C\{1\}$.

$$(Au)(x) = -u''(x) + p(x)u'(x) + q(x)u(x) \quad \text{für } 0 < x < 1,$$

$$(Au)(0) = -u'(0) \cos \varphi + u(0) \sin \varphi,$$

$$(Au)(1) = u'(1) \cos \psi + u(1) \sin \psi,$$

$$s = \begin{pmatrix} r \\ \alpha \\ \beta \end{pmatrix} \text{ mit } r \in C[0,1],\ \alpha,\beta \in \mathbb{R}.$$

Bekanntlich (man vgl. Protter und Weinberger, Kapitel 1) ist A invers-isoton, und aus der Theorie der gewöhnlichen Differentialgleichungen weiß man, daß für gegebenes $s \in S$ die Aufgabe $Au = s$ genau eine Lösung $u \in U$ hat. $Au = s$ ist die Kurzschreibweise für das Zwei-Punkt-Randwertproblem

$$-u'' + pu' + qu = r, \quad -u'(0) \cos \varphi + u(0) \sin \varphi = \alpha,$$

$$u'(1) \cos \psi + u(1) \sin \psi = \beta .$$

Je nach Wahl von φ und ψ ist hierin das Dirichletsche, das Neumannsche und das gemischte Randwertproblem enthalten.

Die Nützlichkeit der etwas kompliziert vereinbarten Struktur der Räume U und S wird klar, wenn man gekoppelte Systeme gewöhnlicher und partieller Differentialgleichungen mit Rand- und Übergangsbedingungen betrachtet, wie sie bei manchen Multidiffusionsprozessen mit absorbierenden Wänden auftreten.

§ 2 Diskretisierung, Konsistenz, Stabilität, Konvergenz

Vollständigkeitshalber und zur Klärung der Bezeichnungsweise skizzieren wir einige wohlbekannte Zusammenhänge.

Wir überziehen $\bar{\Omega}$ mit einem Gitter $\gamma \subset \bar{\Omega}$ endlich vieler Gitterpunkte und lassen γ eine gegebene gerichtete Menge Γ durchlaufen. Wegen $\bar{\Omega} \subset \mathbb{R}^m$ können wir mit der Maximumnorm des $\mathbb{R}^m$ den Feinheitsgrad [2)]

$$\bar{h}(\gamma) = 2 \max_{x \in \bar{\Omega}} \min_{y \in \gamma} \|y-x\|$$

definieren und in Γ die Relation $\succeq$ erklären vermittels

$$\gamma_1 \succeq \gamma_2 \iff \bar{h}(\gamma_1) \leq \bar{h}(\gamma_2) \quad .$$

Die Gerichtetheit von Γ bedeutet also Verfeinerbarkeit des Gitters, und wenn γ die Menge Γ durchläuft, gehe $\bar{h}(\gamma) \to 0$. Der Grenzübergang lim ist im folgenden stets im Sinne von $\bar{h}(\gamma) \to 0$ zu verstehen. Stets sei im folgenden $\gamma \in \Gamma$.

Vermittels des Restriktionsoperators $R_\gamma : U \to V_\gamma$, erklärt durch

$$(R_\gamma u)(x) = u(x) \quad \text{für } u \in U, \quad x \in \gamma,$$

führen wir den endlich-dimensionalen reellen Raum V_γ ein und versehen ihn mit der Maximumnorm und der natürlichen Halbordnung. Den Raum V_γ können wir identifizieren mit einem $\mathbb{R}^{\nu(\gamma)}$. Wir bemerken, daß $\lim \|R_\gamma u\| = \|u\|$ für $u \in U$. Stets ist $\|R_\gamma u\| \leq \|u\|$.

Zum linearen Operator $A : D(A) \subset U \to S$ sei gegeben eine Folge linearer Näherungsoperatoren $A_\gamma : V_\gamma \to V_\gamma$ und eine Folge von Mittelungsoperatoren $M_\gamma : S \to V_\gamma$. Zur Operatorgleichung

(2.1) $\quad Au = s, \quad s \in S$ gegeben, $u \in U$ gesucht,

betrachten wir die Folge der Näherungsgleichungen

(2.2) $\quad A_\gamma u_\gamma = M_\gamma s, \quad s \in S$ gegeben, $\quad u_\gamma \in V_\gamma$ gesucht.

Wir bemerken, daß wir die Räume U_γ und S_γ, die wir hätten einführen können, mit V_γ identifizieren. (2.2) ist ein Gleichungssystem mit $\nu(\gamma)$ reellen Unbekannten. Es definiert ein Differenzenverfahren $\mathcal{D} = \mathcal{D}(\Gamma, A)$.

Definition 2.1: Das Verfahren $\mathcal{D}(\Gamma, A)$ heißt konsistent, wenn

(2.3) $$\lim \|A_\gamma R_\gamma u - M_\gamma A u\| = 0 \qquad \forall\, u \in D(A) \quad .$$

Definition 2.2: Das Verfahren $\mathcal{D}(\Gamma, A)$ heißt stabil, wenn positive Konstanten K und h_o existieren derart, daß für $\bar{h}(\gamma) < h_o$

$$\|v_\gamma\| \leq K \|A_\gamma v_\gamma\| \qquad \forall\, v_\gamma \in V_\gamma \quad .$$

Anmerkung: Da die A_γ mit Matrizen identifizierbar sind, folgt aus der Stabilität die Eindeutigkeit und damit die Existenz der Näherungslösungen u_γ von (2.2). Man beachte, daß die Stabilität nur eine Eigenschaft der Folge $\{A_\gamma\}_{\gamma \in \Gamma}$ ist.

Definition 2.3: Das Verfahren $\mathcal{S}(\Gamma,A)$ heißt diskret konvergent, wenn aus der Existenz einer Lösung $u \in U$ von (2.1) die Lösbarkeit von (2.2) für $\bar{h}(\gamma) < h_o$, h_o passend, und

$$\lim \| R_\gamma u - u_\gamma \| = 0$$

folgt.

Satz 2.1: Wenn das Verfahren $\mathcal{S}(\Gamma,A)$ konsistent und stabil ist, so ist es auch diskret konvergent, und im Falle der Existenz einer Lösung $u \in U$ von (2.1) ist für $\bar{h}(\gamma) < h_o$

$$\| R_\gamma u - u_\gamma \| \leq K \| A_\gamma R_\gamma u - M_\gamma A u \| = K \| A_\gamma R_\gamma u - M_\gamma s \| .$$

Beweis: In der Definition 2.2 der Stabilität setze man $v_\gamma = R_\gamma u - u_\gamma$ und beachte (2.2) und (2.1).

Anmerkung: Aus Satz 2.1 und der Stetigkeit der Funktion $u \in U$ folgt: Wenn $\mathcal{S}(\Gamma,A)$ stabil ist, so hat (2.1) höchstens eine Lösung $u \in U$.

Satz 2.2: Es seien die Voraussetzungen des Satzes 2.1 erfüllt, und es sei $A_\gamma \bar{u}_\gamma = \bar{s}_\gamma$. Dann ist für alle $\gamma \in \Gamma$ mit $\bar{h}(\gamma) < h_o$

$$\| R_\gamma u - \bar{u}_\gamma \| \leq K(\| A_\gamma R_\gamma u - M_\gamma A u \| + \| M_\gamma s - \bar{s}_\gamma \|).$$

Beweis mittels Dreiecksungleichung.

Anmerkungen: Die Konvergenzgeschwindigkeit in Satz 2.1 ist bestimmt durch den Abbruchfehler $\| A_\gamma R_\gamma u - M_\gamma A u \|$ des Verfahrens $\mathcal{S}(\Gamma,A)$. Die in Satz 2.2 auftretenden $M_\gamma s - \bar{s}_\gamma$ können beispielsweise interpretiert werden als durch Rundungsfehler auf einer Rechenanlage auftretende Störungen.

§ 3 Invers-isotone Diskretisierung

Man kann das Verfahren $\mathcal{S}(\Gamma,A)$ invers isoton nennen, wenn für alle $\gamma \in \Gamma$ und $s \geq 0$ aus $A_\gamma v_\gamma = M_\gamma s$ folgt $v_\gamma \geq 0$. Da wir diese Eigenschaft im folgenden jedoch nicht benutzen, definieren wir nur starke und schwache Invers-Isotonie des Verfahrens.

Definition 3.1: Wir nennen das Verfahren $\mathcal{S}(\Gamma,A)$ stark invers-isoton, wenn alle A_γ invers-isoton und alle M_γ isoton sind; wir nennen es schwach invers-isoton, wenn alle A_γ invers-isoton sind ($\gamma \in \Gamma$).

Wir suchen ein hinreichendes Stabilitätskriterium. Hierzu definieren wir als Einsfunktion $\eta \in S$ dasjenige Element von S, dessen Komponenten überall auf den jeweiligen Definitionsbereichen $\equiv 1$ sind (also $\eta(x) \equiv 1$ komponenten- und punktweise). Analog sei der Einsvektor

$\eta_\gamma \in V_\gamma$ definiert als der Vektor, dessen $\nu(\gamma)$ Komponenten alle = 1 sind.

Satz 3.1 (Majorisierungsprinzip): Das Verfahren $\mathcal{J}(\Gamma,A)$ sei schwach invers-isoton. Zu jedem $\gamma \in \Gamma$ existiere ein $w_\gamma \in V_\gamma$ mit $\|w_\gamma\| \leq K$ und $A_\gamma w_\gamma \geq \eta_\gamma$ für $\bar{h}(\gamma) < h_o$. Dabei seien K und h_o geeignete positive Konstanten. Dann ist das Verfahren $\mathcal{J}(\Gamma,A)$ stabil.

Beweis: Sei $v_\gamma \in V_\gamma$ beliebig und $y_\gamma = \|A_\gamma v_\gamma\| w_\gamma$. Dann ist für $\bar{h}(\gamma) < h_o$

$$A_\gamma y_\gamma = \|A_\gamma v_\gamma\| A_\gamma w_\gamma \geq \|A_\gamma v_\gamma\| \eta_\gamma \geq |A_\gamma v_\gamma| ,$$

also $|v_\gamma| \leq y_\gamma$ nach dem Einschließungsprinzip (man vgl. Satz 1.3) und

$$\|v_\gamma\| \leq \|y_\gamma\| \leq K\|A_\gamma v_\gamma\| .$$

Korollar zu Satz 3.1 (lokale Eingrenzung): Die Voraussetzungen des Satzes 3.1 seien erfüllt, und (2.1) habe eine Lösung $u \in U$. Dann ist mit den Lösungen u_γ von (2.2)

$$|u_\gamma - R_\gamma u| \leq \|A_\gamma R_\gamma u - M_\gamma A u\| w_\gamma \quad . \tag{3.1}$$

Beweis: In der beim Beweis des Satzes 3.1 auftretenden Ungleichung $|v_\gamma| \leq y_\gamma$ setze man $v_\gamma = R_\gamma u - u_\gamma$ und beachte (2.1) und (2.2).

Definition 3.2: Die Folge $\{M_\gamma\}_{\gamma \in \Gamma}$ heißt eins-konsistent, wenn $\lim\|\eta_\gamma - M_\gamma \eta\| = 0$ ist.

Definition 3.3: Das Verfahren $\mathcal{J}(\Gamma,A)$ nennen wir ausgeglichen, wenn die Folge $\{M_\gamma\}_{\gamma \in \Gamma}$ eins-konsistent ist.[3)]

Für spätere Bezugnahme formulieren wir zwei Voraussetzungen.

(A1) Das Verfahren $\mathcal{J}(\Gamma,A)$ sei konsistent, schwach invers-isoton und ausgeglichen, und es existiere ein $w \in U$ mit $Aw = \eta$.

(A2) Das Verfahren $\mathcal{J}(\Gamma,A)$ sei konsistent, stark invers-isoton und ausgeglichen, und es existiere ein $w \in U$ mit $Aw \geq \eta$.

Satz 3.2 (Stabilitätskriterium): Es gelte (A1) oder (A2). Dann ist das Verfahren $\mathcal{J}(\Gamma,A)$ stabil. Mit beliebig wählbarem, nach erfolgter Wahl aber festem $a \in (0,1)$ sind mit $w_\gamma = (1/a) R_\gamma w$ und $K = \|w\|/a$ die Voraussetzungen des Satzes 3.1 erfüllt, und nach Satz 2.1 ist das Verfahren diskret konvergent.

Beweis: Mit $\delta_\gamma = A_\gamma R_\gamma w - M_\gamma Aw$ ist $\lim\|\delta_\gamma\| = 0$ wegen der Konsistenz und $A_\gamma R_\gamma w = \delta_\gamma + M_\gamma Aw \geq -\|\delta_\gamma\| n_\gamma + M_\gamma \eta$, und das ist $\geq a n_\gamma$ für $\bar{h}(\gamma) < h_o$, h_o passend.

Kommentar: Oft weiß man aus der Analysis des zu behandelnden Randwertproblems, daß $Aw = \eta$ eine Lösung $w \in U$ hat. Trifft dies zu, so ist das Verfahren $\mathcal{G}(\Gamma,A)$ stabil, wenn es konsistent ist, die A_γ invers-isoton sind und die Folge $\{M_\gamma\}$ eins-konsistent ist. Manchmal kann man ein w mit $Aw \geq \eta$ raten. Dann hat man im Falle eines ausgeglichenen konsistenten stark invers-isotonen Schemas $\mathcal{G}(\Gamma,A)$ mit (3.1) wenigstens eine qualitative lokale Einschließung. Wenn es gelingt, den Abbruchfehler $\|A_\gamma R_\gamma u - M_\gamma Au\| = \|A_\gamma R_\gamma u - M_\gamma s\|$ vermittels der bekannten "Daten" s abzuschätzen (schon Gerschgorin hat das an einem Spezialfall durchgeführt), so hat man auch eine numerische Fehlerabschätzung.

§ 4 Die lineare Zweipunktrandwertaufgabe zweiter Ordnung

Es liege vor die als Beispiel (B) am Ende des § 1 angeschriebene Zweipunktrandwertaufgabe. Zur Abkürzung sei

$$P = \max\{|p(x)| \quad |0 \leq x \leq 1\}, \qquad Q = \max\{q(x) \quad |0 \leq x \leq 1\} \quad .$$

Mit den Maschenweiten $h = 1/N$, $N \in \mathbb{N}$, verwenden wir die Gitter

$$\gamma(h) = \{jh \mid j = 0,1,2,\ldots,N=1/h\} \; ,$$

und es sei $\check{\gamma}(h) = \{jh \mid j = 1,2,\ldots,N-1\}$, $\dot{\gamma}(h) = \{0, Nh=1\}$.
Es sei aus später ersichtlichen Gründen $2N > P$, also

$$Ph < 2 \qquad (4.1)$$

und im Sonderfall $\varphi = \psi = 0$

$$h < h_o, \qquad h_o \text{ so, daß } \gamma(h) \cap \{x|q(x) > 0\} \text{ nichtleer ist.} \qquad (4.1')$$

Wir definieren jetzt $\Gamma = \{\gamma(h) \mid h = 1/N,\ N \in \mathbb{N},\ (4.1),\ \text{ggf. } (4.1')\}$.

Weil $\gamma(h)$ durch die Maschenweite h eindeutig bestimmt ist, ersetzen wir an den entsprechenden Stellen den Index γ durch den Index h.

Um zu diskretisieren, ersetzen wir in der Differentialgleichung für alle $x \in \gamma(h)$ die Ableitungen von u durch zentrale Differenzenquotienten von u_h. Bei $x \in \dot{\gamma}(h)$ diskretisieren wir ebenso die erste Ableitung in der Randbedingung. Hierzu benötigen wir noch die externen Punkte $-h$ und $1+h$ und dort die Werte $u_h(-h)$ und $u_h(1+h)$. Weil wir für jeden der Punkte 0 und 1 aber zwei Gleichungen haben, können wir die externen Werte $u_h(-h)$ und $u_h(1+h)$ eliminieren, so daß auch für die Punkte $x \in \dot{\gamma}(h)$ jeweils eine Gleichung übrigbleibt. Wir erhalten so ein ausgeglichenes Schema $A_h u_h = M_h s$ mit

$$(4.2)\quad \begin{cases} (A_h u_h)(x) = \frac{1}{h^2}\{(-1-\frac{hp(x)}{2})u_h(x-h) + (2+h^2 q(x))u_h(x) + (-1+\frac{hp(x)}{2})u_h(x+h)\} \\ (M_h s)(x) = r(x) \qquad \text{für } x \in \check{\gamma}(h) \quad , \end{cases}$$

$$(4.3)\quad \begin{cases} (A_h u_h)(0) = \{\frac{(2+h^2 q(0))\cos\varphi}{h(2+hp(0))} + \sin\varphi\}u_h(0) - \frac{2\cos\varphi}{h(2+hp(0))}u_h(h) \\ (M_h s)(0) = \alpha + \frac{h\cos\varphi}{2+hp(0)}r(0) \end{cases}$$

$$(4.4)\quad \begin{cases} (A_h u_h)(1) = -\frac{2\cos\psi}{h(2-hp(1))}u_h(1-h) + \{\frac{(2+h^2 q(1))\cos\psi}{h(2-hp(1))} + \sin\psi\}u_h(1) \\ (M_h u_h)(1) = \beta + \frac{h\cos\psi}{2-hp(1)}r(1) \quad . \end{cases}$$

Daß wir zur Herleitung dieses Schemas Werte von u_h an Stellen benutzt haben, an denen u gar nicht definiert ist, braucht uns nach durchgeführter Elimination nicht mehr zu stören.

Im Sonderfall $\varphi = \psi = \pi/2$ hat man die übliche Diskretisierung der Dirichletschen Zweipunktrandwertaufgabe. Durch Ersetzung der Inhomogenitäten α,β,r durch 1 erkennt man die Einskonsistenz der Folge $\{M_h\}$. Elementare Rechnung ergibt Konsistenz des Schemas für $u \in C^2[0,1] = U = D(A)$, während man schärfer

$$\|A_h R_h u - M_h A u\| = O(h^2), \qquad h \to 0$$

für $u \in C^4[0,1]$ erhält. Die Isotonie der M_h ist evident.

Die Operatoren A_h sind als Matrizen irreduzibel mit positiver dominanter Diagonale, während alle ihre Nichtdiagonal-Elemente nichtpositiv sind. Es sind nun zwei Fälle zu unterscheiden:

(a) $\qquad \varphi \neq 0 \quad \text{oder} \quad \psi \neq 0$

(b) $\qquad \varphi = \psi = 0 \quad \text{und} \quad q(x) \not\equiv 0 \quad .$

Im Falle (a) ist A_h in der ersten oder letzten Zeile stark diagonaldominant, weil (4.1) gilt. Im Falle (b) hat man in mindestens einer Zeile starke Diagonaldominanz, weil (4.1') gilt. Bekanntlich (man vgl. z.B. Collatz 1964, Seite 297) ist dann A_h invers-isoton. Da es ferner ein $w \in U$ mit $Aw = \eta$ gibt, sind alle Voraussetzungen des Satzes 3.2 erfüllt. Es gilt der folgende Satz.

Satz 4.1: Das auf Γ durch (4.2), (4.3), (4.4) für die Zweipunktrandwertaufgabe (B) definierte Schema $\mathcal{S}(\Gamma,A)$ ist stark invers-isoton und ausgeglichen, konsistent und stabil (also auch konvergent). Wenn $Au = s$ eine Lösung $u \in C^4[0,1]$ hat, so ist $\|u_h - R_h u\| = O(h^2)$.

Für einige Sonderfälle des Problems (B) geben wir nun Vergleichsfunktionen w mit $Aw \geq \eta$ an, da man oft die Lösung von $Aw = \eta$ nicht kennt. Die in diesen Funktionen w auftretenden Konstanten B und C sind jeweils positiv und genügend groß zu wählen - in Abhängigkeit von p, q, φ, ψ. Wir verzichten auf die Durchführung der hierhergehörenden Détailrechnungen.

1. Fall: $\varphi, \psi \in (0, \frac{\pi}{2}]$: $w(x) = B - \cosh(C(x - \frac{1}{2}))$.

2. Fall: $\varphi \in (0, \frac{\pi}{2}]$, $\psi = 0$: $w(x) = B - \cosh(C(x - 2))$.

3. Fall: $\varphi = \psi = 0$, $q(x) \geq \hat{q} > 0$: $w(x) = B + C(x - \frac{1}{2})^2$.

4. Fall: $p(x) \equiv 0$, $\varphi = \psi = \pi/2$: $w(x) = \frac{9}{8} - \frac{1}{2}(x - \frac{1}{2})^2$.

Wir analysieren für den 4. Fall noch eine andere invers-isotone Diskretisierung, nämlich die von Bramble und Hubbard 1964 angegebene: Es sei $h = \frac{1}{N}$, $N \in \mathbb{N}$,

(4.5) $$\frac{1}{12h^2}\{u_h(x-2h)-16u_h(x-h)+(30+12q(x)h^2)u_h(x)-16u_h(x+h) +u_h(x+2h)\}= r(x)$$

für $x = 2h, 3h, \ldots, 1-2h$,

(4.6) $$\frac{1}{h^2}\{-u_h(x-h) + (2+q(x)h^2)u_h(x) - u_h(x+h)\} = r(x) \text{ für } x = h \text{ und } x = 1-h,$$

(4.7) $$u_h(0) = \alpha, \quad u_h(1) = \beta \quad .$$

Für $u \in C^6[0,1]$ ist hier der Differentialausdruck für $x = 2h, 3h, \ldots, 1-2h$ in der Ordnung 4, für $x = h$ und $x = 1-h$ nur in der Ordnung 2 approximiert, und die Frage ist, wie schnell nun u_h gegen u konvergiert. Bramble und Hubbard haben gezeigt, daß die Systemmatrix für $h < h_o$, h_o passend, invers-isoton ist. Sei also von jetzt ab $h < h_o$. Nimmt man als Gitter $\tilde{\gamma}(h) = \{0,h,2h,\ldots,1\}$ und nimmt für

$x = h, 2h, \ldots, 1-h$ als $(\tilde{A}_h u_h)(x)$ den in (4.5) bzw. (4.6) links stehenden Ausdruck und $(\tilde{M}_h s)(x) = r(x)$, während man $(\tilde{A}_h u_h)(0) = u_h(0)$ und $(\tilde{M}_h s)(0) = \alpha$, $(\tilde{A}_h u_h)(1) = u_h(1)$, $(\tilde{M}_h s)(1) = \beta$ nimmt, so ist das Verfahren $\tilde{\mathcal{F}}(\Gamma, A)$ konsistent, wegen der ungenaueren Approximation bei $x = h$ und $x = 1-h$ ist allerdings selbst für $u \in C^6[0,1]$ im allgemeinen nur $\|A_h R_h u - M_h A u\| = O(h^2)$, und nicht $O(h^4)$. Ferner ist die Bedingung (A2) zu Satz 3.2 erfüllt. Also ist das Verfahren stabil und diskret konvergent, und für $u \in C^4[0,1]$ hat man die Grenzbeziehung $\|u_h - R_h u\| = O(h^2)$, die sich anscheinend auch für $u \in C^6[0,1]$ nicht verbessern läßt. Bramble und Hubbard haben 1964 jedoch nachgewiesen, daß $\|u_h - R_h u\| = O(h^4)$ für $u \in C^6[0,1]$ gilt, und mit einer etwas anderen Methode hat Ciarlet 1970 ebenfalls die schärfere Abschätzung bewiesen.

Um dieses Resultat zu bekommen, verwenden wir die Methode der Gitterverkleinerung, die sich auch auf andere Randwertaufgaben mit schlechterer Approximation in Randnähe anwenden läßt und mittels deren Hilfe man die getrennte Abschätzung über Teilsummen der diskreten Greenschen Funktion in verschiedenen Teilmengen des Gitters vermeiden kann (man vgl. z.B. Ciarlet 1970, der die Methode der getrennten Abschätzungen beschreibt).

Wir setzen $\gamma(h) = \{h, 2h, \ldots, 1-h\}$ und $(A_h u_h)(x)$ gleich dem in (4.5) links stehenden Ausdruck, $(M_h s)(x) = r(x)$ für $x \in \{2h, 3h, \ldots, 1-2h\}$. Für $x = h$ und $x = 1-h$ setzen wir in (4.6) die nach (4.7) vorgegebenen Werte α und β für $u_h(0)$ und $u_h(1)$ ein und schaffen sie nach rechts. Weil sie dann noch den Faktor $\frac{1}{h^2}$ haben, ist das Schema nicht ausgeglichen. Wir multiplizieren mit h^2 und haben

$$(A_h u_h)(h) = (2 + q(h)h^2)u_h(h) - u_h(2h),$$

$$(A_h u_h)(1-h) = -u_h(1-2h) + (2 + q(1-h)h^2)u_h(1-h),$$

$$(M_h s)(h) = \alpha + h^2 r(h), \quad (M_h s)(1-h) = \beta + h^2 r(1-h).$$

Auf dem verkleinerten Gitter $\gamma(h)$ ist jetzt das Schema ausgeglichen und, wegen der erfolgten Multiplikation mit h^2, haben wir jetzt auch für $x = h$ und $x = 1-h$ Konsistenz in der Ordnung 4, wenn $u \in C^6[0,1]$. Nach Satz 3.2 ist das Verfahren stabil, und wegen

$$\|A_h R_h u - M_h A u\| = O(h^4) \qquad \text{für } u \in C^6[0,1]$$

haben wir nach Satz 2.1

$$\|u_h - R_h u\| = O(h^4),$$

wenn die Lösung $u \in C^6[0,1]$ ist. Die schlechtere Approximation des Differentialoperators in Randnähe verringert also nicht die Konvergenzordnung gegenüber der besseren Approximation für weiter innen gelegene Gitterpunkte.

Das Beispiel lehrt, daß das gleiche System von Differenzengleichungen manchmal auf verschiedene Weise als ausgeglichenes Schema geschrieben werden kann, je nachdem, welche Punkte man zum Gitter $\gamma(h)$ zusammenfaßt. In solchen Fällen, in denen in Randnähe die Konsistenzordnung schlechter wird, kommt man manchmal durch Verkleinerung des Gitters auf bessere Konsistenzordnung.

§ 5 Das Dirichlet-Problem für die Poisson-Gleichung

Die Punktmenge $\Omega \subset \mathbb{R}^2$ sei beschränkt, offen und zusammenhängend. Mit

$\Delta = \frac{\partial^2}{\partial x^2} + \frac{\partial^2}{\partial y^2}$ sei für $u \in C^2(\bar{\Omega}) = U = D(A)$

$$(5.1) \quad \begin{cases} (Au)(x,y) = -(\Delta u)(x,y) & \text{für } (x,y) \in \Omega \quad , \\ (Au)(x,y) = u(x,y) & \text{für } (x,y) \in \partial\Omega \quad , \end{cases}$$

$$(5.2) \quad s(x,y) = \begin{cases} f(x,y) & \text{für } (x,y) \in \Omega, \quad f \in C(\bar{\Omega}) \quad , \\ g(x,y) & \text{für } (x,y) \in \partial\Omega, \quad g \in C(\partial\Omega) \quad . \end{cases}$$

Für $s \in S = A(U)$ hat die Aufgabe

$$(5.3) \qquad Au = s$$

bekanntlich genau eine Lösung $u \in U$.

Wir betrachten die Diskretisierung von Shortley und Weller (1938), von der Bramble und Hubbard 1962 und Ciarlet 1970 nachgewiesen haben, daß sie in der Ordnung 2 konvergiert, falls $u \in C^4(\bar{\Omega})$. Um nicht die Summe der diskreten Greenschen Funktion in Randnähe gesondert abschätzen zu müssen, wenden wir, ähnlich wie in § 4, die Methode der Gitterverkleinerung an.

Für gegebene Maschenweite h ziehen wir in x-Richtung und in y-Richtung äquidistante Gitterlinien mit dem Abstand h, deren in Ω liegende Schnittpunkte die Menge $\gamma(h)$ bilden. Die Menge $(\partial\Omega)_h$ bestehe aus den Schnittpunkten der Gitterlinien mit $\partial\Omega$.

Es sei $\check{\gamma}(h) \subset \gamma(h)$ die Menge der Gitterpunkte, die mitsamt ihren vier jeweiligen längs Gitterlinien benachbarten Gitterpunkten in $\gamma(h)$ liegen, und es sei $\dot{\gamma}(h) = \gamma(h) - \check{\gamma}(h)$.

Für einen beliebigen Punkt $P = (x,y) \in \gamma(h)$ seien P_1, P_2, P_3, P_4 die längs Gitterlinien benachbarten Punkte von $\gamma(h) \cup (\partial\Omega)_h$ gemäß

$$P_1 = (x+h_1,y), \quad P_2 = (x,y+h_2), \quad P_3 = (x-h_3,y), \quad P_4 = (x,y-h_4) \quad .$$

Im Falle $P \in \check{\gamma}(h)$ ist $h_1 = h_2 = h_3 = h_4 = h$, stets ist $0 < h_j \leq h$.

Wir wollen ein ausgeglichenes Schema aufstellen. Nach Shortley und Weller sei

$$(5.4) \qquad (D_h u_h)(P) = -\frac{2}{h_1(h_1+h_3)} u_h(P_1) - \frac{2}{h_2(h_2+h_4)} u_h(P_2)$$

$$- \frac{2}{h_3(h_3+h_1)} u_h(P_3) - \frac{2}{h_4(h_4+h_2)} u_h(P_4) + \left(\frac{2}{h_1 h_3} + \frac{2}{h_2 h_4}\right) u_h(P).$$

Wir setzen

$$(5.5) \qquad (A_h u_h)(P) = (D_h u_h)(P), \quad (M_h s)(P) = f(P) \qquad \text{für } P \in \check{\gamma}(h) \ .$$

Für $u \in C^4(\bar{\Omega})$ ist bekanntlich

$$\max\{|(A_h R_h u)(P) - (M_h \ Au)(P)| \, \big| \, P \in \check{\gamma}(h)\} = O(h^2).$$

Würde man auf $\dot{\gamma}(h)$ ebenfalls A_h und M_h wie in (5.5) definieren, so wäre dort i.a., auch für $u \in C^4(\bar{\Omega})$, die Konsistenzordnung nur 1, und mit der naheliegenden Definition $(A_h u_h)(P) = u_h(P)$, $(M_h s)(P) = g(P)$ für $P \in (\partial\Omega)_h$ ergäbe sich auch die Konvergenz von u_h gegen u nur in der Ordnung 1. Wir gehen deshalb so vor:

Für $P \in \dot{\gamma}(h)$ setzen wir $(D_h u_h)(P) = f(P)$ und ersetzen die links auftretenden Randwerte $u(P_j)$ durch die entsprechenden vorgegebenen $g(P_j)$ (für $P_j \in (\partial\Omega)_h$) und schaffen sie dann nach rechts. Das so entstehende Schema ist nicht ausgeglichen, weil im Nenner rechts noch quadratische Poylnome der h_ν stehen. Ausgleichung auf Eins-Konsistenz bewirkt also eine Erhöhung der Konsistenzordnung um 2, so daß das entstehende Schema die Konsistenzordnung 3 auf $\dot{\gamma}(h)$ hat.

Sind beispielsweise $P_1, P_2 \in (\partial\Omega)_h$, $P_3, P_4 \in \gamma(h)$, so dividiert man

die unausgeglichene Differenzengleichung durch $\frac{2}{h_1(h_1+h_3)} + \frac{2}{h_2(h_2+h_4)}$,
und dies bedeutet eine Multiplikation mit dem $O(h^2)$-Ausdruck

$$\alpha(h) = \frac{h_1 h_2 (h_1+h_3)(h_2+h_4)}{2h_1(h_1+h_3)+2h_2(h_2+h_4)} .$$

Für einen solchen Punkt P ist nun

$$(A_h u_h)(P) = \alpha(h)\left\{ -\frac{2u_h(P_3)}{h_3(h_3+h_1)} - \frac{2u_h(P_4)}{h_4(h_4+h_2)} + \left(\frac{2}{h_1 h_3} + \frac{2}{h_2 h_4}\right) u_h(P) \right\} ,$$

$$(M_h s)(P) = \frac{2\alpha(h)}{h_1(h_1+h_3)} g(P_1) + \frac{2\alpha(h)}{h_2(h_2+h_4)} g(P_2) + \alpha(h) f(P) .$$

Nun sei $h < h_o$, h_o passend, so daß die dem Operator A_h entsprechende Matrix irreduzibel ist. Man überlegt sich jetzt leicht, daß die Voraussetzung (A2) des Satzes 3.2 anwendbar ist, das Verfahren also stabil ist. Weil jetzt $\|A_h R_h u - M_h A u\| = O(h^2)$ für $u \in C^4(\bar{\Omega})$, gilt auch, falls die Lösung $u \in C^4(\bar{\Omega})$, die Grenzbeziehung $\|u_h - R_h u\| = O(h^2)$. Eine geeignete Vergleichsfunktion w ist gegeben durch

$$w(x,y) = \frac{1}{2} c^2 + 1 - \frac{1}{2} \{(x-a)^2 + (y-b)^2\} \quad \text{für } (x,y) \in \bar{\Omega} .$$

Hierbei ist (a,b) ein beliebiger, aber fixierter Punkt von Ω, und c ist der Durchmesser von $\bar{\Omega}$.

Schlußbemerkung

Mit der in dieser Arbeit dargestellten Methode lassen sich auch inversisotone Diskretisierungen elliptischer Differentialgleichungen mit gemischten Randbedingungen (man vgl. Bramble und Hubbard 1965), Thuraisamy 1969) und parabolischer Anfangs-Randwert-Aufgaben (mit Dirichletschen oder gemischten seitlichen Randbedingungen) bequem analysieren. Wegen der erforderlichen Schreibarbeit sei hier auf die Durchführung verzichtet.

Literatur

I. BABUŠKA, M. PRÁGER, E. VITÁSEK: Numerical processes in differential equations. Übersetzung aus dem Tschechischen. John Wiley and Sons, London 1966.

J.H. BRAMBLE and B.E. HUBBARD: On the formulation of finite difference analogues of the Dirichlet problem for Poisson's equation. Numerische Mathematik 4, 313 - 327 (1969).

J.H. BRAMBLE and B.E. HUBBARD: On a finite difference analogue of an elliptic boundary value problem which is neither diagonally dominant nor of non-negative type. Journal of Mathematics and Physics 43, 117 - 132 (1964).

J.H. BRAMBLE and B.E. HUBBARD: Approximations of solutions of mixed boundary value problems for Poisson's equation by finite differences. Journal of the ACM 12, 114 - 123 (1965).

J.H. BRAMBLE, B.E. HUBBARD, Vidar THOMÉE: Convergence estimates for essentially positive type discrete Dirichlet problems. Mathematics of Computation 23, 695 - 709 (1969).

P.G. CIARLET: Discrete maximum principles for finite difference operators. Aequationes mathematicae 4, 338 - 352 (1970).

L. COLLATZ: Bemerkungen zur Fehlerabschätzung für das Differenzenverfahren bei partiellen Differentialgleichungen. ZAMM 13, 56 - 57 (1933).

L. COLLATZ: Aufgaben monotoner Art. Arch. math. 3, 366 - 376 (1952).

L. COLLATZ: Funktionalanalysis und numerische Mathematik. Springer-Verlag, Berlin 1964.

G.E. FORSYTHE and W.R. WASOW: Finite difference methods for partial differential equations. John Wiley and Sons, New York 1960.

S. GERSCHGORIN: Fehlerabschätzung für das Differenzenverfahren zur Lösung partieller Differentialgleichungen. ZAMM 10, 373 - 382 (1930).

R. GORENFLO: Differenzenschemata monotoner Art für schwach gekoppelte Systeme parabolischer Differentialgleichungen mit gemischten Randbedingungen. Computing 8, 343 - 362 (1971).

K. HAINER: Lösung des Dirichletproblems und Konvergenz der Differenzenapproximation für elliptische Differentialgleichungen zweiter Ordnung. Mathematische Zeitschrift: demnächst. Manuskript 1972.

W. KOLAR: Über allgemeine monotone Differenzenverfahren zur Lösung des ersten Randwertproblems bei parabolischen Differentialgleichungen. Dissertation RWTH Aachen 1970. Bericht Jül-672-MA der Kernforschungsanlage Jülich GmbH, Juli 1970.

J.M. ORTEGA and W.C. RHEINBOLDT: Iterative solution of nonlinear equations in several variables. Academic Press, New York 1970.

M.H. PROTTER and H.F. WEINBERGER: Maximum principles in differential equations. Prentice Hall, Englewood Cliffs, N.J., 1967.

G. SHORTLEY and R. WELLER: The numerical solution of Laplace's equation. J.Appl.Phys. 9, 334 - 348 (1938).

V. THURAISAMY: Approximate solutions for mixed boundary value problems by finite difference methods. Mathematics of computation 23, 373 - 386 (1969).

V. THURAISAMY: Monotone type discrete analogue for the mixed boundary value problem. Mathematics of Computation 23, 387 - 394 (1969).

W. WALTER: Differential- und Integral-Ungleichungen. Springer-Verlag, Berlin 1964.

Anmerkungen:

1) Wir schließen uns der Terminologie von Ortega und Rheinboldt (1970) an. Monotone Art nach L. Collatz (1952) = Invers-Isotonie.

2) Wenn $\bar{\Omega} = [a,b] \subset \mathbb{R}$, $\|x\| = |x|$ für $x \in \mathbb{R}$ und $a,b \in \gamma$, so ist $\bar{h}(\gamma)$ der maximale Abstand zweier benachbarter Gitterpunkte; ist zusätzlich γ ein äquidistantes Gitter, so ist $\bar{h}(\gamma)$ die Maschenweite.

3) Bramble, Hubbard und Thomée (1969) verwenden in Randnähe eine Schreibweise des Verfahrens, die bis auf einen von 0 und ∞ weg-beschränkten Faktor ausgeglichen ist.

APPROXIMATE SOLUTIONS OF FUNCTIONAL DIFFERENTIAL EQUATIONS

By Myron S. Henry

1. INTRODUCTION

Consider the functional differential system

$$\ddot{x}(t) = f(t,x(g(t)))$$
$$x(0) = c_0\ ,\ \dot{x}(0) = c_1. \tag{1.1}$$

The questions of existence and uniqueness of solutions to initial value problems with deviating arguments have been considered in several recent papers, see, for example, [8,9]. However, numerical techniques for problems of this type are not well developed. In fact, the functional argument $g(t)$ and the lack of an initial segment requirement for $x(t)$ complicate spline (see [1,2,7] for $g(t) = t$) and finite difference techniques. In this paper, an alternative that results in a uniform approximate solution to some forms of (1.1) is discussed. Suppose that (1.1) has the unique solution $y(t)$ on $I = [-\tau,\tau]$, and that

$$Fx = \ddot{x}(t) - f(t,x(g(t))). \tag{1.2}$$

Then on I let $\mathcal{P} = \{p(t)\}$ be the set of all polynomials of degree at most k of the form

$$p(t) = c_0 + c_1t + a_2t^2 + \ldots + a_kt^k.$$

If the problem

$$\inf_{p \in \mathcal{P}} \max_{I} |Fp| = \inf_{p \in \mathcal{P}} \|Fp\| \tag{1.3}$$

has solution $p_k(t)$, then we show that for certain functions $f(t,x)$ and $g(t)$

$$\lim_{k \to \infty} \|p_k^{(j)}(t) - y^{(j)}(t)\| = 0,\ j = 0,1,2. \tag{1.4}$$

If (1.2) is a linear operator, then the existence of $p_k(t)$ is guaranteed and computational techniques are discussed. Best approximation problems of this type have been considered in [1,2,3,4,5,6], but in all of these references the initial value problems involved only $g(t) = t$.

2. THE EXISTENCE OF SOLUTIONS

In this section we present an existence theorem for the initial value problem (1.1). The tools of this proof will be utilized in succeeding sections of this paper, and the proof parallels that given in [9].

Suppose that $g(t)$ in (1.1) is continuous on I, and that $g(I) \subseteq I$. Define

$$\|f\|_1 = \max(|f|, |f'|).$$

Let S be the set of all twice continuously differentiable functions $u(t)$ on I that satisfy the initial conditions of (1.1), and that satisfy

$$\|u(t) - (c_0+c_1t)\|_1 < L \max(2|t|, t^2)$$

for some constant L. Suppose that $f \in C[I\times R]$, and that

$$|f(t,x_1) - f(t,x_2)| \leq K|y_1-y_2|$$

for (t,x_1) and (t,x_2) in $I\times R$.

Theorem 1. Let $g(t)$, $f(t,x)$ be as described above. If $\max_I |g(t)| \leq m$, and if $K \max(2m,m^2) < 2$, then (1.1) has a unique solution in S on I.

Proof: We first note that the pair (S,ρ), where

$$\rho(u_1,u_2) = \inf\{L: \|u_1-u_2\|_1 \leq L \max(2|t|,t^2) \quad \forall t \in I\} \tag{2.1}$$

is a complete metric space.

Let

$$Tu = c_0 + c_1t + \int_0^t (t-s)f(s,u(g(s)))ds . \tag{2.2}$$

Then $u(t)$ satisfies (1.1) if and only if $u \in S$ and $u(t) = Tu(t)$. Thus we show that T is a contraction operator on S. For let $u(t) \in S$ and $u_0(t) = c_0 + c_1t$.

Then

$$(Tu)(0) = c_0, \quad (Tu)'(0) = c_1.$$

Also,

$$|Tu-u_0| \leq \left|\int_0^t (t-s)f(s,u(g(s)))ds\right| \leq \frac{M}{2} t^2 , \tag{2.3}$$

where M is a bound for $f(t,u(g(t)))$ on I. Equation (2.2) also implies that

$$|(Tu-u_0)'| \leq \frac{M}{2} (2|t|). \tag{2.4}$$

Inequalities (2.3) and (2.4) imply that

$$\|Tu-u_0\|_1 \leq \frac{M}{2}\max(2|t|,t^2),$$

and hence $TS \subseteq S$.

Now let $u_1,u_2 \in S$, and let $M_S = \{L: \|u_1-u_2\|_1 \leq L\max(2|t|,t^2), t \in I\}$

and

$N_S = \{L: \|Tu_1-Tu_2\|_1 \leq L\max(2|t|,t^2),\ t \in I\}$.

Suppose that $L_0 \in M_S$. Then direct calculations similar to those in (2.3) and (2.4) show that

$$\|Tu_1-Tu_2\|_1 \leq L_0\ \{\frac{K}{2}\max(2m,m^2)\}\ \max(2|t|,t^2) \quad \text{for all } t \in I.$$

This implies that

$$L_0\ \{\frac{K}{2}\max(2m,m^2)\} \ \in N_S,$$

and since L_0 was an arbitrary element of M_S we conclude that

$$\frac{K}{2}\max(2m,m^2)M_S \subseteq N_S.$$

Thus

$$\inf\ N_S \leq \inf\{\ \frac{K}{2}\max(2m,m^2)M_S\},$$

which implies that

$$\rho(Tu_1,Tu_2) \leq \frac{K}{2}\max(2m,m^2)\ \rho(u_1,u_2).$$

Thus if $\frac{K}{2}\max(2m,m^2) < 1$, then T is a contraction operator.

3. THE APPROXIMATION PROBLEM

Suppose that the conditions of Theorem 1 are satisfied. Then the approximation problem can be considered in two parts. First, show that (1.3) has a solution $p_k(t) \in \boldsymbol{P}$ for each k, and second, show that (1.4) is satisfied.

Suppose that (1.3) holds for some $p_k(t) \in \boldsymbol{P}$, $k = 2,3,\ldots$.

Let

$$Fp_k(t) = \epsilon_k(t) \ , \tag{3.1}$$

and let

$$\max_I |\epsilon_k(t)| = \epsilon_k \ . \tag{3.2}$$

Then the Weierstrass Theorem implies that

$$\lim_{k\to\infty} \epsilon_k = 0. \tag{3.3}$$

Also (3.1) implies that

$$p_k(t) = c_0 + c_1 t + \int_0^t (t-s)f(s,p_k(g(s)))ds + \int_0^t (t-s)\epsilon_k(s)ds. \tag{3.4}$$

Define

$$T_k u = c_0 + c_1 t + \int_0^t (t-s) f(s,u(g(s)))ds + \int_0^t (t-s)\epsilon_k(s)ds \tag{3.5}$$

for $u \in S$. We note that $T_k S \subseteq S$ and that $p_k \in S$ for all k.

Then (2.2), (3.3), and (3.5) imply that

$$\lim_{k \to \infty} \|T_k u - Tu\|_1 = 0$$

uniformly for $t \in I$ and $u \in S$. From (2.2) and (3.5) we conclude that

$$\|T_k p_k - Tp_k\|_1 \leq \epsilon_{\frac{k}{2}} \max(2|t|, t^2),$$

and hence

$$\rho(T_k p_k, Tp_k) \leq \epsilon_{\frac{k}{2}} . \tag{3.6}$$

Since $\rho(T_k u_1, T_k u_2) = \rho(Tu_1, Tu_2)$, T_k is a contraction operator on S. Thus p_k is the unique fixed point of T_k in S. Therefore

$$\rho(p_k, y) = \rho(T_k p_k, Ty) \leq \rho(T_k p_k, Tp_k) + \rho(Tp_k, Ty)$$

$$\leq \rho(T_k p_k, Tp_k) + \alpha\, p(\rho_k, y),$$

where $0 < \alpha < 1$. Hence (3.6) implies that

$$\rho(p_k, y) \leq \frac{1}{1-\alpha} \cdot \frac{\epsilon_k}{2} .$$

From this inequality and (2.1) it is immediate that (1.4) holds for $j = 0,1$. We have proven the following theorem.

THEOREM 2. _If the hypotheses of Theorem 1 are satisfied, and if_ $p_k(t)$ _is the solution to_ (1.3) _for_ $k = 2,3,\ldots,$ _then_

$$\|y^{(j)}(t) - p_k^{(j)}(t)\|_1 \leq \frac{\epsilon_k}{2} \frac{1}{1-\alpha} \max(2|t|, t^2),\ j = 0,1.$$

COROLLARY. _If_ $\{p_k(t)\}$ _are as in Theorem_ 2, _then_

$$|y^{(2)}(t) - p_k^{(2)}(t)| \leq M\epsilon_k$$

for all $t \in I$, _where_ M _is a nonnegative constant_.

Proof: From $Fy = 0$ and $Fp_k = \epsilon_k(t)$ we conclude that $p_k''(t) - y''(t) = f(t,p_k(g(t))) - f(t,y(g(t))) + \epsilon_k(t)$. Hence the Lipschitz condition for f and Theorem 2 imply the result.

4. COMPUTATIONS

In this section we consider the initial value problem

$$\ddot{x}(t) + a(t)\, x(g(t)) = h(t)$$

$$x(0) = c_0 , \ \dot{x}(0) = c_1. \tag{4.1}$$

We assume that if $A = \max_I |a(t)|$, then $A \max(2m, m^2) < 2$.

Define

$$Lx = \ddot{x}(t) + a(t)x(g(t))). \tag{4.2}$$

If $p(t) = c_0 + c_1 t + a_2 t^2 + \dots + a_k t^k$,

then

$$Lp(t) = \sum_{i=2}^{k} a_i\{i(i-1)t^{i-2} + a(t)g^i(t)\} + a(t)(c_0 + c_1 g(t)).$$

Let

$$\phi_i(t) = i(i-1)t^{i-2} + a(t)g^i(t), \tag{4.3}$$

$$i = 2, \dots, k.$$

Thus showing there exists a $p_k(t) \in \mathcal{P}$ that minimizes (1.3) amounts to showing that

$$\inf_{(a_2,\dots,a_k) \in R^{k-1}} \ \sup_I \left| \sum_{i=2}^{k} a_i\phi_i(t) - [h(t) - a(t)(c_0 + c_1 t)] \right| \tag{4.4}$$

is attained for some vector $(a_2^*, \dots, a_k^*)$. If $\{\phi_i(t)\}_{i=2}^k$ is a linearly independent set on I, this is immediate.

THEOREM 3. _The set_ $\{\phi_i(t)\}_{i=2}^k$ _defined in_ (4.3) _is a linear independent set of functions on_ I.

Proof: The theorem follows from the uniqueness of the zero solution to the system

$\ddot{x}(t) + a(t)x(g(t)) = 0$, $x(0) = 0$, $\dot{x}(0) = 0$, on I and the linear independence of $\{t^i\}_{i=2}^k$.

Thus finding the best approximation to the solution of (4.1) in the sense of (1.3) has been reduced to the linear approximation problem (4.4). In general it may be quite difficult to show that $\{\phi_i(t)\}_{i=2}^k$ is a Tchebycheff set on I. However, the second algorithm of Remes may still yield a best approximation, even though $\{\phi_i(t)\}_{i=2}^k$ are only linearly independent, see [1,2]. Also, other algorithms may be applicable, see, for example, [10]. The two examples that follow demonstrate the above theory.

Example 1.
$$\ddot{x}(t) + tx(t/2) = \sin t(t - 8\cos t)$$
$$x(0) = 0, \quad \dot{x}(0) = 2 \tag{4.5}$$

The existence theorem guarantees a unique solution to this initial value problem on $I = [-1,1]$. Identity (4.3) yields

$$\phi_i(t) = i(i-1)t^{i-2} + \frac{t^{i+1}}{2^i} \,.$$

If $i = 5$, then the best approximation to the solution of (4.5) in the sense of (4.4) is

$$p_5(t) = 2t + .000139t^2 - 1.2855706t^3 - .000296t^4 + .206669t^5$$

A comparison between $p_5(t)$ and the solution $x(t) = \sin 2t$ is given in table 1.

Table 1

t	$E_5(t) = \sin 2t - p_5(t)$
-1.0	.11958541235E - 01
-.9	.11089354717E - 01
-.8	.99680497550E - 02
-.7	.83375035339E - 02
-.6	.63366427877E - 02
-.5	.42748884710E - 02
-.4	.24690414570E - 02
-.3	.11392250082E - 02
-.2	.35814535466E - 03
-.1	.45806090752E - 04
0	0
.1	-.48524413808E - 04
.2	-.36830811522E - 03
.3	-.11594267257E - 03
.4	-.24983239926E - 02
.5	-.43073199642E - 02
.6	-.63598965930E - 02
.7	-.83314371575E - 02
.8	-.99032837915E - 02
.9	-.10925851786E - 01
1.0	-.11644184917E - 01

If i = 7, then

$$p_7(t) = 2t + .000001t^2 - 1.331688t^3 - .000003t^4 + .262648t^5 - .000003t^6 - .21462t^7$$

Table 2 compares $p_7(t)$ with $\sin 2t$.

Table 2

t	$E_7(t) = \sin 2t - p_7(t)$
-1.0	.20491747896E - 03
-.9	.18013603450E - 03
-.8	.16707062476E - 03
-.7	.15768469031E - 03
-.6	.13926206268E - 03
-.5	.10814710611E - 03
-.4	.70208190548E - 04
-.3	.35457336668E - 04
-.2	.11904466937E - 04
-.1	.15990718178E - 05
0	0
.1	-.16122516375E - 05
.2	-.11950075298E - 04
.3	-.35531298749E - 04
.4	-.70260922982E - 04
.5	-.10805137803E - 03
.6	-.13876692121E - 03
.7	-.15635146897E - 03
.8	-.16418705310E - 03
.9	-.17460446753E - 03
1.0	-.19511326463E - 03

Example 2.

$$\ddot{x}(t) + (\sin t)x(\sin t) = 6|t| + \sin^4 t \tag{4.6}$$

$$x(0) = 0, \quad \dot{x}(0) = 0.$$

Again the existence theorem guarantees a unique solution to (4.6) on I = [-1.1]. In this example, (4.3) yields

$$\phi_i(t) = i(i-1)t^{i-2} + (\sin t)^{i+1} .$$

Table 3 is for

$$p_6(t) = .199364t^2 + .003045t^3 + .984168t^4 - .024780t^5 - .203431t^6.$$

Table 3

t	$E_6(t) = \|t\|^3 - p_6(t)$
-1.0	-.18352213322E - 02
-.9	.17502091035E - 01
-.8	.28059512199E - 01
-.7	.29826234198E - 01
-.6	.24903025676E - 01
-.5	.16433439325E - 01
-.4	.76814897800E - 02
-.3	.12558019592E - 02
-.2	.15197701453E - 02
-.1	-.10890542312E - 02
0	0
.1	-.10946490236E - 02
.2	-.15526341626E - 02
.3	.12117914640E - 02
.4	.77991959560E - 02
.5	.17220880559E - 01
.6	.27441262036E - 01
.7	.36066725495E - 01
.8	.41180947653E - 01
.9	.42326690558E - 01
1.0	.41634065754E - 01

5. CONCLUSIONS

Some computational difficulties arose in the above examples. Specifically, the ϵ_k of (3.2) should be monotone in k, but this did not always occur. However, the empirical data did support the conclusion of equation (3.3). Although the theory only guarantees that the error is of the form

$$\frac{\epsilon_k}{2} \frac{1}{1-\alpha} \max(2|t|, t^2),$$ in general the errors were much smaller than this estimate. This phenomena is not totally unexpected since the theory also involves approximating $\dot{x}(t)$ and $\ddot{x}(t)$, see equation (1.3). Further computational analysis is necessary. Also the use of different approximating functions in the theory would appear appropriate.

With appropriate modifications, the above analysis can be extended to initial value problems of the form

$$x^{(n)}(t) + f(t,x(g(t))) = 0, \; x^{(i)}(0) = c_i, \quad i = 0,\dots,n-1.$$

As previously indicated in the introduction, it appears that the functional argument $g(t)$ hinders the development of the spline theory given in [1,2,7] for $g(t) = t$. If such a development were possible, then Tchebycheff sets of arbitrarily large order would be unnecessary in the computations.

REFERENCES

1. Allinger, G., Spline approximate solutions to linear initial value problems, Thesis (Ph.D.), University of Utah, Salt Lake City, Utah, 1972.

2. Allinger, G., and S. E. Weinstein, Spline approximate solutions to initial value problems, to appear.

3. Bacopoulous, A., and A. G. Kartsatos, On polynomials approximating solutions of nonlinear differential equations, Pacific J. Math., Vol.40, No.1, 1972, 1-5.

4. Henry, M. S., Best approximate solutions of nonlinear differential equations, J. Approx. Theory, 1(1970), 59-65.

5. Huffstutler, R. G., and F. M. Stein, The approximate solution of certain nonlinear differential equations, Proc. Amer. Math. Soc., Vol. 19, 1968, 998-1002.

6. ________________, The approximate solution of $\dot{y} = F(x,y)$, Pacific J. Math.(1968), 283-289.

7. Loscalzo, F. R. and T. D. Talbot, Spline function approximations for solutions of ordinary differential equations, SIAM J. Numer. Anal., Vol.4, No.3, 1967, 433-445.

8. Oberg, R. J., On the local existence of solutions of certain functional differential equations, Proc. Amer. Math. Soc., Vol. 20, No. 2, 1969, 295-302.

9. Ryder, G. H., Solutions of functional differential equations, Amer. Math. Monthly, Vol. 76, No. 9, 1969, 1031-1033.

10. Scott, P. D., and J. S. Thorp, A descent algorithm for linear continuous Chebyshev approximation, Journal of Approx. Theory, Vol. 6, No. 3, 1972, 231-241.

NUMERICAL CALCULATION OF PRIMARY BIFURCATION POINTS OF THE HAMMERSTEIN OPERATOR

By Jörg Hertling

§ 1 INTRODUCTION

Under appropriate hypotheses the investigation of primary bifurcation points of Hammerstein operators which have the null element as eigenelement leads to an eigenvalue problem for a compact linear operator. For the numerical treatment of this linear eigenvalue problem we use a Ritz-Galerkin approach. In particular we use the subspace of L-splines in order to derive error bounds for the bifurcation points.

§ 2 BACKGROUND

We consider the Hammerstein operator

$$(2.1) \qquad \underline{T}u = \underline{A}\underline{h}u = \int_0^1 K(x,y)g(u(y),y)dy$$

on the Banach space $C[0,1]$ into itself. Let the linear operator $\underline{A}$ be generated by a kernel $K(x,y)$ which is bounded and continuous on $[0,1]\times[0,1]$. This implies that $\underline{A}$ is completely continuous on $C[0,1]$. The function $g(u,y)$ which defines the Nemytskii operator $\underline{h}$ may be continuous in the strip $(-\infty,+\infty)\times[0,1]$, uniformly in u with respect to y. Let $\underline{T}\theta=\theta$ where θ denotes the null element.

Plotting the norm of the eigenelements $\|u\|$ of such a nonlinear odd operator versus the eigenvalues λ one observes that the norms of the eigenelements branch away from the trivial solution $u=\theta$ at the eigenvalues $\mu_i^{(o)}$ of the linearized operator $\underline{T}'(\theta)$. For a proof see e. g. [1] or [7]. We want to approximate numerically these "primery bifurcation points" $\mu_i^{(o)}$.

The following set of assumptions which has been pointed out in [7] is suitable for this problem:

(i) If λ_o belongs to the point spectrum $P\sigma(\underline{T}'(u_o))$ then $\underline{T}'(u_o)$ may be a compact linear operator. It is well known [6] that if $\underline{T}$ is completely continuous on $C[0,1]$, then the Fréchet derivative $\underline{T}'(u)$ is also completely continuous for $u \in C[0,1]$.

(ii) $\underline{T}$ may be an odd three times Fréchet differentiable operator:

$$\underline{T}'(u)v \equiv d\underline{T}(u;v) \equiv \underline{A}g_u'v = \int_0^1 K(x,y)g_u'(u(y),y)v(y)dy,$$

$$\underline{T}''(u)v_1v_2 \equiv d^2\underline{T}(u;v_1,v_2) \equiv \underline{A}g_u''v_1v_2,$$

$$\underline{T}'''(u)v_1v_2v_3 \equiv d^3\underline{T}(u;v_1,v_2,v_3) \equiv \underline{A}g_u'''v_1v_2v_3.$$

Condition (i) is satisfied if $g_u'(u,y)$ exists and is integrable. $\underline{T}$ is odd if we assume $g(-u,y)=-g(u,y)$.

Assuming now (i) with $\lambda_o \in P\sigma(\underline{T}'(\theta))$ and (ii) it has been proved in [7] that there exist two nontrivial branches of eigenfunctions $u^{\pm}(\lambda)$ which bifurcate from the trivial solution $u=\theta$ at the bifurcation point $\lambda=\lambda_o$. These branches exist at least in a small neighborhood of λ_o and $\|u^{\pm}(\lambda)\| \to 0$ as $\lambda \to \lambda_o$.

Alternatively one may only assume (i), $T\theta=\theta$ and $d^2\underline{T}(\theta;h_1,h_2) \neq \theta$. Then there exist two nontrivial branches of eigenfunctions $u^{\pm}(\lambda)$ which bifurcate from the trivial solution $u=\theta$ at the bifurcation point $\lambda=\lambda_o$. Again these branches exist at least in a small neighborhood of λ_o and $\|u^{\pm}\| \to 0$ as $\lambda \to \lambda_o$.

More conditions are needed in order to make sure that the eigenvalues of the linearized problem are of unit multiplicity and unit Riesz index. We introduce some definitions. An $n \times n$ matrix $A=(a_{ik})$ is a completely non-negative (resp. completely positive) matrix if all of its minors of any order are non-negative (resp. positive). An $n \times n$ matrix is an oscillation matrix if it is a completely non-negative matrix and there exists a positive integer $\varkappa$ such that $A^{\varkappa}$ is a completely positive matrix. A continuous kernel $K(x,y)$, $0 \le x,y \le 1$, is an oscillation kernel if for any set of n points $x_1, x_2, \ldots, x_n$, where $0 \le x_i \le 1$, one of which is internal, the matrix $(K(x_i,x_k))$ is an oscillation matrix.

Now we assume

(iii) $K(x,y)$ is a symmetric oscillation kernel,

(iv) $g_u'(u,y) > 0$.

According to [7] these two conditions ensure that the linearized problem

$$(2.2) \qquad \mu h(x) = \int_0^1 K(x,y) g_u'(u(y),y) h(y) dy$$

has a sequence $\{\mu_n\}$ of positive and simple eigenvalues $\mu_0 > \mu_1 > \ldots > 0$ and a corresponding sequence $\{h_n\}$ of continuous eigenfunctions such that $h_n(x)$ has precisely n zeros of odd order on $(0,1)$, $n=0,1,2,\ldots$. We are interested in the sequence $\{\mu_n^{(o)}\}$ of eigenvalues for the linearized equation (2.2) with $u=\theta$. Let us now use the setwise imbedding $\mathbb{C}[0,1] \subset \mathbb{L}^2[0,1]$ and introduce the inner product. Since $\underline{A}$ is symmetric and has positive eigenvalues there exists a unique positive definite square root $\underline{A}^{1/2}$. Thus we may write (2.2) with $u=\theta$ as

$$(2.3) \qquad \mu h = \underline{A}^{1/2} \underline{g_u'} \underline{A}^{1/2} h \equiv \int_0^1 H(x,z) g_u'(\theta,z) \int_0^1 H(z,y) h(y) dy dz.$$

Finally, denoting the Rayleigh quotient by

$$(2.4) \qquad R[h] = \frac{(\underline{A}^{1/2} \underline{g_u'} \underline{A}^{1/2} h, h)}{(h,h)}, \qquad h \neq \theta$$

we may use Courant's and Poincaré's minimax principles and characterize

$$(2.5) \qquad u_k^{(o)} = \min_{\substack{v_1,\ldots,v_{k-1} \\ \text{lin. ind.}}} \Big(\max_{h \perp v_1,\ldots,v_{k-1}} R[h] \Big),$$

(2.6) $$\mu_k^{(o)} = \max_{\substack{v_1,\dots,v_k \\ \text{lin. ind.}}} \left(\min_{c_1,\dots,c_k} R[\sum_{i=1}^{k} c_i v_i] \right).$$

If $\{h_i\}_{i=1}^{\infty}$ is the sequence of eigenfunctions corresponding to $\{\mu_i^{(o)}\}_{i=1}^{\infty}$ then we can write equivalently

(2.7) $$\mu_k^{(o)} = \max_{h \perp h_1,\dots,h_{k-1}} R[h],$$

(2.8) $$\mu_k^{(o)} = \min_{c_1,\dots,c_k} R[\sum_{i=1}^{k} c_i h_i].$$

§ 3 ERROR BOUNDS FOR BIFURCATION POINTS

Obviously we can apply a Rayleigh-Ritz procedure in order to calculate approximate eigenvalues of (2.3). This procedure consits in choosing a finite dimensional subspace $\mathbb{L}_M^2$ of $\mathbb{L}^2[0,1]$ (resp. $\mathbb{C}[0,1]$) and calculating the extremal points of the Rayleigh quotient over $\mathbb{L}_M^2$.

We let $\mathbb{S}_M$ denote the subspace which is spanned by the first M orthogonal eigenfunctions $\{h_i\}_{i=1}^{M}$ of the Rayleigh quotient (2.6) with eigenvalues $\{\mu_i^{(o)}\}_{i=1}^{M}$. P may be a linear approximation scheme which maps $\mathbb{S}_M$ linearly into the subspace $\mathbb{L}_M^2$. We write

(3.1) $$Ph_i = \tilde{h}_i = h_i - \epsilon_i .$$

Application of the Rayleigh-Ritz method to $\mathbb{L}_M^2$ yields a sequence of approximate eigenvalues (primary bifurcation points) $\hat{\mu}_1^{(o)} \geq \hat{\mu}_2^{(o)} \geq \dots > 0$ where $\mu_i^{(o)} \geq \hat{\mu}_i^{(o)}$. This follows from (2.6).

Now we consider a subspace $\mathbb{L}_{M,k}^2 \subset \mathbb{L}_M^2$ with $\mathbb{L}_{M,k}^2 = P[\mathbb{S}_k]$. We decompose each $\epsilon = h - Ph$, $h \in \mathbb{S}_k$ into its component $\epsilon' \in \mathbb{S}_k$ and its component $\epsilon^{\perp} \in \mathbb{S}_k^{\perp}$, where $\mathbb{S}_k^{\perp}$ is the orthocomplement of $\mathbb{S}_k$. Then $\tilde{h}' = h - \epsilon'$ is the orthogonal projection of $\tilde{h}$ into $\mathbb{S}_k$. Since $\mathbb{S}_k$ and $\mathbb{S}_k^{\perp}$ are complementary eigensubspaces for the problem (2.3) we have for $u \in \mathbb{S}_k$ and $v \in \mathbb{S}_k^{\perp}$ the following relations

(3.2) $$(u+v,u+v) = (u,u) + (v,v),$$

(3.3) $$(\underline{A}^{1/2}g_{\underline{u}}'\underline{A}^{1/2}(u+v),u+v) = (\underline{A}^{1/2}g_{\underline{u}}'\underline{A}^{1/2}u,u) + (\underline{A}^{1/2}g_{\underline{u}}'\underline{A}^{1/2}v,v)$$

and from this

(3.4) $$(\epsilon^{\perp},\epsilon^{\perp}) \le (\epsilon,\epsilon),$$

(3.5) $$(\underline{A}^{1/2}g_{\underline{u}}'\underline{A}^{1/2}\epsilon^{\perp},\epsilon^{\perp}) \le (\underline{A}^{1/2}g_{\underline{u}}'\underline{A}^{1/2}\epsilon,\epsilon).$$

The argumentation is now analoguously to the argumentation in [2]. For all $\tilde{h}\in\mathbb{L}^2_{M,k}$, $\tilde{h}\neq\theta$

(3.6) $$R[\tilde{h}] = \frac{(\underline{A}^{1/2}g_{\underline{u}}'\underline{A}^{1/2}\tilde{h},\tilde{h})}{(\tilde{h},\tilde{h})} = \frac{(\underline{A}^{1/2}g_{\underline{u}}'\underline{A}^{1/2}\tilde{h}',\tilde{h}')}{(\tilde{h},\tilde{h})} - \frac{(\underline{A}^{1/2}g_{\underline{u}}'\underline{A}^{1/2}\epsilon^{\perp},\epsilon^{\perp})}{(\tilde{h},\tilde{h})} \ge \mu_k^{(o)} - \frac{(\underline{A}^{1/2}g_{\underline{u}}'\underline{A}^{1/2}\epsilon^{\perp},\epsilon^{\perp})}{(\tilde{h},\tilde{h})}$$

since

(3.7) $$(\tilde{h}',\tilde{h}') = (\tilde{h},\tilde{h}) + (\epsilon^{\perp},\epsilon^{\perp}) \ge (\tilde{h},\tilde{h})$$

and

(3.8) $$R[\tilde{h}'] \ge \mu_k^{(o)}$$

according to (2.6). From (3.5) follows

(3.9) $$R[\tilde{h}] \ge \mu_k^{(o)} - \frac{(\underline{A}^{1/2}g_{\underline{u}}'\underline{A}^{1/2}\epsilon,\epsilon)}{(\tilde{h},\tilde{h})}, \qquad h \neq \theta.$$

We use these results in order to prove the following

<u>Theorem 1:</u> With the conditions which we have imposed on the eigenvalue problem (2.3) let $\{h_i\}_{i=1}^M$ be the first M orthonormalized eigenfunctions. Let $\{\tilde{h}_i\}_{i=1}^M$ be a set of functions such that

(3.10) $$\sum_{i=1}^M \|\tilde{h}_i - h_i\|^2 < 1.$$

Then the functions $\{\tilde{h}_i\}_{i=1}^M$ are linearly independent and with $\epsilon_i=h_i-\tilde{h}_i$, $1\le i\le M$ we have

(3.11) $$\mu_k^{(o)} \ge \hat{\mu}_k^{(o)} \ge \mu_k^{(o)} - \frac{\sum_{i=1}^k (\underline{A}^{1/2}g_{\underline{u}}'\underline{A}^{1/2}\epsilon_i,\epsilon_i)}{(1-(\sum_{i=1}^k \|\epsilon_i\|^2)^{1/2})^2}, \qquad 1 \le k \le M$$

where

$$(3.12)\qquad \hat{\mu}_k^{(o)} = \min_{c_1,\dots,c_k} R[\sum_{i=1}^{k} c_i\tilde{h}_i], \qquad 1 \le k \le M$$

is the k-th approximate eigenvalue obtained by the Rayleigh-Ritz method for the space spanned by $\{\tilde{h}_i\}_{i=1}^{M}$.

Proof: The triangle inequality, Schwarz' inequality and the orthonormality of the h_i yields with $h=\sum_{i=1}^{k} c_i h_i$

$$(3.13)\qquad \|\epsilon\|^2 = \|\sum_{i=1}^{k} c_i\epsilon_i\|^2 \le (\sum_{i=1}^{k}|c_i|\|\epsilon_i\|)^2 \le$$
$$\le (\sum_{i=1}^{k} c_i^2)(\sum_{i=1}^{k}\|\epsilon_i\|^2) = \|h\|^2\sum_{i=1}^{k}\|\epsilon_i\|^2.$$

Analoguously one shows

$$(3.14)\qquad (\underline{A}^{1/2}g_u'\underline{A}^{1/2}\epsilon,\epsilon) \le (\sum_{i=1}^{k}|c_i|(\underline{A}^{1/2}g_u'\underline{A}^{1/2}\epsilon_i,\epsilon_i)^{1/2})^2 \le$$
$$\le \|h\|^2\sum_{i=1}^{k}(\underline{A}^{1/2}g_u'\underline{A}^{1/2}\epsilon_i,\epsilon_i).$$

Then, from (3.13)

$$(3.15)\qquad \|\tilde{h}\|^2 = \|h-\epsilon\|^2 \ge (\|h\|-\|\epsilon\|)^2 \ge (1-(\sum_{i=1}^{k}\|\epsilon_i\|^2)^{1/2})^2\|h\|^2.$$

This shows that $\|h\|>0$ implies $\|Ph\|>0$, P is nonsingular on $\mathbb{S}_k$ and $\mathbb{L}^2_{M,k}$ is k-dimensional. From (2.6), (3.9), (3.14) and (3.15) we obtain finally

$$(3.16)\qquad 0 \le \mu_k^{(o)} - \hat{\mu}_k^{(o)} \le \sup\frac{(\underline{A}^{1/2}g_u'\underline{A}^{1/2}\epsilon,\epsilon)}{(\tilde{h},\tilde{h})} \le$$
$$\le \frac{\sum_{i=1}^{k}(\underline{A}^{1/2}g_u'\underline{A}^{1/2}\epsilon_i,\epsilon_i)}{(1-(\sum_{i=1}^{k}\|\epsilon_i\|^2)^{1/2})^2}.$$

Q. E. D.

If one satisfies condition (3.10) one obtains a general convergence criterion for the eigenvalues. Changing slightly Theorem 3 of [3] yields the following

Theorem 2: With the conditions which we have imposed on the eigenvalue problem (2.3) let $\{\mathbb{L}^2_{M_j}\}_{j=1}^{\infty}$ be a sequence of subspaces of $\mathbb{L}^2$ with $\dim \mathbb{L}^2_{M_j}=M_j\ge k$ for all $j\ge 1$. Let $\hat{\mu}_{k,j}$ be the k-th approximate eigenvalue obtained from applying the Rayleigh-Ritz method to the subspace $\mathbb{L}^2_{M_j}$, $j \geq 1$. If $\{h_i\}_{i=1}^{k}$ are the first k eigenfunctions of (2.3) and

$$(3.17) \qquad \lim_{j\to\infty} \{ \inf_{\omega\in\mathbb{L}_{M_j}} \|\omega - h_i\| \} = 0 \qquad \text{for each } 1 \leq i \leq k,$$

then the sequence $\{\hat{\mu}_{k,j}^{(o)}\}_{j=1}^{\infty}$ converges to $\mu_k^{(o)}$ from below. Moreover there exists a positive integer j_o such that for each $j \geq j_o$ there exist k functions $\{\tilde{h}_{i,j}\}_{i=1}^{k}$ in $\mathbb{L}^2_{M_j}$ which together with $\{h_j\}_{j=1}^{\infty}$ satisfie (3.10) and consequently

$$(3.18) \qquad \mu_k^{(o)} \geq \hat{\mu}_{k,j}^{(o)} \geq \mu_k^{(o)} - \frac{\sum_{i=1}^{\infty} (\underline{A}^{1/2} g_{\underline{u}}' \underline{A}^{1/2}(\tilde{h}_{i,j}-h_i), \tilde{h}_{i,j}-h_i)}{(1-(\sum_{i=1}^{k} \|\tilde{h}_{i,j}-h_i\|^2)^{1/2})^2}$$

for $j \geq j_o$.

§ 4 APPLICATION OF SPLINE SUBSPACES

As an important example of a finite dimensional subspace $\mathbb{L}^2_M$ we consider now the L-spline space as discussed in [8]. For generalizations of this concept see [5] and [4]. Coupling the interpolation estimate for L-splines with the previous Theorem yields the rate of convergence.

L-splines may be defined as follows: Let π be a partition of the interval $I=[0,1]$

$$(4.1) \qquad \pi: \; 0=x_0 < x_1 < \ldots < x_M=1.$$

If $\bar{\pi} = \max_{0\leq i\leq M-1} (x_{i+1}-x_i)$ and $\underline{\pi} = \min_{0\leq i\leq M-1} (x_{i+1}-x_i)$, let $\mathcal{P}_\lambda(I)$ denote all partitions π of the interval I such that $\underline{\pi} \leq \lambda\bar{\pi}$, where λ is a constant. One introduces the m-th order differential operator

$$(4.2) \qquad Lu(x) = \sum_{j=0}^{m} c_j(x) D^j u(x)$$

where $c_j \in C^j[0,1]$, $0\leq j\leq m$, with $c_m(x) \geq \delta > 0$ for all $x\in[0,1]$. Then, if z is a fixed positive integer $1\leq z\leq m$, we denote as L-spline space $Sp(L,\pi,z)$ the collection of all real-valued functions w on $[0,1]$ such that

(4.3)
$$L^*L\omega(x) = 0 \qquad \text{for all } x\in(a,b)-\{x_i\}_{i=1}^{M-1} \text{ with}$$
$$D^k\omega(x_{i-}) = D^k\omega(x_{i+}) \qquad \text{for } 0\le k\le 2m-1-z,\ 0<i<M.$$

L^* denotes the formal adjoint of L.

For convenience we use now the Sobolew norm

(4.4)
$$\|u\|_{W_2^1[0,1]} = \sum_{|\alpha|\le 1} \|D^\alpha u\|_{\mathbb{L}^2[0,1]}.$$

The interpolation of a function $f\in\mathbb{C}^{m-1}[0,1]$ by $s\in Sp(L,\pi,z)$ is unique and defined by

(4.5)
$$D^j(f-s)(x_i) = 0, \qquad j=0,1,\dots,z-1 \text{ if } 0<i<M,$$
$$D^j(f-s)(x_o) = D^j(f-s)(x_M) = 0, \qquad j=0,1,\dots,m-1.$$

The following result has been proved in [8]: Let $f\in W_2^\sigma[0,1]$ with $m\le\sigma\le 2m$. If s is the unique element in $Sp(L,\pi,z)$ interpolating f in the sense of (5.5), then

(4.6)
$$\|D^l(f-s)\|_{\mathbb{L}^2[0,1]} \le \|f-s\|_{W_2^1[0,1]} \le C\bar{\pi}^{\sigma-l}\|f\|_{W_2^\sigma[0,1]}$$

for any $0\le l\le m$. One can apply Sobolew's Lemma here in order to obtain for $0\le m\le l-1$

(4.7)
$$\max_{x\in[0,1],\ i\le m} |D^i(f(x)-s(x))| \le \overline{C}\|f-s\|_{W_2^1[0,1]} \le$$
$$\le C\overline{C}\bar{\pi}^{\sigma-l}\|f\|_{W_2^\sigma[0,1]}$$

where the constant $\overline{C}$ does not depend on f-s. Together with Theorem 2 this yields the following

<u>Theorem 3:</u> Let $\{\wp_{\lambda,j}[I]\}_{j=1}^\infty$, be a squence of partitions of [0,1] such that $\underline{\pi}_j\le\lambda\bar{\pi}_j$ and $\lim_{j\to\infty}\bar{\pi}_j=0$. With the conditions which we have imposed on the eigenvalue problem (2.3) let $\hat{\mu}_{k,j}^{(o)}$ be the k-th approximate eigenvalue of (2.3) (bifurcation point of (2.1)), obtained by applying the Rayleigh-Ritz method to the subspace $\mathbb{L}^2_{M_j} = Sp(L,\pi,z)$ of $\mathbb{L}^2$. If the eigenfunctions $\{h_i\}_{i=1}^\infty$ of (2.3) are of

class $W_2^{\sigma}[0,1]$ with $m \le \sigma \le 2m$, then there exists a constant $\bar{\bar{C}}$, dependent on k and m but independent of j, and a positive integer j_o such that

(4.8) $$\mu_k^{(o)} \ge \hat{\mu}_{k,j}^{(o)} \ge \mu_k^{(o)} - \bar{\bar{C}}(\bar{\pi}_j)^{2\sigma} \qquad \text{for all } j \ge j_o.$$

BIBLIOGRAPHY

[1] Berger, M. S.: A bifurcation theory for nonlinear elliptic partial differential equations and related systems.
Keller, J. B.; Antmann, S. (ed.): Bifurcation theory and nonlinear eigenvalue problems.
W. A. Benjamin, Inc., New York - Amsterdam, 1969.

[2] Birkhoff, G.; de Boor, C.; Swartz, B.; Wendroff, B.: Rayleigh-Ritz approximation by piecewise cubic polynomials.
SIAM J. Numer. Anal. 3 (1966), 188-203.

[3] Ciarlet, P. G.; Schultz, M. H.; Varga, R. S.: Numerical methods of high-order accuracy for nonlinear boundary value problems. III. Eigenvalue problems.
Numer. Math. 12 (1968), 120-133.

[4] Jerome, J.; Pierce, J.: On spline functions determined by singular self-adjoint differential operators.
J. Approx. Theory 5 (1972), 15-40.

[5] Jerome, J. W.; Varga, R. S.: Generalizations of spline functions and applications to nonlinear boundary value and eigenvalue problems.
Greville, T. N. E. (ed.): Theory and applications of spline functions.
Academic Press, New York - London, 1969.

[6] Krasnosel'skii, M. A.: Topological methods in the theory of nonlinear integral equations.
Pergamon Press, Oxford - London - New York - Paris, 1964.

[7] Pimbley, G. H. Jr.: Eigenfunction branches of nonlinear operators and their bifurcations.
Lecture Notes in Mathematics 104.
Springer-Verlag, Berlin - Heidelberg - New York, 1969.

[8] Schultz, M. H.; Varga, R. S.: L-splines.
Numer. Math. 10 (1967), 345-369.

ONE-STEP METHODS WITH ADAPTIVE STABILITY FUNCTIONS FOR THE INTEGRATION OF DIFFERENTIAL EQUATIONS

By P.J. van der Houwen

Paper for the conference on "Numerische, insbesondere Approximationstheoretische Behandlung von Funktionalgleichungen".
Oberwolfach, december, 3-12, 1972.

1. Introduction

Let

(1.1) $$\frac{dy}{dx} = f(y)$$

represent a vector differential equation in which f(y) belongs to a class of sufficient differentiability. Furthermore, let $J(y)$ represent the Jacobian matrix of system (1.1).

We shall consider integration formulas of the type

(1.2) $$y_{n+1} = y_n + \Theta_0 h_n f(y_n) + \Theta_1 h_n f(y_n + \Lambda h_n f(y_n)),$$

where $h_n = x_{n+1} - x_n$ denotes the step length and Θ_0, Θ_1 and Λ are polynomial or rational functions of $h_n J(y_n)$.

A number of formulas of type (1.2) are known in the literature; we mention the formulas of

Rosenbrock [1963]: $$\Theta_0(z) = 0,\ \Theta_1(z) = \frac{1}{1 - (1 - \frac{1}{2}\sqrt{2})z},$$
$$\Lambda(z) = \tfrac{1}{2}(\sqrt{2} - 1)\ \Theta_1(z);$$

Calahan [1968]: $$\Theta_0(z) = \frac{3}{4(1-(\frac{1}{2} + \frac{1}{6}\sqrt{3})z)},\ \Theta_1(z) = \frac{1}{3}\Theta_0(z),$$
$$\Lambda(z) = -\frac{8}{9}\sqrt{3}\ \Theta_0(z);$$

Liniger and Willoughby [1970]: ($F^{(2)}$ formula with a single Newton step) $$\Theta_0(z) = \frac{1 + (\frac{1}{2} + \alpha_1)z}{1 + \alpha_1 z + \alpha_2 z^2},$$
$$\Theta_1(z) = 0,\ \alpha_1 \geq -\frac{1}{2},\ \alpha_2 \geq -\frac{1 + 2\alpha_1}{4}.$$

The Rosenbrock and Liniger-Willoughby formulas are second order exact; the Calahan formula is third order exact. They are all A-stable in the sense of Dahlquist [1963]. The parameters α_1 and α_2 occurring in the Liniger-Willoughby method can be used for exponential fitting of the stability function (see section 5). Exponentially fitted stability functions have proved to be useful for the integration of stiff differential equations (cf. Liniger and Willoughby [1970]).

In this paper first to fourth order accurate integration formulas will be given of which the stability functions can be freely chosen, provided that they are compatible with the order of accuracy. Thus, when we are dealing with a stiff equation, an exponentially fitted stability function may be used. But also other types of differential equations can take advantage of the possibility to adapt the stability function to the problem under consideration. For instance, systems which originate from partial discretization of parabolie differential equations (by the method of lines) require stability functions with a large real stability boundary. Hyperbolie equations require large imaginary stability boundaries, etc. A survey of stability polynomials appropriate for the integration of respectively stiff, parabolic and hyperbolic equations may be found in van der Houwen [1972 b].

2. Consistency conditions

Formula (1.2) is said to have an order of accuracy p when the expansion of y_{n+1} in powers of h_n agrees with p+1 terms with the Taylor expansion for the local solution y(x) of (1.1) through the point (x_n, y_n). The requirement that (1.2) has an order of accuracy p leads to a set of relations between the derivatives of the functions $\Theta_0(z)$, $\Theta_1(z)$ and $\Lambda(z)$ at z=0. These relations will be called consistency conditions. By writing

$$\theta_0 = \theta_0(0),\ \theta_0' = \frac{d}{dz}\,\theta_0(z)|_{z=0},\ \theta_0'' = \frac{d^2}{dz^2}\,\theta_0(z)|_{z=0},\ \ldots ,$$

(and similar notations for $\Theta_1(z)$ and $\Lambda(z)$),we arrive at the consistency conditions listed in table 2.1.

TABLE 2.1 Consistency conditions for formula (1.2)

p	Consistency conditions
$p \geq 1$	$\theta_0 + \theta_1 = 1;$
$p \geq 2$	$\theta_0' + \theta_1' + \theta_1\lambda = 1/2;$
$p \geq 3$	$\theta_0'' + \theta_1'' + 2(\theta_1\lambda' + \theta_1'\lambda) = 1/3,$ $\theta_1\lambda^2 = 1/3;$
$p \geq 4$	$\theta_0''' + \theta_1''' + 3(\theta_1\lambda'' + 2\theta_1'\lambda' + \theta_1''\lambda) = 1/4,$ $\theta_1\ \lambda\lambda' = 1/8,$ $\theta_1'\ \lambda^2 = 1/12,$ $\theta_1\ \lambda^3 = 1/4.$

3. Stability functions

Suppose that formula (1.2) is applied to the scalar differential equation

$$\frac{dy}{dx} = \delta y.$$

Then, a relation of the type

$$y_{n+1} = R(h_n\delta)y_n \tag{3.1}$$

is obtained, where

$$R(z) = 1 + z[\Theta_0(z) + \Theta_1(z)] + z^2\Theta_1(z)\,\Lambda\,(z). \tag{3.2}$$

R(z) is called the stability function of (1.2). From (3.1) it is easily seen that the stability function R(z) of a p-th order accurate method satisfies the derivative conditions

$$\frac{d^j}{dz^j} R(z)|_{z=0} = 1,\ j = 0,1,\ \dots\ , p. \tag{3.3}$$

Such a stability function will be called consistent of order p.

When an integration formula is required of which the stability function is a priori given, one cannot expect an order of accuracy which is higher than the order of consistency of the stability function. On the other hand it is desirable that the order of accuracy is not lower than the order of consistency of R(z). Let us denote an arbitrary stability function which is consistent of order p by $R^{(p)}_{m_1,m_2}(z)$ (see table 3.1). Then we have to solve the problem to construct an integration formula with prescribed stability function $R^{(p)}_{m_1,m_2}(z)$ and with order of accuracy p. In the following section formulas are presented which satisfy this requirement for $p = 1, 2, 3, 4$.

TABLE 3.1 p-th order consistent stability functions

$$R^{(p)}_{m_1,m_2}(z) = \frac{1 + \beta_1 z + \beta_2 z^2 + \ldots + \beta_{m_1} z^{m_1}}{1 + \alpha_1 z + \alpha_2 z^2 + \ldots + \alpha_{m_2} z^{m_2}}$$

p	Coefficients β_j
$p \geq 1$	$\beta_1 = 1 + \alpha_1$
$p \geq 2$	$\beta_2 = \frac{1}{2} + \alpha_1 + \alpha_2$
$p \geq 3$	$\beta_3 = \frac{1}{6} + \frac{1}{2}\alpha_1 + \alpha_2 + \alpha_3$
$p \geq 4$	$\beta_4 = \frac{1}{24} + \frac{1}{6}\alpha_1 + \frac{1}{2}\alpha_2 + \alpha_3 + \alpha_4$

4. Integration formulas with prescribed stability function

In this section integration methods of order p, p = 1,2,3,4 are given of which the stability function is given by

$$R(z) = R^{(p)}_{m_1,m_2}(z),$$

where $R^{(p)}_{m_1,m_2}(z)$ is defined in table 3.1.

First order methods

$$\theta_0(z) = \frac{R^{(1)}_{m_1,m_2}(z) - 1}{z},$$

(4.1)

$$\theta_1(z) \equiv 0.$$

Second order methods

$$\theta_0(z) = \frac{R^{(2)}_{m_1,m_2}(z) - 1}{z},$$

(4.2)

$$\theta_1(z) \equiv 0.$$

Third order methods

$$\theta_0(z) = \frac{1}{4}, \quad \theta_1(z) = \frac{3}{4},$$

(4.3)

$$\Lambda(z) = \frac{4}{3}\,\frac{R^{(3)}_{m_1,m_2}(z) - 1 - z}{z^2}$$

Fourth methods

$$\Theta_0(z) = \frac{R^{(4)}_{m_1,m_2}(z) - 1}{z} - \frac{(1+\frac{3}{4}z+\frac{9}{32}z^2)(\frac{16}{27}+\frac{4}{27}(4\alpha_1+1)z)}{1+\alpha_1 z+\alpha_2 z^2+\ldots+\alpha_{m_2} z^{m_2}},$$

(4.4) $$\Theta_1(z) = \frac{\frac{16}{27}+\frac{4}{27}(4\alpha_1+1)z}{1+\alpha_1 z+\alpha_2 z^2+\ldots+\alpha_{m_2} z^{m_2}},$$

$$\Lambda(z) = \frac{3}{4}+\frac{9}{32}z.$$

The formulas presented above are not the only possibilities. Many other formulas can be constructed; for instance, in van der Houwen [1972a] some third order methods are given which do not fit into class (4.3). However, these methods have a more complicated structure than the formulas given here.

5. Numerical example

An example of a highly stiff, non-linear system is given by a problem of Robertson [1967]:

$$y_1' = -.04y_1 + 10^4 y_2 y_3,$$
$$y_2' = .04y_1 - 10^4 y_2 y_3 - 3.10^7 y_2^2 ,$$
$$y_3' = 3 .10^7 y_2^2 ,$$

$$y_1(0) = 1,\ y_2(0) = y_3(0) = 0.$$

A first analysis of this system reveals that

$$y_1' + y_2' + y_3' = 0,$$

so that

$$y_1 + y_2 + y_3 = \text{constant}.$$

By virtue of the initial conditions the constant equals 1; hence the system can be reduced to two first order equations. By eliminating y_1 we obtain

$$\text{(5.1)} \qquad \begin{aligned} y_2' &= .04 - .04(y_2 + y_3) - 10^4 y_2 y_3 - 3.10^7 y_2^2 , \\ y_3' &= 3.10^7 y_2^2 . \end{aligned}$$

The Jacobian matrix is given by

$$J = \begin{pmatrix} -.04 - 10^4 y_3 - 6.10^7 y_2 & -.04 - 10^4 y_2 \\ 6.\ 10^7 y_2 & 0 \end{pmatrix}$$

By Gerschgorin's theorem the eigenvalues δ_0 and δ_1 of J are situated in the regions

$$|\delta| < |.04 + 10^4 y_2| , |\delta + .04 + 10^4 y_3 + 6.10^7 y_2| < |.04 + 10^4 y_2| .$$

From this it follows that initially both eigenvalues are small ($|\delta_0| \le .04$, $|\delta_1| \le .04$); but as soon as y_3 differs from zero one eigenvalue (δ_1) decreases to the asymptotic value $- 10^4$, while the other one (δ_0) remains in the neighbourhood of the origin. Clearly, we are dealing with a rather stiff system.

Problem (5.1) has been solved by the third order method (4.3) with stability function

$$(5.2) \qquad R(z) = R^{(3)}_{2,2}(z) = \frac{1 + (1 + \alpha_1)z + (\frac{1}{3} + \frac{1}{2}\alpha_1)z^2}{1 + \alpha_1 z - (\frac{1}{6} + \frac{1}{2}\alpha_1)z^2} ;$$

The free parameter α_1 was chosen in such a way that $R(z)$ is exponentially fitted at $z_1 = h_n \delta_1$, i.e.

$$(5.3) \qquad R(z_1) = \exp(z_1).$$

Here, δ_1 is the large negative eigenvalue of the Jacobian matrix. It was proved by Liniger and Willoughby [1970] that with this value of α_1, the function $R^{(3)}_{2,2}(z)$ is A-stable, hence no stability problems are to be expected.

In table 5.2 the results of this experiment are given in terms of the correct number of digits, i.e. the value of

$$- {}^{10}\log|1 - \frac{y_n}{y(x_n)}|.$$

Since no analytical solution $y(x)$ is available we used as reference solution the results of method (4.3) with relatively small h_n; in fact, we took

$$h_n = 10^{-3}, \quad n < 400, \quad 0 \le x \le .4,$$

$$h_n = 2_{10}{}^{-2}, \quad n \ge 400, \quad .4 \le x \le 10.$$

This experiment produced the values listed in table 5.2. In order to evaluate the merits of method (4.3) we have also listed in table 5.2 the results of Calahan's method.

TABLE 5.1 Reference solution for Robertson's problem

x	y_2	y_3
.4	$.338639538_{10}^{-4}$	$.147940228_{10}^{-1}$
10	$.162340057_{10}^{-4}$	.158613844

TABLE 5.2 Results produced by method (4.3) and Calahan's method for Robertson's problem

x	method (4.3),(5.2),(5.3) $h_n=10^{-3}$, $n<3$, $h_n=10^{-1}$, $n\geq 3$			Calahan's method $h_n=5_{10}^{-3}$, $n<200$, $h_n=2_{10}^{-2}$, $n\geq 200$		
	y_2	y_3	N	y_2	y_3	N
.4	3.8	5.4	7	-.8	.8	80
10	2.8	4.3	102	.2	.3	560

which were obtained by Lapidus and Seinfeld [1971]. Clearly, the new method is both more efficient and more accurate than Calahan's method.

References

Calahan, D.A. [1968]: A stable, accurate method of numerical integration for non-linear systems, Proc. IEEE 56, 744.

Dahlquist, G.G. [1963]: A special stability problem for linear multistep problems, BIT 3, 27.

Houwen, P.J. van der [1972a]: Explicit and semi-implicit Runge-Kutta formulas for the integration of stiff equations, Report TW 132/72, Mathematisch Centrum, Amsterdam.
[1972b]: Explicit Runge-Kutta formulas with increased stability boundaries, Numerische Mathematik (to appear).

Lapidus, L. and J.H. Seinfeld [1971]: Numerical solution of ordinary differential equations, Academic Press, New York.

Liniger, W. and R.A. Willoughby [1970]: Efficient integration methods for stiff systems of ordinary differential equations, SIAM J., Numer. Anal. 7, 47.

Robertson, H.H. [1967]: The solution of a set of reaction rate equations in "Numerical Analysis", (J. Walsh, ed.), Thompson Book Co., Washington.

Rosenbrock, H.H. [1963]: Some general implicit processes for the numerical solution of differential equations, Comput. J. 5, 329.

MCC-VERFAHREN

(NUMERISCHE ČEBYŠEV-ENTWICKLUNG EINER STAMMFUNKTION)

Von F. Locher und K. Zeller

1. Einleitung

Bei der Lösung von Funktionalgleichungen kann man Reihenentwicklungen nach Čebyšev-Polynomen 1. bzw. 2. Art ansetzen (gute C- bzw. L-Approximation beim Abschneiden). Benützt wurde dies vor allem bei Differentialgleichungen; als Musterbeispiel diente die Bildung einer Stammfunktion: CLENSHAW-CURTIS [2] (CC-Verfahren) bzw. FILIPPI [3] [4] (Modifikation MCC). Über den gesamten Fragenkomplex gibt es zahlreiche Untersuchungen, u.a. von LANCZOS, IMHOF, ELLIOT, COOPER, CHAWLA, HAVIE, OLIVER, SCHMIDT; siehe insbesondere WRIGHT [13], O'HARA-SMITH [11], BULIRSCH-STOER [1], FOX-PARKER [5], GENTLEMAN [6], SALZER [12].

FILIPPI [4] erläutert die Vorzüge von MCC u.a. durch folgende Aussage:

Lemma 1 (FILIPPI; Vergleich unter einer Lipschitzbedingung).
Bei einer Funktion $f \in \mathrm{Lip}\,\alpha$ (mit $0 < \alpha \leq 1$) liefert MCC für die Stammfunktion Approximationen mit einem Fehler $O(n^{-\alpha-1})$, während bei CC nur eine Fehlerordnung $O(n^{-\alpha})$ garantiert werden kann.

Durch praktische Beispiele bestätigt er das, z.T. auch für andere Funktionsklassen. Auch WRIGHT [13] berichtet von günstigen Erfahrungen mit MCC, während O'HARA-SMITH [11] auf Vorzüge von CC bei bestimmter Integration hinweisen.

Diese Vergleichsuntersuchungen führen wir fort. Der grundsätzliche Unterschied zwischen CC und MCC läßt sich durch drei Reihenglieder beschreiben; zu berücksichtigen ist aber auch der (u.U. kompensierende) Einfluß der Koeffizientenabweichungen. Darauf basieren eingehendere Erörterungen der Verfahren: Pro und Kontra (verschiedene Argumente); Kombination von CC mit MCC (Ausgleich); problemorientierte Anpassung der Koeffizienten-Numerik.

2. Grundlagen

Mit f,g,h bezeichnen wir jeweils eine Funktion zusammen mit einer Reihenentwicklung (die in $[-1,1]$ gleichmäßig konvergiere):

$$(1)\qquad f = \sum_{k=o}^{\infty} a_k T_k\,, \qquad g = \sum_{k=o}^{\infty} b_k U_k\,, \qquad h = \sum_{k=o}^{\infty} c_k T_k\,.$$

Dabei gelte

$$(2)\qquad f(x) = g(x)\,, \qquad h(x) = \int_{-1}^{x} f(t)\,dt \qquad (-1 \leq x \leq 1);$$

also stellen f und g dieselbe Funktion dar, ferner h die zugehörige Stammfunktion mit $h(-1) = o$. Dann sind die Koeffizienten in bekannter Weise aneinander gekoppelt:

<u>Lemma 2</u> (Koeffizientenformeln). Bestehen die Beziehungen (2), so gelten für die Koeffizienten in (1) die Gleichungen

$$(3)\qquad a_k - a_{k+2} = 2b_k = 2(k+1)c_{k+1} \qquad (k \geq 1),$$

$$(4)\qquad 2a_o - a_2 = 2b_o = 2c_1\,,$$

$$(5)\qquad \sum_{k\neq 1} \frac{(-1)^{k+1}}{k^2-1}\, a_k - \tfrac{1}{4} a_1 = \sum_{k=o}^{\infty} \frac{(-1)^k}{k+1}\, b_k = c_o\,.$$

<u>Beweis.</u> Man erhält (3) und (4) aus geläufigen Integrationsformeln für Čebyšev-Polynome, während (5) aus $h(-1) = o$ folgt; vgl. [5] S. 54, 55, 59.

3. Abschneiden

In (1) bilden wir die Teilsummen

$$(6)\qquad f_n = \sum_{k=o}^{n} a_k T_k\,, \qquad g_n = \sum_{k=o}^{n} b_k U_k\,, \qquad h_{n+1} = \sum_{k=o}^{n+1} c_k T_k$$

und wenden auf erstere (gesamt und gliedweise) den Integrationsoperator S an,

$$S:\ u \longrightarrow V \qquad \text{mit} \qquad V(x) := \int_{-1}^{x} u(t)\,dt\,.$$

Es ist also

$$(7) \qquad Sf_n = \sum_{k=o}^{n} a_k \, ST_k \,, \qquad Sg_n = \sum_{k=o}^{n} b_k \, SU_k \,.$$

Wir erkennen

$$(8) \qquad Sg_n - \text{const.} = \sum_{k=o}^{n} \frac{1}{k+1} b_k T_{k+1} = \sum_{k=1}^{n+1} c_k T_k \,.$$

Abgesehen von der Integrationskonstante (die in Satz 4 erörtert wird) liefert der MCC-Ansatz mit den g_k die Teilsumme h_{n+1}, welche die Funktion h fast optimal approximiert:

$$(9) \qquad \left| h(x) - h_{n+1}(x) \right| \leq E_{n+1}(h)\,(1 + L_{n+1}) \qquad (n = o,1,\ldots; \; -1 \leq x \leq 1);$$

dabei ist $E_n(h)$ die Čebyšev-Abweichung und L_n die Lebesgue-Konstante (Norm des Fourier-Teilsummen-Operators), siehe [10] S. 54 - 55. Laut [7] gilt

$$(1o) \qquad L_n = (4/\pi^2)\log n + c_n \quad \text{mit} \quad o \leq c_n \leq 3 \qquad (n = 1,2,\ldots).$$

Rechnungen zeigen, daß die c_n im Bereich $1 \leq n \leq 1000$ abnehmen; es ergeben sich die gerundeten Werte

$$c_1 = 1{,}44 \;; \quad c_2 = 1{,}36 \;; \quad c_{10} = 1{,}29 \;; \quad c_{1000} = 1{,}27 \,.$$

In den meisten Fällen wird die obige Schranke recht grob sein; bei Übergang zu einer Abschätzung mittels f' oder Var f stößt man auf die Gibbs-Konstante.

Wir fragen nun, wie stark Sf_n von dem ziemlich günstigen Approximationsausdruck Sg_n abweicht.

Satz 3 (Vergleich von CC und MCC). Für $n \geq 1$ gilt

$$(11) \qquad Sf_n - Sg_n = \text{const.} + \frac{1}{2n} a_{n+1} T_n + \frac{1}{2n+2} a_{n+2} T_{n+1} \,.$$

Beweis. Man modifiziert die Formeln aus Lemma 2.

Satz 3 deutet (wegen der Fast-Optimalität von h_{n+1} bzw. Sg_n) auf gewisse Vorzüge von MCC hin. In manchen Fällen gilt

$$a_{n+1} \sim E_n(f) \sim nE_{n+1}(h) \;;$$

dann ist der Unterschied nicht sehr bedeutend. Typisch für einen größeren Unterschied ist der Fall

$a_k = 1/k^2$, also $b_k \sim 1/k^3$; CC-Fehler $\sim 1/n^2$, MCC-Fehler $\sim 1/n^3$.

Bei Integration verbessert sich die Approximationsgüte hier nicht um den erwarteten Faktor $1/n$, sondern um $1/n^2$; dies ist durch die Sonderrolle der Randpunkte bei algebraischer Approximation zu erklären (s. [10] S. 65, vgl. auch S. 61). Viel mehr als den Gewinnfaktor $1/n$ kann man allgemein nicht erwarten; denn nach einem Satz von DINI 1867 ([8] S. 302) ist

$$\sum b_k/a_k \quad \text{divergent} \quad (\text{falls } b_k > 0).$$

Für einzelne n kann natürlich der Unterschied sehr groß sein.

4. Integrationskonstanten

Bei unseren Betrachtungen spielen die Integrationskonstanten eine überraschend große Rolle. Wir formulieren daher

Satz 4 (Integrationskonstanten bei CC und MCC). Das konstante Glied in $h_{n+1}-Sf_n$ (und ebenso in $h-Sf_n$) ist für $n \geq 1$

$$\sum_{k=n+1}^{\infty} \frac{(-1)^{k+1}}{k^2-1} a_k \; ; \tag{12}$$

das konstante Glied in $h_{n+1}-Sg_n$ (und ebenso in $h-Sg_n$) ist für $n \geq 0$

$$\sum_{k=n+1}^{\infty} \frac{(-1)^k}{k+1} b_k = (-1)^{n+1}\left\{\frac{a_{n+1}}{2(n+2)} - \frac{a_{n+2}}{2(n+3)}\right\} + \sum_{k=n+3}^{\infty} \frac{(-1)^{k+1}}{k^2-1} a_k. \tag{13}$$

Beweis. Man erhält die Beziehungen aus Lemma 2 Formel (5).

Satz 4 deutet auf einen Vorteil von CC hin, der den in Satz 3 beschriebenen Vorteil von MCC ungefähr kompensieren kann. Der Unterschied zwischen (12) und (13), also das konstante Glied in Sf_n-Sg_n wird durch

$$(-1)^{n+1}\left\{\frac{a_{n+1}}{2n} - \frac{a_{n+2}}{2(n+1)}\right\} \tag{14}$$

wiedergegeben. Man vergleiche Lemma 7 und die anschließenden Bemerkungen.

5. Bestimmtes Integral

Auch bei der Berechnung des bestimmten Integrals bietet CC gewisse Vorteile. Bei CC kommt man auf

$$(15) \quad \int_{-1}^{1} f(t)\,dt = 2a_o - a_2 + \sum_{k=2,4,..} \frac{a_k - a_{k+2}}{k+1}$$

$$= 2a_o + \sum_{k=2,4,...} a_k \left(\frac{1}{k+1} - \frac{1}{k+3}\right),$$

bei CC auf

$$(16) \quad \int_{-1}^{1} g(t)\,dt = \sum_{k=2,4,..} \frac{2b_k}{k} = \sum_{k=o,2,..} 2c_{k+1}$$

$$= \sum_{k=o,2,..} \frac{a_k - a_{k+2}}{k+1} \quad .$$

Bei nicht zu langsam fallenden a_k deutet (15) wegen der kleineren Gewichte $\frac{1}{k+1} - \frac{1}{k+3}$ (statt $\frac{1}{k+1}$ in (16)) auf ein günstigeres Fehlerverhalten von CC beim Abschneiden hin; dies steht im Einklang mit bekannten Fehlerformeln (vgl. [9]). Wir formulieren

Lemma 5 (Bestimmtes Integral). Der Wert von $Sf_n - Sg_n$ an der Stelle 1 (also der Unterschied der Näherungen für das bestimmte Integral von -1 bis 1) ist

$$\frac{a_{n+1}}{n} \quad \text{(n ungerade)}, \qquad \frac{a_{n+2}}{n+2} \quad \text{(n gerade)} .$$

Beweis. Aus den obengenannten Formeln leiten wir die jeweiligen Reste her; Subtraktion liefert die Behauptung.

6. Koeffizienten

Näherungen $\tilde{a}_k$, $\tilde{b}_k$, $\tilde{c}_k$ für die Koeffizienten aus (1) berechnet man gewöhnlich mit Hilfe diskreter Euler-Fourier-Formeln (Interpolation). Beim CC-Verfahren laut [2] benützt man als Knoten die Extremalstellen von T_m (m=n) im Intervall $[-1,1]$. Die Ausnahmen von der diskreten Orthogonalität für $k > m$ führen zu

Lemma 6 (Koeffizientenfehler). Die durch Interpolation an den Extremalstellen von T_m in $[-1,1]$ bestimmten $\tilde{a}_k$ erfüllen

$$(17)\quad \begin{aligned} \tilde{a}_k &= a_k + a_{2m-k} + a_{2m+k} + a_{4m-k} + a_{4m+k} + \dots \qquad (k=1,2,\dots,m-1), \\ \tilde{a}_o &= a_o + a_{2m} + a_{4m} + \dots , \\ \tilde{a}_m &= a_m + a_{3m} + a_{5m} + \dots \end{aligned}$$

Beweis. Siehe [5] S. 67.

Insbesondere haben wir

$$(18)\quad \tilde{a}_{n-1} = a_{n-1} + a_{n+1} + \dots \qquad (\text{falls } m=n).$$

Das Hauptrestglied $+a_{n+1}$ verstärkt die in Lemma 2 beschriebene Abweichung. Das läßt sich umgehen, wenn man m größer wählt oder die Nullstellen von T_m $(m \geqslant n+1)$ als Knoten benützt. In diesem Fall gilt (vgl. [5] S. 67)

$$(19)\quad \tilde{a}_k = a_k - a_{2m+2-k} - a_{2m+2+k} + a_{4m+4-k} + a_{4m+4+k} --++ \dots \qquad (k = o,1,\dots,m+1) .$$

Beim MCC-Verfahren [4] nimmt man als Knoten die Extremalstellen von T_m $(m=n+1)$ im offenen Intervall $(-1,1)$. Die Berechnung ist etwas aufwendiger als bei CC, weil ein Zusatzfaktor hereinkommt. Hier wird der Zusammenhang zwischen den numerischen und den exakten Čebyšev-Koeffizienten durch

$$(2o)\quad \tilde{a}_k = a_k - a_{2m-k} + a_{2m+k} - a_{4m-k} + a_{4m+k} -+ \dots$$

wiedergegeben (vgl. [4] S. 3o).

7. Bemerkungen

Bei überschlägigen Rechnungen wird man a_{n+1} als Hauptursache des Fehlers nehmen. Wir formulieren daher

Lemma 7 (Hauptglied in den Fehlerformeln). Der Koeffizient a_{n+1} tritt in den Fehlerformeln für $n \geqslant 1$ in folgender Gestalt auf

$$(21)\quad Sf_n - Sg_n \text{ (ohne Konstante)}: \quad \frac{1}{2n} a_{n+1} \quad ,$$

(22) $h_{n+1} - Sf_n$ (konstantes Glied): $\frac{(-1)^n}{(n+1)^2-1} a_{n+1}$,

(23) $h_{n+1} - Sg_n$ (konstantes Glied): $\frac{(-1)^n}{2(n+1)} a_{n+1}$,

(24) $\tilde{a}_{n-1}$: a_{n+1} .

Beweis. Wir greifen in Satz 3, Satz 4 und Lemma 6 die Glieder mit a_{n+1} heraus.

Zum Vergleich ist auch Lemma 5 heranzuziehen. Wir vermerken, daß (22) einen Vorteil von CC bei der Integrationskonstanten signalisiert, während sich der Einfluß von a_{n+1} in (21) und (24) verstärkt. Per Saldo verbleibt also ein Vorteil für MCC. Jedoch legen die Betrachtungen nahe, gewisse Kombinationen von CC und MCC in Verbindung mit einer angepaßten Koeffizientenformel zu verwenden. Die Wahl hängt natürlich ab von den vorhandenen Informationen. Es kann günstig sein, eine hohe Knotenzahl m und verhältnismäßig wenige (dafür genaue) Koeffizienten a_k und b_k zu verwenden. In anderen Fällen kann man den Kompensationseffekt ungenauer Koeffizienten ausnützen. Meistens aber empfiehlt es sich, bei MCC die Integrationskonstante präziser zu bestimmen.

Literatur

1. BULIRSCH, R., STOER, J.: Darstellung von Funktionen in Rechenautomaten. In: Mathematische Hilfsmittel des Ingenieurs; herausgegeben von R. Sauer und I. Szabó; Teil III, S. 352 - 446. Berlin-Heidelberg-New York: Springer 1968.
2. CLENSHAW, C.W., CURTIS, A.R.: A Method for Numerical Integration on an Automatic Computer. Numer. Math. 2, 197 - 2o5 (196o).
3. FILIPPI, S.: Angenäherte Tschebyscheff-Approximation einer Stammfunktion - eine Modifikation des Verfahrens von Clenshaw und Curtis. Numer. Math. 6, 32o - 328 (1964).
4. FILIPPI, S.: Untersuchung über die Fourier-Tschebyscheff-Approximation von Stammfunktionen. Köln und Opladen: Westdeutscher Verlag 197o.
5. FOX, L., PARKER, I.B.: Chebyshev Polynomials in Numerical Analysis. Oxford: University Press 1968.

6. GENTLEMAN, M.W.: Implementing Clenshaw-Curtis Quadrature. I, II. Comm. ACM 15, 337 - 346 (1972).

7. GRONWALL, T.: Uber die Lebesgueschen Konstanten bei den Fourierschen Reihen. Math. Ann. 72, 244 - 261 (1912).

8. KNOPP, K.: Theorie und Anwendung der unendlichen Reihen. 5. Auflage. Berlin-Heidelberg-New York: Springer 1964.

9. LOCHER, F.: Fehlerabschätzungen für das Quadraturverfahren von Clenshaw und Curtis. Computing 4, 3o4 - 315 (1969).

1o. LORENTZ, G.: Approximation of Functions. New York: Holt, Rinehart and Winston 1966.

11. O'HARA, H., SMITH, F.J.: Error estimation in the Clenshaw-Curtis quadrature formula. Comput. J. 11, 213 - 219 (1968).

12. SALZER, H.E.: Lagrangian interpolation at the Chebyshev points $x_{n,\nu} = \cos(\nu\pi/n)$, $\nu = o(1)n$; some unnoted advantages. Comput. J. 15, 156 - 159 (1972).

13. WRIGHT, K.: Series methods for integration. Comput. J. 9, 191 - 199 (1966).

EINE VARIANTE DES ZWEISCHRITT-LAX-WENDROFF-VERFAHRENS

Von Hans-Werner Meuer

Zusammenfassung

Zur numerischen Behandlung des Anfangswertproblems in beliebig vielen Ortsveränderlichen für Systeme linearer hyperbolischer Differentialgleichungen 1. Ordnung wird in Abhängigkeit eines reellen Parameters eine Schar von Differenzenverfahren 2. Ordnung untersucht. Als Spezialfall ist die von Richtmyer angegebene Version des Lax-Wendroff-Verfahrens enthalten. Es wird ein Kriterium angegeben, das die Konvergenz der Verfahren garantiert und für eine gewisse Klasse auch physikalisch interessanter Probleme sogar notwendig für die Konvergenz ist. Der Zusammenhang zur Courant-Friedrichs-Lewy-Bedingung wird hergestellt.

1. Einführung

Betrachtet werde das Cauchy Problem für ein System partieller Differentialgleichungen 1. Ordnung:

$$(1)\qquad \begin{aligned} & u_t = P(x,\tfrac{\partial}{\partial x})\, u := \sum_{j=1}^{s} A_j(x)\, u_{x_j} + B(x)\, u\ , \quad -\infty < x_j < \infty,\ 0 \leq t \leq T, \\ & u(x,0) = f(x)\quad . \end{aligned}$$

Hier ist $(x,t) := (x_1,\ldots,x_s,t)^T$ ein Punkt des reellen Euklidischen Raumes R^{s+1}, $u = u(x,t) := (u_1(x,t),\ldots,u_n(x,t))^T$ die gesuchte reellwertige Vektorfunktion mit n Komponenten, und $A_1(x),\dots,A_s(x)$, $B(x)$ sind vorgegebene reellwertige n x n Matrizen. u^T bezeichnet den zu u transponierten Vektor. Bezüglich der Anfangsbedingung $f(x)$ setzen wir voraus: $f(x) \in L_2(R^s)$. Dabei verstehen wir unter $L_2(R^s)$ den Hilbertraum der im R^s quadratintegrierbaren Funktionen u,v,... der unabhängigen Veränderlichen x mit dem Skalarprodukt

$$(u,v) := \int_{R^s} u^T v \, dx \text{ und der Norm } \| u \|^2 := (u,u) .$$

Für Operatoren L in $L_2(R^2)$ sei die Norm durch

$$\| L \| := \sup_{\|u\|=1} \|Lu\| \text{ eingeführt.}$$

Dagegen werde die Euklidische Voktornorm mit $|\cdot|$ bezeichnet, wobei $|v|^2 := v_1^2+\dots+v_s^2$ und für eine n x n Matrix M die zugeordnete Matrixnorm mit

$$|M| := \sup_{|u|=1} |Mu| \text{ definiert ist.}$$

M^T bezeichnet die Transponierte und $\rho(M)$ den Spektralradius von M.
Dem Differentialoperator $P(x,\frac{\partial}{\partial x})$ in (1) ordnen wir die n x n Matrix

$$(2) \qquad \hat{P}(x,\omega) := i \sum_{j=1}^{s} A_j(x)\, \omega_j \,, \quad -\infty < \omega_j < \infty \quad \text{zu; dabei ist}$$

ω ein s-dimensionaler reeller Vektor.
Wir setzen voraus, daß (1) hyperbolisch im folgenden Sinne ist: Es existiert eine nichtsinguläre Matrix $R(x,\omega)$, so daß gilt $R\hat{P}R^{-1} = iD$, mit $|R|$, $|R^{-1}| \leq K$ gleichmäßig beschränkt in x und ω, und D ist eine reelle Diagonalmatrix.
Für das folgende machen wir die vereinfachende Voraussetzung, daß in (1) $B(x) \equiv 0$ ist. Der allgemeinere Fall wird in [8] behandelt.

Zur numerischen Behandlung von (1) überziehen wir den R^{s+1} mit einem Gitter R_Δ^{s+1}. Dabei seien Δt die Schrittweite in t-Richtung und $\Delta x_j = \frac{\Delta t}{\lambda_j}$ die Schrittweiten in x_j-Richtung mit den fest vorgegebenen Schrittweitenverhältnissen λ_j.
Geeignete Diskretisierung von (1) führt nun zu einer Differenzenapproximation von (1):

$$(3) \qquad \begin{aligned} v(x,t+\Delta t) &= S(x,\Delta t)\, v(x,t) \\ v(x,0) &= f(x) . \end{aligned}$$

Dabei sei $S(x,\Delta t)$ ein expliziter Differenzenoperator, und durch das Einschritt-Verfahren (3) lassen sich die Werte der Gitterfunktion $v(x,t)$, beginnend zur Zeit $t = 0$,

auf den Schichten Δt, $2\Delta t,\ldots,T$ bestimmen.
Bekannte Differenzenoperatoren S zur numerischen Lösung von (1) sind z.B. der Friedrichs-Operator S_F und der Lax-Wendroff-Operator S_{LW}:

$$(4)\qquad S_F := I + \sum_{j=1}^{s} \frac{\lambda_j}{2} A_j D_j^{(1)} + \frac{1}{2s} \sum_{j=1}^{s} D_j^{(2)} \quad ,$$

$$(5)\qquad S_{LW} := I + \sum_{j=1}^{s} \frac{\lambda_j}{2} A_j D_j^{(1)} + \sum_{j=1}^{s} \frac{\lambda_j^2}{2} A_j D_j^{(2)} + \sum_{j<k} \frac{\lambda_j \lambda_k}{8} (A_j A_k + A_k A_j) D_j^{(1)} D_k^{(2)} \quad .$$

Dabei bedeuten

$$D_j^{(1)} := T_j^{+1} - T_j^{-1} \quad , \quad D_j^{(2)} := T_j^{+1} - 2I + T_j^{-1}$$

mit dem Identitätsoperator I und den Translationsoperatoren $T_j^{\pm 1}$ in x_j-Richtung:

$$T_j^{\pm 1}\, g(x) := g(x_1,\ldots,x_j \pm \Delta x_j, x_{j+1},\ldots,x_s) \quad .$$

S_F approximiert (1) von 1. Ordnung, S_{LW} dagegen ist ein Verfahren 2. Ordnung. Beide Verfahren sind bedingt konvergent, d.h. konvergieren nur dann gegen die exakte Lösung im L_2-Sinne, falls gewisse Bedingungen an die Maschenweite des Orts-Zeit-Gitters erfüllt sind. Im Falle symmetrischer Koeffizientenmatrizen A_j lauten typische Konvergenzbedingungen:

$$\max_j \lambda_j \, \rho(A_j) \le \frac{1}{s} \qquad \text{für } S_F \qquad \text{(vgl. z.B. [5])}$$

$$\max_j \lambda_j \, \rho(A_j) \le \frac{1}{s\sqrt{s}} \qquad \text{für } S_{LW} \qquad \text{(vgl. z.B. [7] und [3]).}$$

2. Das verallgemeinerte Zweischritt-Lax-Wendroff-Verfahren

Für eine Ortsveränderliche (s=1) und konstante Koeffizientenmatrix $A := A_1$ läßt sich S_{LW}, wie R. D. Richtmyer 1962 (vgl. [10]) feststellte, als Zweischritt-, besser Zwischenschrittverfahren S_{LWR} formulieren.
Ausgehend von Werten $v(x,t)$ auf der Schicht $t = n\cdot\Delta t$ werden zunächst provisorische Werte $v(x,t+\Delta t)$ auf der Schicht $(n+1)\Delta t$ mit Hilfe von S_F berechnet:

$$v(x,t+\Delta t) = S_F\, v(x,t) = v(x,t) + \frac{\lambda}{2}\, A(v(x+\Delta x,t) - v(x-\Delta x,t)) .$$

Die endgültigen Werte auf der Schicht $(n+2)\Delta t$ erhält man mit Hilfe eines anderen bekannten Verfahrens, des schon 1928 in der Arbeit von Courant, Friedrichs und Lewy (vgl. [1]) erwähnten 'leap-frog'-Verfahrens:

$$v(x,t+2\Delta t) = v(x,t) + \lambda A(v(x+\Delta x,t+\Delta t) - v(x-\Delta x,t+\Delta t)).$$

Die Identität von S_{LW} und S_{LWR} für $s = 1$ läßt sich leicht nachweisen, wenn man beachtet, daß ein Zeitschritt bei S_{LWR} $2\Delta t$ ist. Allerdings bietet S_{LWR} gegenüber S_{LW} rechentechnische Vorteile, da die Matrix A^2 nicht explizit gebildet werden muß. Für $s > 1$ sind beide Verfahren nicht identisch. Wie Yamaguti (vgl. [18]) nachgewiesen hat, sind die Unterschiede erheblich im nichtsymmetrischen, jedoch hyperbolischen Falle. Es lassen sich nämlich Beispiele angeben, für die S_{LW} niemals stabil und damit auch nicht konvergent ist, wie klein auch die Maschenweite λ gewählt ist. Für S_{LWR} ist dies nicht der Fall (vgl. [16]).
Richtmyer hat für $s = 2$ im Falle einer linearisierten Euler'schen Gleichung aus der Hydrodynamik eine Stabilitätsbedingung für S_{LWR} angegeben (vgl. [10] und [11]). Für diese Gleichung im Falle dreier Ortsveränderlicher wurde die Richtmyer'sche Bedingung kürzlich von Rubin und Preiser (vgl. [12], [2]) verallgemeinert. J. C. Wilson +) (vgl. [17]) gibt eine s-dimensionale Verallgemeinerung, die man aber auch schon bei Wendroff (vgl. [16]) findet.

Wir kommen jetzt zur Formulierung der hier untersuchten Variante des Richtmyer-Verfahrens:
Sei ω ein reeller Parameter. Dann werden zunächst analog wie bei S_{LWR} provisorische Zwischenwerte gebildet:

$$\text{(6a)} \quad \begin{aligned} v(x,t+\Delta t) &= S_F(\omega)\, v(x,t) \qquad \text{mit} \\ S_F(\omega) &:= I + \sum_{j=1}^{s} \frac{\lambda_j}{2} A_j D_j^{(1)} + \frac{\omega}{2s} \sum_{j=1}^{s} D_j^{(2)} . \end{aligned}$$

Die endgültigen Werte erhält man wie bei S_{LWR} durch das 'leap-frog'-Verfahren:

+) Auf diese Arbeit hat mich in Oberwolfach freundlicherweise J. Ll. Morris aufmerksam gemacht. Ihm möchte ich an dieser Stelle auch für einige sehr anregende Diskussionen herzlich danken.

(6b) $$v(x,t+2\Delta t) = v(x,t) + \sum_{j=1}^{s} \lambda_j A_j D_j^{(1)} v(x,t+\Delta t) .$$

Bemerkung:

Das im 1. Schritt benutzte $S_F(\omega)$ resultiert in folgender Approximation der Differentialgleichung (1) (e_j j-ter Einheitsvektor):

(7) $$\frac{v(x,t+\Delta t) - v(x,t)}{\Delta t} = \sum_{j=1}^{s} A_j \frac{v(x+\Delta x_j e_j,t) - v(x-\Delta x_j e_j,t)}{2\Delta x_j}$$

$$+ \frac{\omega}{2s} \sum_{j=1}^{s} \frac{\Delta x_j}{\lambda_j} \frac{v(x+\Delta x_j e_j,t) - 2v(x,t) + v(x-\Delta x_j e_j,t)}{\Delta x_j^2} .$$

Durch Addition einer Pseudo-Viskosität, Approximation von $\frac{\omega}{2s} \sum_{j=1}^{s} \frac{\Delta x_j}{\lambda_j} u_{x_j x_j}$,

wird aus einem instabilen Verfahren ein stabiles Verfahren. Für $\omega = 1$ hat man das Friedrichs-Verfahren, im Falle s = 1 und s = 2 wurden in praktischen Problemen schon öfters $\omega \neq 1$ verwendet (vgl. z.B. [9], [13], [15]).

Im folgenden machen wir die vereinfachende Voraussetzung, daß alle Schrittweitenverhältnisse gleich sind $\lambda := \lambda_1 = \ldots = \lambda_s$ und verweisen für den allgemeinen Fall auf [8].

Wegen (6a) und (6b) lautet der verallgemeinerte Lax-Wendroff-Richtmyer-Operator $S_{LWR}(\omega)$:

(8) $$S_{LWR}(\omega) = I + [\lambda \sum_{j=1}^{s} A_j D_j^{(1)}] \cdot [I + \frac{\lambda}{2} \sum_{j=1}^{s} A_j D_j^{(1)} + \frac{\omega}{2s} \sum_{j=1}^{s} D_j^{(2)}] .$$

$S_{LWR}(\omega)$ ist für jede Wahl des reellen Parameters ω eine Approximation 2. Ordnung von (1).

Es gilt

Satz 1 Im Falle konstanter Koeffizienten A_j ist $S_{LWR}(\omega)$ für $0 < \omega \leq 1$ ein konvergentes Verfahren, falls gilt

(9) $$\lambda\rho(\sum_{j=1}^{s} A_j \alpha_j) \leq \sqrt{\frac{\omega}{s}} \quad \text{für alle } |\alpha| = 1 .$$

Für $\omega \leq 0$ und $\omega > 1$ ist $S_{LWR}(\omega)$ nicht konvergent, unabhängig wie klein $\lambda > 0$ gewählt wurde.

Beweis:

Wir bilden die Amplifikations-Matrix $\hat{S}_{LWR}(\omega,\xi)$ zu $S_{LWR}(\omega)$ durch die Substitution $T_j^{\pm 1} \rightarrow e^{\pm i\xi_j}$; $\hat{S}_{LWR}$ ist ein Polynom in

$$Q := \lambda\sum_{j=1}^{s} A_j \sin\xi_j, \text{ nämlich}$$

$$\hat{S}_{LWR} = E + 2i(1-\omega+ \frac{\omega}{s}\sum_{j=1}^{s} \cos\xi_j)\, Q - 2Q^2 \quad . \tag{10}$$

Für einen Eigenwert g_μ von $\hat{S}_{LWR}$ läßt sich unter der Voraussetzung (9) nach einfacher Rechnung zeigen:

$$|g_\mu|^2 \leq 1 - 4\kappa_\mu^2 \frac{\omega(1-\omega)}{s} \sum_{j=1}^{s} (1-\cos\xi_j)^2 \;; \kappa_\mu \text{ ist Eigenwert von } \lambda\sum_j A_j \sin\xi_j \;. \tag{11}$$

Für $0 < \omega \leq 1$ ist aber damit Teil 1 der Behauptung bewiesen, denn (11) besagt, daß die von-Neumann'sche Bedingung (vgl. [11]) erfüllt ist, und die von Neumann'sche Bedingung ist wegen der speziellen Gestalt der Amplifikationsmatrix und der Hyperbolizität von (1) hinreichend für Lax-Richtmyer-Stabilität des Differenzenoperators $S_{LWR}(\omega)$. Das sieht man so:

$$|\hat{S}_{LWR}^n| = |R^{-1}D^nR| \leq |R^{-1}|\; |D^n|\; |R| \leq K\rho^n(\hat{S}_{LWR}) \leq K \;. \tag{12}$$

Wegen der Hyperbolizität von (1) existiert nämlich ein R, das Q und wegen (10) auch $\hat{S}_{LWR}$ auf Hauptachsen transformiert. (12) besagt wegen der konstanten Koeffizienten A_j, daß $\hat{S}_{LWR}$ Lax-Richtmyer-stabil ist (vgl. [11]), die Konvergenz folgt aus dem Äquivalenzsatz von Lax (vgl. [11]).

Es bleibt zu zeigen, daß für $\omega \leq 0$ bzw. $\omega > 1$ es für jede Wahl von $\lambda > 0$ $\bar{\xi}_j$, $j=1, \ldots, s$ gibt, so daß die von-Neumann'sche Bedingung verletzt ist.

Wir führen hier nur den Fall $\omega = 0$ vor, die anderen Fälle gehen analog (vgl. [8]).

Zunächst folgt aus $\omega = 0$ für den Eigenwert g_μ von $\hat{S}_{LWR}$:

$$|g_\mu|^2 = 1 + 4\kappa_\mu^4 \geq 1 \quad , \text{ wieder mit } \kappa_\mu \text{ Eigenwert von } \lambda\sum_j A_j \sin\xi_j \;. \tag{13}$$

(13) gilt für beliebige $\lambda > 0$ und $0 \leq |\xi_j| \leq \pi$. Es kann aber nicht immer das Gleichheitszeichen gelten, denn das würde bedeuten, daß alle Eigenwerte von $A_1, \ldots, A_s$ Null wären. Wir hätten, wegen der Hyperbolizität von (1), $A_1 = \ldots = A_s = 0$ und damit nur das Problem $u_t = 0$ vorliegen.

<u>Satz 2</u> Im Falle symmetrischer und konstanter A_j ist $S_{LWR}(\omega)$ für $0 < \omega \leq 1$ ein konvergentes Verfahren, falls gilt

(14) $$\lambda \max_j \rho(A_j) \leq \frac{\sqrt{\omega}}{s} .$$

Für $A := A_1 = \ldots = A_s$ ist (14) äquivalent (9) und auch notwendig für Konvergenz.

<u>Beweis:</u>
Zunächst läßt sich wegen der Symmetrie der A_j zeigen:

(15) $$\sup_{|\alpha|=1} \rho(\sum_{j=1}^{s} A_j\alpha_j) \leq \sqrt{s} \max_j \rho(A_j) \text{ (vgl. [8]).}$$

Wegen (14) folgt aber (9) und damit wegen Satz 1 die Konvergenz.
In (15) tritt das Gleichheitszeichen ein, falls $A := A_1 = \ldots = A_s$ ist; also sind in diesem Spezialfall die Bedingungen (14) und (9) äquivalent.
Es bleibt zu zeigen, daß die von-Neumann'sche Bedingung verletzt ist, falls im Spezialfall (14) verletzt ist. Das sieht man wie folgt:
Speziell für die Wahl $\xi_o := \xi_1 = \ldots = \xi_s$ gilt für die Eigenwerte g_μ von $\hat{S}_{LWR}$ unter der Voraussetzung (14)

(16) $$|g_\mu|^2 = 1 + 4K_\mu^2 \, (K_\mu^2 + 2\omega(\omega-1) + 2\omega(1-\omega)\cos\xi_o - \omega^2\sin^2\xi_o)$$
mit K_μ Eigenwert von $s\lambda A\sin\xi_o$.

Sei nun (14) verletzt, d.h. es gibt μ_o mit

$$K_{\mu_o}^2 > \omega\sin^2\xi_o \quad \text{bzw.} \quad K_{\mu_o}^2 = \omega\sin^2\xi_o + \varepsilon_o s^2\sin^2\xi_o \quad \text{mit} \quad \varepsilon_o > 0 \quad .$$

Wegen (16) folgt aber

$$|g_{\mu_o}|^2 = 1 + 4K_{\mu_o}^2 \, [\xi_o^2 \, (\varepsilon_o s^2 - (\frac{\omega(1-\omega)}{4} + \frac{\varepsilon_o s^2}{3}) \, \xi_o^2) + O(\xi_o^6)] \quad .$$

Hieraus ergibt sich

$$|g_{\mu_o}|^2 > 1 \quad \text{für} \quad 0 < \xi_o \leq \overline{\xi}_o \, , \quad \text{falls } \overline{\xi}_o \text{ genügend klein gewählt wird,}$$

denn dann ist auch $K_{\mu_o}^2 > 0$. Damit ist unter der Annahme, daß (14) verletzt ist, für die spezielle Wahl

$\xi_o := \xi_1 = \ldots = \xi_s$ und ξ_o genügend klein, auch die von-Neumann'sche

Bedingung verletzt und Satz 2 vollständig bewiesen.

Bemerkung:

1) Ersetzt man (14) durch die für $s > 1$ einschränkendere Bedingung

$$(14a) \quad \lambda \max_j \rho(A_j) \leq \frac{\sqrt{\omega}}{s\sqrt{s}} ,$$

so folgt trivialerweise die Bedingung (9) und damit Konvergenz. Für $\omega = 1$ und $s = 2$ ist (14a) als Stabilitätsbedingung für S_{LWR} von Gourlay und Morris (vgl. [4]) benutzt, aber als zu einschränkend bei dem dortigen numerischen Beispiel empfunden worden. Wir werden dasselbe Beispiel im Abschnitt 5 numerisch behandeln.

2) In einem gewissen Sinne ist die Bedingung (9) und damit Bedingung (14) in der Klasse der symmetrischen Matrizen $A_1, A_2, \ldots, A_s$ bestmöglich, denn für $A := A_1 = \ldots = A_s$ läßt sie sich nicht verbessern.

3) Als Unterschied zwischen S_{LW} und $S_{LWR}(\omega=1)$ bezüglich ihrer Stabilität ergibt sich damit:
Für $A := A_1 = \ldots = A_s$ ist

$\lambda\rho(A) \leq \frac{1}{s\sqrt{s}}$ notwendig und hinreichend bei S_{LW} (vgl. [7]),

$\lambda\rho(A) \leq \frac{1}{s}$ notwendig und hinreichend bei S_{LWR} .

Satz 3 Im Falle symmetrischer und variabler $A_j(x)$ ist $S_{LWR}(\omega,x)$ für $0 < \omega \leq 1$ ein konvergentes Verfahren, falls gilt

$$(a) \quad \left| \frac{\partial^\nu A_j(x)}{\partial x_\mu^\nu} \right| \leq K \quad \text{für} \quad \begin{aligned} j &= 1,\ldots,s \\ \mu &= 1,\ldots,s \\ \nu &= 0,1,2 \end{aligned}$$

$$(b) \quad \lambda\rho\Big(\sum_{j=1}^{s} A_j(x)\alpha_j\Big) \leq \sqrt{\frac{\omega}{s}} .$$

Der Beweis ergibt sich unmittelbar durch Anwendung eines Satzes von Lax und Nirenberg (vgl. [6]). Für Einzelheiten sei auf [8] verwiesen.

Bemerkung:

1) Wegen Satz 2 läßt sich in Satz 3 die Bedingung (b) durch die einschränkendere, aber handlichere Bedingung

$$(b') \qquad \lambda \max_j \rho(A_j) \leq \frac{\sqrt{\omega}}{s}$$

ersetzen.

2) Den regulär hyperbolischen Fall für variable Koeffizienten $A_j(x)$ hat Yamaguti mit Hilfe von Pseudo-Differenzenschemata betrachtet (vgl. [19]). Für das Richtmyer-Verfahren $\omega = 1$ erhält er die in Satz 3 unter (b) angegebene hinreichende Stabilitätsbedingung.

3. Anwendung auf eine Klasse skalarer Differentialgleichungen 2. Ordnung

Gegeben sei das skalare hyperbolische Anfangswertproblem 2. Ordnung

$$(17) \qquad \begin{aligned} u_{tt} &= \sum_{j=1}^{s} c_j^2 u_{x_j x_j} + \sum_{j=1}^{s} d_j u_{x_j} + e u_t \quad , \quad -\infty < x_j < \infty \, , \; 0 \leq t \leq T \, , \\ u(x,0) &= f(x) \, , \\ u_t(x,0) &= g(x) \, . \end{aligned}$$

Wir transformieren die Aufgabe (17), die eine Anzahl von Problemen der mathematischen Physik wie etwa die Wellengleichung in s Ortsveränderlichen umfaßt, in ein System 1. Ordnung der Form (1):

$$(18) \qquad \begin{aligned} W_t &= \sum_{j=1}^{s} B_j W_{x_j} + DW \quad , \\ W(x,0) &= F(x) \end{aligned}$$

mit

$$W := (u_t, c_1 u_{x_1}, \dots, c_s u_{x_s})^T \, ,$$

$$F := (g, c_1 f_{x_1}, \dots, c_s f_{x_s})^T$$

und

$$B_j := c_j \cdot \begin{pmatrix} 0 & \cdots & 1 & 0 & \cdots & 0 \\ \vdots & & & & & \\ 1 & & & & & \\ \vdots & & & & \mathbf{0} & \\ 0 & & & & & \end{pmatrix} \quad \begin{matrix} \leftarrow (j+1)\text{-te Spalte (Spalte der 1 in der ersten Zeile)} \\ \leftarrow (j+1)\text{-te Zeile (Zeile der 1 in der ersten Spalte).} \end{matrix}$$

Die Matrix D hat die Gestalt

$$D := \begin{pmatrix} e & \frac{d_1}{c_1} & \cdots & \frac{d_s}{c_s} \\ & & \mathbf{0} & \end{pmatrix}$$

Es gilt

<u>Satz 4</u> Im Falle des auf ein System 1. Ordnung (18) transformierten skalaren hyperbolischen Anfangswertproblems (17) ist $S_{LWR}(\omega)$ <u>genau dann</u> stabil für $0 < \omega \leq 1$, falls gilt:

(19) $$\lambda \max_j \rho(B_j) \leq \sqrt{\frac{\omega}{s}} \; .$$

Die Bedingung (19) ist äquivalent der Bedingung (9).

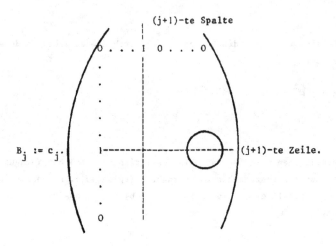

Beweis:

1. Schritt: (19) ist hinreichend für Stabilität.

Wegen $\rho(B_j) = c_j$ gilt

$$\rho^2(\lambda \sum_{j=1}^{s} B_j \alpha_j) = \lambda^2 \sum_{j=1}^{s} \rho^2(B_j)\, \alpha_j^2 \leq \frac{\omega}{s} \sum_{j=1}^{s} \alpha_j^2 \quad ,$$

und damit ist der 1. Schritt gezeigt.

2. Schritt: (19) ist notwendig für Stabilität.

(19) sei verletzt, d.h. o.B.d.A. gelte $\lambda\rho(B_1) > \sqrt{\frac{\omega}{s}}$, $0 < \omega \leq 1$. Also folgt

$$(20) \qquad \lambda^2 c_1^2 = \frac{\omega}{s\varepsilon_o} \qquad \text{mit } 0 < \varepsilon_o < 1 .$$

Wir betrachten jetzt die Amplifikations-Matrix $\hat{S}_{LWR}$ und werden zeigen, daß sich $\overline{\xi}_1,\dots,\overline{\xi}_s$ so wählen lassen, daß die von-Neumann-Bedingung verletzt ist, falls (20) gilt. Wir wählen $\overline{\xi}_j$ in $0 \leq \overline{\xi}_j \leq \frac{\pi}{2}$ so, daß gilt:

$$(21) \qquad \begin{aligned} &\sin^2\overline{\xi}_1 = 1 - \varepsilon_o , \\ &\sin^2\overline{\xi}_k = 0 \quad \text{für } k = 2,\dots,s . \end{aligned}$$

Wegen (20) läßt sich für die durch (21) bestimmten speziellen $\overline{\xi}_j$ der Spektralradius von $\hat{S}_{LWR}$ angeben:

$$(22) \qquad \rho^2(\hat{S}_{LWR}) = 1 + \frac{4\omega}{s\varepsilon_o}(1-\varepsilon_o)\left[\frac{\omega}{s\varepsilon_o}(1-\varepsilon_o) + \left(1+\frac{\omega}{s}(\sqrt{\varepsilon_o}-1)\right)^2 - 1\right] .$$

Wir betrachten $\rho^2(\hat{S}_{LWR})$ als Funktion von ε:

$$f_\omega(\varepsilon) := \rho^2(\hat{S}_{LWR}) \text{ mit } \varepsilon \text{ statt } \varepsilon_o \text{ in (22).}$$

Wir sind fertig, falls wir $f_\omega(\varepsilon_o) > 1$ für $0 < \omega \leq 1$ gezeigt haben.
Man verifiziert sofort

1) $\qquad f_\omega(1) = 1$

2) $\qquad \lim_{\varepsilon\to o} f_\omega(\varepsilon) = +\infty .$

Es bleibt also wegen $0 < \varepsilon_o < 1$ zu zeigen, daß $f_\omega(\varepsilon)$ in $(0,1]$ monoton fällt. Man rechnet nach:

$$(23) \qquad \frac{s^2}{4\omega^2}\,\frac{\partial f_\omega}{\partial\varepsilon} = -\frac{2}{\varepsilon^3} + \frac{4}{\varepsilon^2} - \frac{1}{\varepsilon\sqrt{\varepsilon}} - \frac{1}{\sqrt{\varepsilon}} - \frac{\omega}{s}\left(\frac{1}{\varepsilon^2} - \frac{1}{\varepsilon\sqrt{\varepsilon}} - \frac{1}{\sqrt{\varepsilon}} + 1\right).$$

Wir machen die Variablensubstitution $\sqrt{\varepsilon} := \frac{1}{1+\eta}$, und damit folgt aus (23) mit Hilfe des binomischen Lehrsatzes

$$\frac{\partial f_\omega}{\partial\varepsilon} < 0 \text{ für } 0 < \omega \leqslant 1 ,$$

und der 2. Schritt ist bewiesen.

3. Schritt: <u>(19) ist äquivalent (9).</u>

Im 1. Schritt wurde bereits gezeigt, daß aus (19) die Bedingung (9) folgt. Aber auch die Umkehrung ist richtig, denn die Gegenannahme führt wegen des 2. Schrittes zu einem Widerspruch.
Damit ist auch der 3. Schritt gezeigt und Satz 4 vollständig bewiesen.

Bemerkung:

In [3] zeigte kürzlich v. Finckenstein, daß sich die hinreichende Stabilitätsbedingung

$$\lambda \max_j \rho(A_j) \leqslant \frac{1}{s\sqrt{s}}$$

für das s-dimensionale Lax-Wendroff-Verfahren S_{LW} im Spezialfall der s-dimensionalen Wellengleichung mit konstanten Koeffizienten

$$u_{tt} = \sum_{j=1}^{s} c_j^2\, u_{x_j x_j}$$

als System 1. Ordnung geschrieben zu

$$\lambda \max_j \rho(A_j) \leqslant \frac{1}{s}$$

abschwächen läßt. Numerische Ergebnisse in [3] ließen sogar vermuten, daß

$$(24) \qquad \lambda \max_j \rho(A_j) \leqslant \frac{1}{\sqrt{s}}$$

hinreichend für Stabilität ist.

Als Spezialfall in Satz 4 ist enthalten, daß (24) notwendig und hinreichend für die Stabilität von S_{LWR} im Falle der s-dimensionalen Wellengleichung ist.

4. Die Courant-Friedrichs-Lewy-Bedingung

Courant, Friedrichs und Lewy haben 1928 (vgl. [1]) festgestellt, daß Differenzenverfahren bei hyperbolischen Anfangswertproblemen nicht immer konvergent sind:

Falls das Verhältnis zwischen Zeit- und Raumschrittweite so groß ist, daß das numerische Abhängigkeitsgebiet D_Δ eines Punktes nicht das Abhängigkeitsgebiet D bezüglich der Differentialgleichung enthält, dann ignoriert ein Differenzenschema mit einem solchen Schrittweitenverhältnis Informationen, von denen die exakte Lösung abhängt. Es kann folglich im allgemeinen nicht gegen die exakte Lösung konvergieren.

Zwar ist es einfach, daß numerische Abhängigkeitsgebiet D_Δ eines Punktes bei einem expliziten Differenzenverfahren zu bestimmen, das Abhängigkeitsgebiet D bezüglich der Differentialgleichung läßt sich jedoch in den meisten Fällen nicht ohne weiteres explizit angeben. Man kann sich aber helfen, in dem man nicht die Abhängigkeitsgebiete D_Δ und D selbst, sondern die Stützfunktionen der konvexen Hüllen der Abhängigkeitsgebiete betrachtet und so zu einer anderen Formulierung der CFL-Bedingung kommt. Auf Einzelheiten soll hier verzichtet werden, es sei auf [8] und die dort angegebene Literatur verwiesen.
Abschließend geben wir ohne Beweis (vgl. [8]) den

Satz 5 Die CFL-Bedingung für das Verfahren $S_{LWR}(\omega)$ lautet

(25) $$\lambda\rho\left(\sum_{j=1}^{s} A_j\alpha_j\right) \leq \max_j |\alpha_j| \quad \text{für alle } |\alpha| = 1 .$$

Falls die Stabilitätsbedingung (9), nämlich

$$\lambda\rho\left(\sum_{j=1}^{s} A_j\alpha_j\right) \leq \sqrt{\frac{\omega}{s}} \quad \text{für alle } |\alpha| = 1 ,$$

erfüllt ist, dann ist auch die CFL-Bedingung (25) erfüllt.
Für $\omega = 1$, das Richtmyer-Verfahren, gilt die Umkehrung in folgenden Fällen:

(a) $A := A_1 = \ldots = A_s$

(b) Problem (18) und $c := c_1 = \ldots = c_s$.+)

Bemerkung:

1) Die CFL-Bedingung für das Problem (18) beim Verfahren $S_{LWR}(\omega)$ lautet:

$$\lambda \sum_{j=1}^{s} c_j^2 \leq 1 \quad .$$

2) Für $A := A_1 = \ldots = A_s$ ist $S_{LWR}(\omega=1)$ optimal stabil im Sinne von Strang (vgl. [14]), falls $\lambda\rho(A) \leq \frac{1}{s}$ erfüllt ist. Von-Neumann-Bedingung und CFL-Bedingung sind in diesem Falle äquivalent. Insbesondere ist für $s = 1$ $S_{LWR}(\omega=1)$ stets optimal stabil, falls $\lambda\rho(A) \leq 1$ erfüllt ist.

3) Es kann kein explizites Differenzenverfahren geben, das bei gleichem numerischem Abhängigkeitsgebiet wie das Richtmyer-Verfahren $S_{LWR}(\omega=1)$ für $A := A_1 = \ldots = A_s$ bei einer weniger einschränkenden Bedingung als $\lambda\rho(A) \leq \frac{1}{s}$ konvergiert.

5. Numerische Beispiele ++)

I) Im R^2 betrachten wir das skalare hyperbolische Anfangswertproblem

$$u_{tt} = \Delta_2 u := u_{xx} + u_{yy}$$

(26) $\quad u(x,y,0) = \sin x + \sin y$

$$u_t(x,y,0) = -(\sin x + \sin y) \sin t \; .$$

Die exakte Lösung von (26) ist $u(x,y,t) = (\sin x + \sin y) \cos t$.

+) Das ist keine Spezialisierung von (a) für das Problem (18), denn für die in (18) auftretenden Matrizen B_j gilt $B_\nu \neq B_\mu$ für $\nu \neq \mu$, auch wenn (b) gilt.

++) Die Rechnungen wurden auf der IBM/370-165 der Kernforschungsanlage Jülich durchgeführt; Fräulein A. Gränz danke ich für die Erstellung der Programme.

Gemäß (18) transformieren wir (26) in ein System 1. Ordnung und behandeln dieses Problem mit Hilfe von $S_{LWR}(\omega)$ für verschiedene ω. In den folgenden Tabellen sind die Ergebnisse der Rechnungen aufgeführt. Die numerische Lösung wurde in

$$G := \{(x,y,t) \mid 0 \leq x \leq 1 \ , \ 0 \leq y \leq 1 \ , \ 0 < t \leq T\}$$

berechnet. Als Randwerte wurde die exakte Lösung verwendet. Im einzelnen bedeuten:

$h := \Delta x = \Delta y$	die Ortsschrittweite
$\lambda := \frac{\Delta t}{h}$	das Schrittweitenverhältnis
n :	die Anzahl der Schritte in t-Richtung
$T := n\Delta t$	der Endwert in t-Richtung
F :	der relative Fehler zwischen Näherungslösung und exakter Lösung; als Fehlermaß wurde die diskrete L_2-Norm verwendet
R :	Resultat der Rechnungen

Durch die Rechnungen wurde die in Satz 4 angegebene notwendige und hinreichende Stabilitätsbedingung (19) auch numerisch sehr gut bestätigt.

h	λ	n	T	F	R
0.04	0.65	380	9.88	7.01 E- 4	stabil
		760	19.76	2.87 E- 4	
	0.70	350	9.80	5.09 E- 4	stabil
		700	19.60	2.02 E- 4	
	$\sqrt{\omega/2}$	350	9.90	4.93 E- 4	stabil
		700	19.80	1.81 E- 4	
	0.75	330	9.90	1.50 E- 3	instabil
		660	19.80	1.00 E+ 1	
	0.80	310	9.92	2.91 E+13	instabil
		620	19.84	1.91 E+33	
0.02	0.65	760	9.88	2.11 E- 4	stabil
		1520	19.76	9.03 E- 5	
	0.70	700	9.80	1.79 E- 4	stabil
		1400	19.60	5.17 E- 5	
	$\sqrt{\omega/2}$	700	9.90	1.47 E- 4	stabil
		1400	19.80	5.06 E- 5	
	0.75	660	9.90	2.68 E+12	instabil
		1320	19.80	3.14 E+32	
	0.80	500	8.00	1.08 E+33	instabil
		540	8.48	>1.0 E+75	

Stabilitätsbedingung : $\lambda \leq \sqrt{\frac{\omega}{2}} = 0.707106\ldots \quad (\omega=1)$

h	λ	n	T	F	R
0.04	0.55	450	9.90	7.37 E- 4	stabil
		900	19.80	5.12 E- 4	
	0.60	410	9.84	5.84 E- 4	stabil
		820	19.68	3.88 E- 4	
	$\sqrt{\omega/2}$	390	9.86	4.96 E- 4	stabil
		780	19.72	2.83 E- 4	
	0.70	350	9.80	3.13 E- 4	?
		700	19.60	2.26 E- 3	
	0.75	330	9.90	1.24 E+ 7	instabil
		660	19.80	2.06 E+20	
0.02	0.55	900	9.90	2.49 E- 4	stabil
		1800	19.80	1.52 E- 4	
	0.60	820	9.84	1.91 E- 4	stabil
		1640	19.68	1.02 E- 4	
	$\sqrt{\omega/2}$	780	9.86	1.58 E- 4	stabil
		1560	19.72	6.67 E- 5	
	0.70	700	9.80	6.72 E+ 1	instabil
		1400	19.60	4.85 E+10	
	0.75	660	9.90	2.30 E+29	instabil
		790	11.85	>1.0 E+75	

Stabilitätsbedingung $\lambda \leq \sqrt{\frac{\omega}{2}} = 0.632455\cdots \quad (\omega=0.8)$

II) Im R^2 betrachten wir das nichtlineare skalare hyperbolische Anfangswertproblem

$$(27)\qquad \begin{aligned} &\frac{\partial u}{\partial t} + \frac{\partial}{\partial x}\left(\frac{1}{4}u^2\right) + \frac{\partial}{\partial y}\left(\frac{1}{4}u^2\right) = z(u,x,y) \\ &u(x,y,0) = 100\, x^2\,(1-x)\, y^2\,(1-y)\,. \end{aligned}$$

Dabei ist $z(u,x,y)$ folgendermaßen definiert:

$$z(u,x,y) = 50\, uxy\, [y(1-y)(2-3x) + x(1-x)(2-3y)]\,.$$

Dieses von Gourlay und Morris (vgl. [4]) angegebene Beispiel hat die zeitunabhängige Lösung

$$u(x,y,t) = 100\, x^2 y^2\,(1-x)(1-y)\,.$$

Wie in I) berechnen wir die numerische Lösung in

$$G := \{(x,y,t) \mid 0 \leq x \leq 1,\ 0 \leq y \leq 1,\ 0 < t \leq T\}\,,$$

mit der exakten Lösung als Randbedingung, also der 0-Randbedingung.

Nach Satz 2 lautet die Stabilitätsbedingung (im linearisierten Sinne) wegen

$$\rho(A_1) = \rho(A_2) = \frac{|u|_{max}}{2}:$$

(28) $\lambda \; |u|_{max} \leq \sqrt{\omega}$.

Die Bedingung (28) ist nach Satz 2 (im linearisierten Sinne) sogar notwendig für Stabilität.
Die Ergebnisse in der folgenden Tabelle bestätigen das recht gut. Es wurden 300 Schritte in t-Richtung gerechnet.

ω	h	Δt	λ	$\lambda\|u\|_{max}$	$\sqrt{\omega}$	Resultat
1	0.02	0.007	0.35	0.768	1	stabil
		0.008	0.4	0.878		stabil
		0.009	0.45	0.988		stabil
		0.01	0.5	1.097		overflow nach 60 Schritten
0.8	0.02	0.007	0.35	0.767	0.894	stabil
		0.008	0.4	0.878		stabil
		0.009	0.45	0.988		overflow nach 82 Schritten

Stabilitätsbedingung $\lambda \; |u|_{max} \leq \sqrt{\omega}$

6. Literaturverzeichnis

[1] Courant, R., Friedrich, K.O., und Lewy, H. (1928): Über die partiellen Differenzengleichungen der mathematischen Physik. Math. Ann. 100, S. 32

[2] Eilon, B. (1972): A note concerning the Two-Step Lax-Wendroff Method in Three Dimensions. Math. Comp. 26., P. 41

[3] Finck v. Finckenstein, K. Graf (1969): On the Numerical Treatment of Hyperbolic Differential Equations with Constant Coefficients, particuarly the n-Dimensional Wave Equation,Conference on the Numerical Solution of Differential Equations, P. 154, Lecture Notes in Mathematics 109, Springer, Berlin

[4] Gourlay, A.R., and Morris, J.Ll. (1968): A Multistep Formulation of the Optimized Lax-Wendroff Method for Nonlinear Hyperbolic Systems in Two Space Variables. Math. Comp. 22, P. 715

[5] John, F. (1967): Lectures on Advanced Numerical Analysis. Gordon and Breach, New York

[6] Lax, P.D., and Nirenberg, L. (1966): On Stability for Difference Schemes; a Sharp Form of Garding's Inequality. Comm. Pure Appl. Math. Vol. 19, P. 473

[7] Lax, P.D., and Wendroff, B. (1964): Difference Schemes for Hyperbolic Equations with High Order of Accuracy. Comm. Pure Appl. Math., Vol. 17, P. 381

[8] Meuer, H.W. (1972): Zur numerischen Behandlung von Systemen hyperbolischer Anfangswertprobleme in beliebig vielen Ortsveränderlichen mit Hilfe von Differenzenverfahren. KFA-Bericht JÜL-861-MA

[9] von Neumann, J., und Richtmyer, R.D. (1950): A Method for the Numerical Calculation of Hydrodynamic Shocks. Journal of Appl. Phys. 21, P. 232

[10] Richtmyer, R.D. (1962): A Survey of Difference Methods for Non-Steady Fluid Dynamics. NCAR Technical Notes 63-2, Boulder Col.

[11] Richtmyer, R.D., and Morton, K.W. (1967): Difference Methods for Initial Value Problems. Wiley (Interscience), New York

[12] Rubin, E.L., and Preiser, S. (1970): Three-dimensional second-order accurate difference schemes for discontinous hydrodynamic flows. Math. Comp. 24, P. 57

[13] Rusanov, V.V. (1962): The Calculation of the Interaction of Non-Stationary Shock Waves and Obstacles. USSR Comp. Math. and Math. Phys. 2, P. 304

[14] Strang, G. (1968): On the Construction and Comparison of Difference Schemes. Siam J. Num. Anal., Vol. 5, P. 506

[15] Tyler, L.D. (1971): Heuristic Analysis of Convective Finite Difference Techniques. Proceedings of the Second International Conference on Numerical Methods in Fluid Dynamics, P. 314, Lecture Notes in Physics 8, Springer, Berlin

[16] Wendroff, B. (1968): Well posed problems and stable difference operators. Siam J. Num. Anal., Vol. 5, P. 71

[17] Wilson, J. C.: Stability of Richtmyer Type Difference Schemes in any Number of Space Variables and their Comparision with Multistep Strang Schemes. to appear

[18] Yamaguti, M. (1967): Some Remarks on the Lax-Wendroff scheme for non-symmetric hyperbolic systems. Math. Comp. 21, P. 611

[19] Yamaguti, M. (1970): On the pseudo difference schemes. Proceedings of the International Conference on Functional Analysis and Related Topics. Tokyo, April 1969, University of Tokyo Press, Tokyo 1970

ASYMPTOTISCHE LÖSUNGEN VON FUNKTIONALGLEICHUNGEN

Von Franz Pittnauer

Aufgabenstellung: Wir beschäftigen uns im folgenden mit einer speziellen Methode zur asymptotischen Lösung von Funktionalgleichungen. Ausgangspunkt unserer Überlegungen ist dabei der folgende

Darstellungssatz: Es sei G ein (offenes) Gebiet der komplexen Zahlenebene, welches $z = o$ als inneren Punkt und $z_o \neq o$ als Randpunkt besitze und welches ganz in der Kreisscheibe $|z| < R$ mit einem Wert $R > o$ liege.

Es sei $F(t_1, t_2, \ldots, t_k)$ holomorph für $|t_\varkappa| < R$ $(\varkappa = 1, 2, \ldots, k)$.

Ferner sei $f_\varkappa$ für $\varkappa = 1, 2, \ldots, k$ in G holomorph und besitze die Eigenschaften

$$|f_\varkappa(z)| < R,$$

$$f_\varkappa(z) \sim \sum_{n=o}^{\infty} b_{n\varkappa} \cdot h^n(z - z_o) \quad \text{für } z \to z_o,\ z \in G.$$

Dabei sei h eine in $z = o$ holomorphe Funktion, welche in diesem Punkt eine einfache Nullstelle besitze.

Dann gilt

$$F[f_1(z), f_2(z), \ldots, f_k(z)] \sim \sum_{n=o}^{\infty} c_n \cdot h^n(z - z_o) \quad \text{für } z \to z_o,\ z \in G.$$

Den einfachen Beweis des Satzes findet man etwa in [1o] S. 21–22, Satz 2.6. Die Werte c_n erhält man mit der Methode der unbestimmten Koeffizienten.

Wir betrachten nun Funktionalgleichungen der Gestalt

$$(1) \qquad F[D_1 g(z),\ D_2 g(z), \ldots, D_k g(z)] = o \quad \text{für } z \in G.$$

Dabei sei g als in G holomorphe Funktion gesucht. Unter den $D_\varkappa$ $(\varkappa = 1, 2, \ldots, k)$ wollen wir geeignete Funktionen oder auch Operatoren verstehen und stets $|D_\varkappa g(z)| < R$ für $z \in G$ verlangen.

Unser Gleichungstyp liefert im Falle, daß F in allen Argumenten linear ist, beispielsweise die folgende Gleichung, wenn wir unter s eine in G holomorphe Funktion mit Werten in G verstehen,

(2) $$g[s(z)] - a\cdot g(z) - b = o.$$

Diese Gleichung reduziert sich für

$a = 1, \quad b \neq o$ auf die sog. Abelsche,

$a \neq o, \quad b = o$ auf die sog. Schrödersche

Funktionalgleichung. Beide Gleichungen sind sehr genau untersucht worden, vgl. etwa Kuczma [7] S. 163-179, 135-162.

Wir wollen nun unter einer asymptotischen Lösung von (1) eine in G holomorphe Funktion g verstehen, welche der Beziehung

(3) $$F[D_1 g(z), D_2 g(z), \ldots, D_k g(z)] \sim O \quad \text{für } z \to z_o,\ z \in G$$

genügt. Im Unterschied zu (1), wo Bedingungen für alle $z \in G$ verlangt waren, verlangen wir in (3) nur noch Bedingungen für $z \to z_o$, $z \in G$. Wir werden deshalb von (3) als von einer "asymptotischen Funktionalgleichung" sprechen.

Eine weitere Berechtigung zu diesem Vorgehen kann man auch dem folgenden Sachverhalt entnehmen.

i) Erfüllt in (2) die Funktion s noch die Bedingung $s(o) = o$, so hat die Abelsche Funktionalgleichung keine Lösung, wie man sofort einsieht.

ii) Erfüllt in (2) die Funktion s noch die Bedingungen $s(o) = o$, $s'(o) = -1$, $s[s(z)] \neq z$, so hat die Schrödersche Funktionalgleichung für $a = -1$ keine in G holomorphe Lösung, wie man Satz 6. 1o bei Kuczma [7] S. 147 entnehmen kann.

Betrachten wir dagegen unsere Aufgabe (3) und verlangen von s noch zusätzlich

$$s(z) \sim z - z_o \quad \text{für } z \to z_o,\ z \in G,$$

so ist diese auch noch in den genannten Ausnahmefällen sinnvoll und wir können leicht eine Reihe von Lösungen explizit angeben, vgl. [1o] S. 151, Aufgaben 1 und 2.

Wenden wir uns nun der Lösung der asymptotischen Funktionalgleichung (3) zu. Um den obigen Darstellungssatz anwenden zu können, ist es einmal nötig, daß wir uns auf in G holomorphe Funktionen g mit

$$g(z) \sim \sum_{n=0}^{\infty} a_n \cdot h^n(z - z_0) \quad \text{für } z \to z_0,\ z \in G \tag{4}$$

beschränken. Zum anderen ist es nötig, daß wir unter Verwendung von (4) auch noch die asymptotischen Reihen nach Potenzen von h aller Funktionen $D_\varkappa g(z)$ ($\varkappa = 1, 2, \dots, k$) für $z \to z_0$, $z \in G$ angeben können, was Einschränkungen für die in (3) zulässigen $D_\varkappa$ bedeutet.
Damit liefert dann der Darstellungssatz über die Methode der unbestimmten Koeffizienten ein Gleichungssystem für die a_n in (4). Dieses ist in vielen Fällen rekursiv lösbar. Schließlich liefert der Existenzsatz von Ritt (vgl. etwa [1o] S. 49, Satz 5. 1) eine große Reihe von Lösungen g.

Das hier beschriebene Verfahren, auf das wir auch in [1o] S. 144-159 hingewiesen haben, unterliegt freilich starken Einschränkungen was den Typ der Gleichung (3) wie auch was den Typ der Lösung (4) anbelangt. Dafür ist es aber einfach zu handhaben und in einer großen Zahl interessanter Fälle anwendbar.
Um einen besseren Einblick in die Möglichkeiten unserer Methode zu gewinnen, betrachten wir im folgenden drei Arten von asymptotischen Funktionalgleichungen des Typs (3) etwas genauer.

Gleichungen erster Art: Als Gleichungen erster Art wollen wir solche Gleichungen (3) bezeichnen, zu deren Lösung der Ansatz (4) mit einer fest vorgegebenen, bekannten, den Bedingungen des Darstellungssatzes genügenden Funktion h zum Ziele führt. Ohne Einschränkung können wir $h(z) = z$ wählen.

Dann folgt aus den Rechenregeln für asymptotische Potenzreihen sofort, daß man für jedes $D_\varkappa$ in (3) unter anderem eine der folgenden einfachen Möglichkeiten wählen darf.

i) $D_\nu g(z) = z$,

ii) $D_\nu g(z) = \int_z^{z_o} g(t) \cdot dt$ längs eines bis auf z_o ganz

in G liegenden, rektifizierbaren Integrationsweges,

iii) $D_\nu g(z) = g^{(n_\nu)}(z)$ mit $n_\nu \geq o$ ganzzahlig.

Wählt man ein $D_\varkappa$ gemäß i), so bedeutet dies lediglich, daß Gleichung (3) explizit von z abhängt. Wählt man alle $D_\varkappa$ gemäß iii) mit $n_\varkappa = o$, so erhält man eine implizite asymptotische Gleichung. Einen zugehörigen Lösungssatz findet man in [1o] S. 144–145, Satz 16. 2.

Besonderes Interesse verdient schließlich noch der Fall, daß man

iv) $D_\nu g(z) = g(z - a_\nu)$ mit $a_\nu \neq o, \infty$

wählt und gleichzeitig unbeschränkte Definitionsgebiete G mit $z_o = \infty$ zuläßt. Hier wird man dann beispielsweise auf die Behandlung nichtlinearer Differenzen-Differentialgleichungen wie etwa

$$F[z^{-1}, g(z), g'(z), g(z-a)] \sim o \quad \text{für } z \to \infty, \ |\arc z| < \alpha < \frac{\pi}{2} \tag{5}$$

geführt.

Geht man nun von für Re z > Re a holomorphen Funktionen g mit

$$g(z) \sim \sum_{n=o}^{\infty} a_n \cdot z^{-n} \quad \text{für } z \to \infty, \ \text{Re } z > \text{Re } a$$

aus und verwendet die Beziehung

$$g(z-a) \sim \sum_{n=o}^{\infty} b_n \cdot z^{-n} \quad \text{für } z \to \infty, \ \text{Re } z > \text{Re } a$$

mit
$$b_n = a^n \cdot \sum_{\nu=o}^{n} \binom{n-1}{\nu-1} \cdot a^{-\nu} \cdot a_\nu \quad (n = o, 1, 2, \ldots),$$

so kann man (5) im Bereich Re z > 2· Re a, $|\arc z| < \alpha$ mit unserer Methode behandeln. Der Wert a_o ergibt sich dabei als eine Lösung der Gleichung $F(o, a_o, o, a_o) = o$.

Es sei noch erwähnt, daß die verschiedenen hier aufgeführten Gleichungstypen in der Literatur auch mit anderen, teils allgemeineren Methoden behandelt wurden. Wir verweisen hier lediglich auf Kuczma [7], Harris-Sibuya [4], Harris-Tolle [6],

Lublin [8], Rozmus-Chmura [12], H. Schmidt [13], Smajdor [14-15], sowie auf Berg [1], Van der Corput [2], Entringer [3], Harris-Sibuya [5], Rickstinš--Muriņa [11], Strodt-Wright [16], Swanson-Schulzer [17] und Wasow [18].

Gleichungen zweiter Art: Als Gleichungen zweiter Art wollen wir solche Gleichungen (3) bezeichnen, zu deren Lösung der Ansatz (4) mit der Funktion $h(z) = g(z)$ zum Ziele führt. Aus technischen Gründen wollen wir dabei von der Lösung g die Zusatzbedingungen $g(o) = o$, $g'(o) \neq o$ fordern.

Anstelle von (4) haben wir hier also den Ansatz

$$(6) \qquad g(z) \sim \sum_{n=o}^{\infty} c_n \cdot g^n(z - z_o) \quad \text{für } z \to z_o, \ z \in G$$

zu verwenden. Wir werden (6) als die "asymptotische Selbstentwicklung" von g für $z \to z_o$, $z \in G$ ansprechen. Bei solchen Aufgaben ist es also nötig, g einmal in seinem asymptotischen Verhalten um $z = z_o$ und zum anderen in seinem regulären Verhalten um $z = o$ zu bestimmen und dies ist - anders als bei Gleichungen erster Art - kein rein lokales Problem mehr.

Aufgaben dieser Art ergeben sich etwa, wenn man für die $D_\varkappa$ in (3) die folgenden Möglichkeit zuläßt.

v) $D_\nu g(z) = g[s_\nu(z)]$ mit einer in G holomorphen Funktion s_ν, welche ihre Werte in G annimmt und der Bedingung $s_\nu(z) \sim z - z_o$ für $z \to z_o$, $z \in G$ genügt.

Wenden wir unsere Methode auf die Gleichung

$$(7) \qquad F[g(z), g(z - z_o)] \sim o \quad \text{für } z \to z_o, \ z \in G$$

an, so erhalten wir den folgenden

Reduktionssatz: Jedem Wert c_o mit $F(c_o, o) = o$, $\left.\frac{\partial}{\partial u} F(u,v)\right|_{\substack{u=c_o \\ v=o}} \neq o$ läßt sich eine Reihe der Gestalt (6) zuordnen derart, daß jede in G holomorphe Funktion g mit $g(o) = o$, $g'(o) \neq o$, welche diese Reihe als asymptotische Selbstentwicklung besitzt, eine Lösung von (7) ist.

Den Beweis des Satzes findet man in [9] bzw. [1o] S. 15o-151.

Als einfache und elementar lösbare Beispiele sei hier auf die Abelsche und Schrödersche Funktionalgleichung verwiesen. Im allgemeinen stehen wir allerdings vor dem Problem, eine Funktion g mit $g(o) = o$, $g'(o) \neq o$ zu finden, welche einer Beziehung (6) mit vorgegebenen Koeffizienten c_n genügt. Dies ist eine offenbar schwierige nichtlineare Aufgabe. Näherungsweise kann sie durch den folgenden Approximationssatz gelöst werden.

Approximationssatz: Es seien die Koeffizienten c_n von (6) gegeben und es sei $c_m \neq o$ für wenigstens ein $m \geq o$. Dann existieren zu jedem $s \geq m + 1$ unendlich viele Polynome P_{2s} mit $P_{2s}(o) = o$, $P'_{2s}(o) \neq o$ und höchstens vom Grade 2s, welche der Beziehung

$$P_{2s}(z) = \sum_{n=o}^{\infty} d_n \cdot P_{2s}^{n} (z - z_o)$$

mit $d_n = c_n$ für $n = o, 1, \ldots, s-1$ genügen.

Für den Beweis des Satzes sowie für weitere Einzelheiten vgl. man [9] bzw. [1o] S. 1o7 - 113.

Gleichungen dritter Art: Als Gleichungen dritter Art wollen wir solche Gleichungen (3) bezeichnen, welche Terme wie etwa

vi) $D_\nu\, g(z) = g[g(z - z_o)]$; $D_\nu\, g(z) = g[g^{(n_\nu)}(z)]$ mit $n_\nu \geq o$ ganzzahlig

enthalten.

Man überlegt sich leicht, daß hier im allgemeinen unsere Methode versagt.

Betrachten wir beispielsweise die sog. asymptotische Funktionalgleichung von Babbage (vgl. Kuczma [7] S. 288)

$$g[g(z)] \sim z \quad \text{für } z \to z_o,\ z \in G. \tag{8}$$

Verwenden wir zur Lösung den Ansatz (4) und ist $a_o = \lim_{z \to z_o} g(z) \neq z_o$, $a_o \in G$, so müssen wir neben (4) auch noch die Taylorreihe von g um $z = a_o$ heranziehen. Damit werden wir aber auf das ungelöste und schwierige Problem geführt, die asymptotischen Potenzreihen einer holomorphen Funktion aus ihrer Taylorreihe zu bestimmen.

Im Falle, daß $a_o = z_o$ ist, können wir dieses Problem freilich umgehen und erhalten mit unserer Methode zwei Klassen von Lösungen. Für die Durchführung der formalen Rechnungen vgl. man Kuczma [7] S. 3o3-3o4.

Schlußbemerkungen: Wie bereits erwähnt, gibt es in der Literatur allgemeinere Methoden zur Lösung asymptotischer Funktionalgleichungen, insbesondere auch solche, die erlauben, den Typ der asymptotischen Entwicklung der gesuchten Funktionen zu bestimmen. Wir verweisen hier lediglich auf die Arbeiten von Berg [1] und Van der Corput [2]. Dabei hat es sich freilich stets um Probleme gehandelt, die eine Betrachtung der gesuchten Funktionen lediglich in der Umgebung eines einzigen Punktes (etwa für $z \to \infty$) erfordern.

Demgegenüber erlaubt unsere Methode nur, Lösungen des Typs (4) zu finden. Allerdings ist die Methode besonders einfach anzuwenden und erlaubt ferner auch noch Aussagen für den Fall, daß die gesuchte Funktion in der Umgebung von zwei verschiedenen Punkten betrachtet werden muß.

Meines Erachtens ist die hier geschilderte Methode in der Literatur, abgesehen von der asymptotischen Lösung gewöhnlicher Differentialgleichungen (vgl. etwa Wasow [18], S. 57, Satz 12. 1), kaum verwendet worden und noch sehr ausbaufähig. Wir würden uns freuen, wenn es uns hier gelungen wäre, weitere einschlägige Arbeiten anzuregen.

Literatur

[1] Berg, L.: Asymptotische Lösungen für Operatorgleichungen. Zeitschr. Ang. Math. Mech. 45(1965) 333 - 352.

[2] Van der Corput, J.G.: Introduction to the method of asymptotic functional relations. Journal d'Analyse Math. 14(1965) 67-78.

[3] Entringer, R.C.: Functions and inverses of asymptotic functions. Amer. Math. Monthly 74(1967) 1o95-1o97.

[4] Harris, W.A. Jr. and Sibuya, J.: General solution of nonlinear difference equations.
Trans. AMS 115(1965) 62–75.

[5] Harris, W.A. Jr. and Sibuya, J.: On asymptotic solutions of systems of nonlinear difference equations.
J. Reine u. Ang. Math. 222(1966) 12o–135.

[6] Harris, W.A. Jr. and Tolle, J.W.: A nonlinear functional equation.
Proc. United States – Japan Seminar on differential and functional equations; Univ. of Minnesota, 1967. New York–Amsterdam: W.A. Benjamin, Inc. 1967, S. 449–454.

[7] Kuczma, M.: Functional equations in a single variable.
Warszawa: Polish Scientific Publ. 1968.

[8] Lublin, M.: On a class of nonlinear difference equations.
Mat. Tidsskr. B(1946) 12o–128.

[9] Pittnauer, F.: Asymptotische Selbstentwicklungen von holomorphen Funktionen.
Math. Zeitschr. 123(1971) 365–372.

[1o] Pittnauer, F.: Vorlesungen über asymptotische Reihen.
Lecture Notes in Math. Nr. 3o1; Berlin–Heidelberg–New York: Springer, 1972.

[11] Rickstiņš, E.J. und Muriņa, M.J.: Über die asymptotische Darstellung impliziter Funktionen. (russ.)
Lat. Mat. Ezegodnik, Riga 1(1965) 23–36.

[12] Rozmus-Chmura, M.: Sur l'equation fonctionnelle $\varphi(x+1) = f[\varphi(x)]$.
Mat. Vesnik 4(1967) 75–78.

[13] Schmidt, H.: Über ganze Lösungen einer Klasse nichtlinearer Differenzengleichungen.
Vortrag, Universität Dortmund, 23. 1o. 72

[14] Smajdor, W.: On the existence and uniqueness of analytic solutions of the functional equation $\varphi(z) = h(z, \varphi[f(z)])$.
Ann. Polon. Math. 19(1967) 37–45.

[15] Smajdor, W.: Analytic solutions of the equation $\varphi(z) = h(z, \varphi[f(z)])$ with right side contracting.
Aequat. Math. 2(1969) 3o-38.

[16] Strodt, W. and Wright, R.K.: Asymptotic behavior of solution and adjunction fields for nonlinear first order differential equations.
Memoirs of the AMS, No. 1o9, Providence, R.I. 1971.

[17] Swanson, C.A. and Schulzer, M.: Asymptotic solutions of equations in Banach space.
Canad. J. Math. 13(1961) 493-5o4.

[18] Wasow, W.: Asymptotic expansions for ordinary differential equations.
New York: Interscience Publ. 1965.

ESTIMATES FOR FREDHOLM EIGENVALUES BASED ON QUASICONFORMAL MAPPING

By Glenn Schober

INTRODUCTION

The Dirichlet problem of potential theory for a smoothly bounded domain can be reduced to solving a linear integral equation with a double layer kernel. In two dimensions the same integral equation arises in constructive methods for conformal mapping. In solving the integral equation by successive approximations, the speed of convergence depends on the smallest positive non-trivial Fredholm eigenvalue. We shall survey some properties and estimates for this eigenvalue emphasizing connections with the theory of quasiconformal mapping.

ITERATIVE PROCEDURES FOR SOME CLASSICAL PROBLEMS

Let C be a smooth (e.g., class C^3) Jordan curve in the complex plane with interior G and exterior $\tilde{G}$. We mention briefly five problems:

(i) Dirichlet problem for G . Given is a continuous function u on C and sought is its harmonic extension to G . The classical approach of C.Neumann, Poincaré and others is to assume a solution in the form

$$(1) \qquad u(z) = \int_C K(z,t)\varphi(t)ds_t , \qquad z \in G ,$$

where $K(z,t) = -\pi^{-1} \frac{\partial}{\partial n_t} \log|z-t|$ is the familiar double layer kernel and φ is an unknown density function. Here n_t is the unit interior normal and the integration is with respect to arc length. The kernel is amenable to practical considerations since it has the simple geometric representations $K(z,t) = \pi^{-1}|z-t|^{-1}\cos(n_t,z-t)$ and $K(z,t)ds_t = \pi^{-1}d_t\arg(t-z)$. From the discontinuity behavior of (1) at the boundary one obtains the linear integral equation

During the preparation of this article the author was a guest of the Forschungsinstitut für Mathematik, ETH, Zürich.

$$\varphi(z) + \int_C K(z,t)\varphi(t)ds_t = u(z) \quad , \quad z \in C \ , \tag{2}$$

for the density φ . To solve (2) by successive approximations, let $\varphi_o = u$ and $\varphi_{n+1} = -\int K\varphi_n + u$ $(n \geq 0)$. Then $(\varphi_{n+1}+\varphi_n)/2$ converges uniformly to the solution φ .

(ii)-(iv) The Dirichlet problem for $\tilde{G}$ and Neumann problems for G and $\tilde{G}$ all have versions of the above with a sign change, transposed kernel, or both.

(v) Conformal mapping. We wish to construct the function f mapping G conformally onto the unit disk, normalized by $f(z_o) = 0$, $f(z_1) = 1$, for fixed $z_o \in G$, $z_1 \in C$. The Cauchy representation $f(z) = (2\pi i)^{-1}\int_C e^{i\arg f(t)}(t-z)^{-1}dt$ reduces the problem to finding the boundary correspondence $\theta(z) = \arg f(z)$, $z \in C$. By integrating $\log f$ over an appropriate contour one finds that θ satisfies the linear integral equation of S.Gershgorin

$$\theta(z) = \int_C K(z,t)\,\theta(t)ds_t - 2\beta(z) \quad , \quad z \in C \ .$$

Here $K(z,t)$ is again the double layer kernel and $\beta(z) = \arg[(z_1-z)/(z_o-z)]$ is a simple geometric quantity. To find θ by successive approximations, let $\theta_o = -2\beta$ and $\theta_{n+1} = \int K\theta_n - 2\beta$ $(n \geq 0)$. Then θ_n converges uniformly to a function $\tilde{\theta}$ as $n \to \infty$ and the solution $\theta(z) = \tilde{\theta}(z)-\tilde{\theta}(z_1)$.

We refer the reader to the book of Gaier [2] and the work of Warschawski [16] for details and further references concerning the above procedures. In all cases the convergence is geometric and explicit error estimates are known. The error is $O(\lambda^{-n})$ where λ is the smallest Fredholm eigenvalue larger than 1 of the kernel $K(z,t)$. It is therefore important to have estimates for λ from below and to know its properties.

THE EIGENVALUE λ

Denote the Dirichlet integral over a set S by $D_S(h) = \iint_S (h_x^2+h_y^2)dxdy$. Then the eigenvalue λ has the characterization [1, 7, 16]

$$(3) \qquad \frac{1}{\lambda} = \sup \frac{|D_G(h) - D_{\tilde{G}}(h)|}{D_G(h) + D_{\tilde{G}}(h)}$$

where the supremum is over all functions h which are continuous in the extended plane and harmonic off C (i.e., in $G \cup \tilde{G}$). We may then use (3) to extend the definition of λ to arbitrary Jordan curves and consider λ as a curve functional. Several properties of λ are evident:

1. $1 \leq \lambda \leq \infty$.
2. λ is invariant under Möbius transformations.
3. $\lambda = \infty$ iff C is a circle [13].

We mention also a useful estimate for λ :

4. If C is convex, then $\lambda \geq [1-(L/2\pi R)]^{-1}$ where L is the length of C and R is the supremum of the radii of all circles which intersect C in at least 3 points. (In case C is smooth, R is the maximum radius of curvature.) This is due to Neumann [5]; but see also [12].

Several other estimates can be found in the article of Warschawski [16]. However, we wish to turn to connections with the theory of quasi-conformal mapping.

QUASICONFORMAL MAPPING

A <u>K-quasiconformal mapping</u> f of a domain G is an orientation preserving homeomorphism of G such that

(a) relative to each standard rectangle in G , f is absolutely continuous on a.e. horizontal and vertical line, and

(b) $\left(\left|\frac{\partial f}{\partial z}\right| + \left|\frac{\partial f}{\partial \bar{z}}\right|\right) / \left(\left|\frac{\partial f}{\partial z}\right| - \left|\frac{\partial f}{\partial \bar{z}}\right|\right) \leq K$ a.e.

The condition (b) controls the distortion. It means that the differential of f at a.e. point maps circles onto ellipses whose major axis to minor axis ratio is at most K . The 1-quasiconformal mappings are just conformal mappings. For us an important property of K-quasi-conformal mappings is the quasi-invariance of Dirichlet integrals:

$$(4) \qquad K^{-1} D_{f(G)}(h) \leq D_G(h \circ f) \leq K D_{f(G)}(h) .$$

We refer the reader to [4] for details and further properties.

Here is the basic idea we wish to emphasize. To study a functional, such as λ , first look at transformations that leave it invariant (e.g., 2). If these are few, next look for transformations which leave the functional, or fundamental parts, quasi-invariant (e.g., (4)).

FURTHER PROPERTIES AND ESTIMATES FOR λ

With the help of quasiconformal mappings it is possible to show that

5. λ is an upper semicontinuous curve functional [14]. That is, if $C_n \to C$ in the sense of Fréchet, then $\limsup_{n\to\infty} \lambda(C_n) \leqslant \lambda(C)$. If, in addition, C_n has tangent vectors which converge appropriately to those for C , then one has continuity [13]. For the latter property quasiconformal mappings play a particularly interesting role. In the absence of conformal mappings of the plane which take C_n onto C , one constructs K_n-quasiconformal mappings with $K_n \to 1$. In the limit, then, estimates such as (4) approach identities.

6. If C is the image of $|z| = 1$ under a plane homeomorphism which is analytic for $r < |z| < R$, K-quasiconformal for $|z| < r < 1$, and $\tilde{K}$-quasiconformal for $|z| > R > 1$, then its eigenvalue

$$\lambda \geqslant (R^2+k\tilde{k}r^2)/(\tilde{k}+kr^2R^2) \tag{5}$$

where $k = (K-1)/(K+1)$ and $\tilde{k} = (\tilde{K}-1)/(\tilde{K}+1)$. This estimate is obtained in [9] by employing the calculus of variations for quasiconformal mappings developed by Schiffer [8]. The functional λ is minimized over the family of such curves C . From this point of view the estimate (5) is necessarily sharp.

One limiting case of (5) is the estimate

$$\lambda \geqslant (R^2+r^2)/(1+r^2R^2)$$

corresponding to $K = \tilde{K} = \infty$ for the class of "uniformly analytic" curves (Schiffer [6]).

A second limiting case of (5) is the estimate

$$\lambda \geqslant (1+k\tilde{k})/(\tilde{k}+k)$$

corresponding to $r = R = 1$ for the class of "quasicircles" (Springer [15]). In particular, if C is the image of $|z| = 1$ under a homeomorphism that is conformal in $|z| > 1$ and K-quasiconformal in $|z| < 1$, then $\tilde{k} = 0$ and

(6) $$\lambda \geqslant 1/k \qquad \text{(Ahlfors [1])}.$$

We shall illustrate the practicality of such estimates by an elementary example. The mapping

$$f(z) = \begin{cases} \frac{1}{4}[(a+b)z+(a-b)z^{-1}] & \text{for } |z| \geqslant 1 \\ \frac{1}{4}[(a+b)z+(a-b)\bar{z}] & \text{for } |z| \leqslant 1 \end{cases} \qquad (a>b>0)$$

is conformal in $|z| > 1$ and (a/b)-quasiconformal in $|z| < 1$. It maps $|z| = 1$ onto an ellipse with major axis a and minor axis b. From (6) the eigenvalue of this ellipse satisfies

$$\lambda \geqslant (a+b)/(a-b) .$$

In fact, $\lambda = (a+b)/(a-b)$ (Schiffer [6]).

More generally, Ahlfors [1] has applied the estimate (6) to curves C which are starlike with respect to an interior point z_o to obtain the useful estimate

$$\lambda \geqslant \inf|\csc(n_t, z_o-t)|$$

over the points $t \in C$ where n_t exists. In particular, if C is a regular n-gon, then $\lambda \geqslant \csc(\pi/n)$.

Still more generally, if the Jordan curve C has the property that $\sigma_{zt} \leqslant c|z-t|$ uniformly for some constant c, where σ_{zt} denotes the shorter arc length between $z,t \in C$, then ([10])

$$\lambda \geqslant (K+1)/(K-1) , \qquad K = 8 \exp\{5\pi(1+c^{10}e^{2\pi})^2\} .$$

However, this estimate is hardly of practical value since it differs little from 1.

OTHER CONNECTIONS

The non-trivial eigenvalues of (2) are also eigenvalues of the complex Hilbert transform over G. For multiply connected domains they play a role in the circular normalization problem. In addition, they can be interpreted in terms of the Grunsky inequalities from univalent function theory. An excellent survey of these connections can be found in Schiffer [7].

We close with an open problem related actually to (6) and the Grunsky inequalities. Let f be analytic and univalent for $|z| < 1$ and

$$\log \frac{f(z)-f(\zeta)}{z-\zeta} = \sum_{m,n=0}^{\infty} c_{mn} z^m \zeta^n \qquad |z|,|\zeta| < 1 .$$

If f has a K-quasiconformal extension to $|z| \geqslant 1$ and $k = (K-1)/(K+1)$, then

$$(7) \qquad \Big|\sum_{m,n=1}^{\infty} \sqrt{mn}\, c_{mn} \xi_m \xi_n\Big| \leqslant k \quad \text{whenever} \quad \sum_{n=1}^{\infty} |\xi_n|^2 \leqslant 1$$

(Kühnau [3]). Is the converse true? That is, if (7) is satisfied, does f have a precisely K-quasiconformal extension to $|z| \geqslant 1$?

REFERENCES

[1] L.Ahlfors, "Remarks on the Neumann-Poincaré integral equation", Pacific J. Math. 2 (1952), 271-280.

[2] D.Gaier, Konstruktive Methoden der konformen Abbildung, Springer Verlag, 1964.

[3] R.Kühnau, "Verzerrungssätze und Koeffizientenbedingungen vom Grunskyschen Typ für quasikonforme Abbildungen", Math. Nachr. 48 (1971), 77-105.

[4] O.Lehto and K.Virtanen, Quasikonforme Abbildungen, Springer Verlag 1965.

[5] C.Neumann, Untersuchungen über das logarithmische und Newton'sche Potential, Leipzig 1877.

[6] M.Schiffer, "The Fredholm eigenvalues of plane domains", Pacific J. Math. 7 (1957), 1187-1225.

[7] M.Schiffer, "Fredholm eigenvalues and conformal mapping", Rend. Mat. 22 (1963), 447-468.

[8] M.Schiffer, "A variational method for univalent quasiconformal mappings", Duke Math. J. 33 (1966), 395-412.

[9] M.Schiffer and G.Schober, "An extremal problem for the Fredholm eigenvalues", Arch. Rational Mech. Anal. 44 (1971), 83-92, and 46 (1972), 394.

[10] G.Schober, "On the Fredholm eigenvalues of plane domains", J. Math. Mech. 16 (1966), 535-541.

[11] G.Schober, "A constructive method for the conformal mapping of domains with corners", J. Math. Mech. 16 (1967), 1095-1115.

[12] G.Schober, "Neumann's lemma", Proc. AMS 19 (1968), 306-311.

[13] G.Schober, "Continuity of curve functionals and a technique involving quasiconformal mapping", Arch. Rational Mech. Anal. 29 (1968), 378-389.

[14] G.Schober, "Semicontinuity of curve functionals", Arch. Rational Mech. Anal. 33 (1969), 374-376.

[15] G.Springer, "Fredholm eigenvalues and quasiconformal mapping", Acta Math. 111 (1964), 121-141.

[16] S.Warschawski, "On the solution of the Lichtenstein-Gershgorin integral equation in conformal mapping: I.Theory", Nat. Bur. Standards Appl. Math. Ser. 42 (1955), 7-29.

DISCRETE CONVERGENCE OF CONTINUOUS MAPPINGS IN METRIC SPACES

By Friedrich Stummel and Jürgen Reinhardt

The purpose of this paper is to prove a number of new, necessary and sufficient conditions for stability and convergence of approximations of continuous mappings in metric spaces. The main results with respect to numerical analysis are the equivalent characterizations of the general stability conditions by stability inequalities and the derivation of fundamental error estimates for the convergence of approximants as well as for the approximate solution of the approximating equations. Thus we extend and generalize methods and results of Ansorge [1], Aubin [3], Lax-Richtmyer [11], Pereyra [12], Spijker [18], Stetter [19], Watt [24]. Another paper Stummel [22] deals with stability criteria, convergence theorems and associated error estimates for the approximation of differentiable mappings in normed spaces.

Section 1 presents the basic theorems and definitions of discrete convergence and discrete limit spaces. In particular, the concepts of convergence, consistency, stability and inverse stability are defined in their most general form for sequences of mappings between sets. For the elementary proofs of these theorems we refer to Stummel [21]. Most approximations of metric or normed spaces in the applications are metric discrete limit spaces in the sense of section 2 or Stummel [21]. This is true, for instance, for the approximations of normed spaces, definable by restriction operators, in Aubin [3], [4], Henrici [8], [9], Spijker [18], Stetter [19], Stummel [2o], Watt [24] and for the approximations defined by projection operators as in Browder [5], Krasnosel'skii et al. [1o], Petryshyn [13], [14], Vainikko [23]. In metric discrete limit spaces, the stability of a sequence of mappings is equivalent to the asymptotic equicontinuity of the sequence (Theorem 2.(1)). As a simple conclusion we obtain the equivalence of discrete convergence to continuous discrete convergence for sequences of mappings (Theorem 2.(3); cf. Reinhardt [15]). Here, the concept of continuous discrete convergence is an obvious generalization of the concept of continuous convergence for sequences of mappings in metric spaces (cf. Rinow [16], § 9).

Section 3 establishes the equivalence of stability and equicontinuity for sequences of continuous mappings in metric discrete limit spaces (Theorem 3.(1)). Correspondingly, one obtains the equivalence of inverse stability and equicontinuity of the sequence of inverses for sequences of continuously invertible mappings (Theorem 3.(2)). Let us remark here, that the concept of stability defined by Aubin [3] requires the uniform equicontinuity of the sequence of mappings (A_ι) on the whole sequence space. As opposed to that, the stability concept of this paper requires only the pointwise equicontinuity of the sequence (A_ι) at every discretely (A_ι)-convergent sequence. In the case of continuous linear mappings in normed spaces, stability is equivalent to boundedness of the sequence (Theorems 3.(4),(5)). Further, equicontinuity and thus stability of a sequence of mappings may be characterized equivalently at each point by inequalities using an associated modulus of continuity (Theorems 4.(1),(2)). A general stability inequality of similar type is studied in Spijker [17]. One finds special inequalities, as sufficient stability conditions, of the type studied here, for instance, in Browder [5], Petryshyn [13], [14] and in the condition (S) of Stetter [19]. Moreover, section 4 establishes very interesting two-sided stability inequalities which characterize equivalently stable, inversely stable sequences or equicontinuous, equicontinuously invertible sequences of mappings (Theorem 4.(3)).

These results yield new equivalent characterizations of the concept of discrete convergence for sequences of continuous mappings by continuous discrete convergence, by consistency and equicontinuity as well as by consistency and stability inequalities (Theorems 5.(1),(2),(3)). From these inequalities one obtains fundamental error estimates for the discretization error and the total error of approximants as well as a-posteriori error estimates for approximate solutions of the approximating equations (5.(5),..., 5.(12)). Essentially, these concepts are obvious generalizations of well-known concepts of the approximation of initial value problems. Here, the two-sided error estimates are particularly interesting since they establish both upper and lower bounds for the errors. A very special form of a similar error estimate is found in Krasnosel'skii et al. [10] and in Vainikko [23]. Finally, section 6 shows for metric discrete limit spaces with discretely convergent metrics that the limit of a discretely convergent sequence of mappings is continuous (Theorem 6.(2)) and that the consistency of a sequence of mappings can be verified already on a dense subspace (Theorem 6.(3)).

1. Discrete Convergence

This section presents the basic theorems and definitions of discrete convergence and discrete limit spaces (cf. Stummel [21]). Here, a set I is called an index sequence if I is directed. This means, I is non-void and partially ordered by a reflexive and transitive binary relation $\geqq$ such that for every pair $\iota, \varkappa$ there exists a λ in I with the property $\lambda \geqq \iota$, $\lambda \geqq \varkappa$. Further, let X be a set, let $(X_\iota)_{\iota \in I}$ be a family of sets $X_\iota, \iota \in I$, and let lim be a mapping with values in X and with domain of definition in the set of all sequences (u_ι) of elements $u_\iota \in X_\iota$, $\iota \in I$. The mapping lim is said to be a discrete convergence. The sequences (u_ι) in the domain of definition of lim are called discretely convergent, the value $u \in X$ of the mapping lim at (u_ι) is the discrete limit of the sequence (u_ι) and accordingly (u_ι) is said to converge discretely to u. For the sake of simplicity, we use the notation $\lim u_\iota = u$ or $u_\iota \rightarrow u$ $(\iota \in I)$ to denote $\lim (u_\iota) = u$. Let $\prod_\iota X_\iota$ be the cartesian product of the family $(X_\iota)_{\iota \in I}$. Thus $\prod_\iota X_\iota$ is the set of all sequences (u_ι) of elements $u_\iota \in X_\iota$, $\iota \in I$. Then the triple $X, \prod_\iota X_\iota, \lim$ is said to be a discrete limit space.

Two sequences $(u_\iota), (v_\iota) \in \prod_\iota X_\iota$ are said to be equivalent if either the two sequences do not converge discretely or if the two sequences converge discretely to the same limit $\lim u_\iota = \lim v_\iota$. Obviously, this constitutes an equivalence relation in $\prod_\iota X_\iota$. Then the inverse mapping $\lim^{-1}$ from X to the family of subsets of equivalent sequences in $\prod_\iota X_\iota$ is defined by

$$\lim^{-1} u = \left\{ (u_\iota) \in \prod_\iota X_\iota \,|\, \lim u_\iota = u \right\}, \qquad u \in X.$$

The mapping lim is surjective if and only if each $u \in X$ is the discrete limit of some sequence $(u'_\iota) \in \prod_\iota X_\iota$, or if for each $u \in X$ the set $\lim^{-1} u$ is non-void. In this case, the discrete limit space $X, \prod_\iota X_\iota, \lim$ is said to be a discrete approximation $\mathcal{A}(X, \prod_\iota X_\iota, \lim)$. Next we consider a sequence of non-void subsets $Z \subseteq X$, $Z_\iota \subseteq X_\iota$, $\iota \in I$. Then $\lim(\prod_\iota Z_\iota, Z)$ denotes the restriction of the mapping lim to sequences (u_ι) with elements $u_\iota \in Z_\iota, \iota \in I$, and discrete limits u in Z. The mapping $\lim(\Pi_\iota Z_\iota, Z)$ is surjective if and only if for each $z \in Z$ there is a discretely convergent sequence of elements $z'_\iota \in Z_\iota, \iota \in I$, such that $\lim z'_\iota = z$.

The fundamental concept in this theory is the concept of discrete convergence of mappings. Let $X, \Pi_\iota X_\iota, \lim^X$ and $Y, \Pi_\iota Y_\iota, \lim^Y$ be two discrete limit spaces with the same index sequence I and let $A, (A_\iota)$ be a sequence of mappings $A: X \to Y$, $A_\iota: X_\iota \to Y_\iota$ $\iota \in I$. For the sake of notational simplicity, we shall use the same symbol lim to denote the discrete convergences $\lim^X$, $\lim^Y$. The sequence (A_ι) is said to <u>converge discretely</u> to the mapping A if $\lim^X$ is surjective and if the following relation holds,

(a) $$\lim u_\iota = u \Longrightarrow \lim A_\iota u_\iota = Au$$

for every discretely convergent sequence (u_ι). This relation may be written $\lim A_\iota = A$ or $A_\iota \to A (\iota \in I)$. Indeed, the discrete convergence of mappings defines a mapping lim or $\to$ from the set of sequences (A_ι) of mappings $A_\iota: X_\iota \to Y_\iota, \iota \in I$, to the set of mappings $A: X \to Y$.

Now we are going to characterize the discrete convergence of mappings by necessary and sufficient stability and consistency conditions. Here, a sequence (A_ι) is said to be <u>stable</u> if the <u>stability condition</u>

(b) $$\lim u_\iota = \lim v_\iota \Longrightarrow \lim A_\iota u_\iota = \lim A_\iota v_\iota$$

holds for every pair of discretely convergent sequences $(u_\iota), (v_\iota)$ such that $(A_\iota u_\iota)$ or $(A_\iota v_\iota)$ converges discretely. The sequence $A, (A_\iota)$ is said to be <u>consistent</u> if, for every $u \in X$, there exists a discretely convergent sequence of elements $(u'_\iota) \in \Pi_\iota X_\iota$ such that the following <u>consistency condition</u> is valid,

(c) $$\lim u'_\iota = u, \ \lim A_\iota u'_\iota = Au.$$

Having made these basic definitions, we are in a position to state the first fundamental theorem (cf. [21], Theorem 2.(1)).

(1) The stability of (A_ι) and the consistency of $A,(A_\iota)$ is a necessary and sufficient condition for the discrete convergence $A_\iota \rightarrow A(\iota \in I)$.

We now consider the important question of the discrete convergence of solutions u_ι of the equations $A_\iota u_\iota = w_\iota, \iota \in I$, to solutions u of the equation $Au = w$. More precisely, we shall state conditions for the validity of the inverse convergence relation

$$(a_{-1}) \qquad \lim A_\iota u_\iota = Au \Longrightarrow \lim u_\iota = u.$$

The basic concept in this context is the inverse stability of the sequence of mappings (A_ι) defined by the inverse stability condition

$$(b_{-1}) \qquad \lim A_\iota u_\iota = \lim A_\iota v_\iota \Longrightarrow \lim u_\iota = \lim v_\iota$$

for every pair of discretely convergent sequences $(A_\iota u_\iota),(A_\iota v_\iota)$ such that (u_ι) or (v_ι) converges discretely. If the inverse convergence relation holds, if A is bijective and $\lim(\Pi_\iota A_\iota X_\iota, Y)$ is surjective, then (A_ι) is necessarily inversely stable. Indeed, for every pair of discretely convergent sequences $(A_\iota u_\iota)$, $(A_\iota v_\iota)$ such that $\lim A_\iota u_\iota = \lim A_\iota v_\iota = w$, $u = A^{-1}w$, the inverse convergence relation implies the discrete convergence of (u_ι), (v_ι) and $\lim u_\iota = \lim v_\iota$. Thus we obtain the following general equivalence theorem (cf. [21], Theorem 2.(2)).

(2) Let A be surjective and let $\lim(\Pi_\iota A_\iota X_\iota, Y)$ be surjective. Then the inverse stability of (A_ι) and the consistency of $A,(A_\iota)$ is a necessary and sufficient condition for the bijectivity of A and the inverse convergence relation of $A,(A_\iota)$.

In this theory, the well-known equivalence theorem of Lax is a simple conclusion of the general equivalence theorem above. Note, that the consistency of $A,(A_\iota)$ and the surjectivity of A implies that $\lim(\Pi_\iota A_\iota X_\iota, Y)$ is surjective. Thus we obtain the following special equivalence theorem.

(3) Let the sequence $A,(A_\iota)$ be consistent and let A be bijective. Then the inverse stability of (A_ι) is a necessary and sufficient condition for the inverse convergence relation of $A,(A_\iota)$.

For bijective mappings $A:X \to Y$, $A_\iota:X_\iota \to Y_\iota$, $\iota \in I$, the inverse convergence relation and the surjectivity of $\lim^Y$ is equivalent to the discrete convergence $A_\iota^{-1} \to A^{-1}$ $(\iota \in I)$. Under these assumptions, the inverse convergence relation may be written in the form

(a_{-1}) $$\lim A_\iota u_\iota = w \Longrightarrow \lim u_\iota = A^{-1}w$$

for every $w \in Y$ and every sequence of elements $u_\iota \in X_\iota$, $\iota \in I$. In this case, the general equivalence theorem now has the following form (cf. [21], Theorem 2.(3)).

(4) Let A be surjective and let $A_\iota:X_\iota \to Y_\iota$ be bijective for each $\iota \in I$. Then the inverse stability of (A_ι) and the consistency of $A,(A_\iota)$ is a necessary and sufficient condition for the bijectivity of A and the discrete convergence $A_\iota^{-1} \to A^{-1}$ $(\iota \in I)$.

2. Continuous Discrete Convergence

In metric discrete limit spaces, the stability of a sequence of mappings is equivalent to the asymptotic equicontinuity of the sequence. In this way, we obtain as a first important conclusion the equivalence of discrete convergence and continuous discrete convergence for sequences of mappings (cf. Reinhardt [15]). Here, the concept of continuous discrete convergence is an obvious generalization of the concept of continuous convergence of sequences of mappings in metric spaces (cf. Rinow [17], §9). In the following, we assume that the index sequence I is denumerably infinite and linearly ordered. Thus the index sequence I has the form $\{\iota_1,\iota_2,\iota_3,\dots\}$, $\iota_1 < \iota_2 < \iota_3 < \dots$, where $\iota < \iota'$ denotes the relation $\iota' \geq \iota$ and $\iota' \neq \iota$.

A discrete limit space $X, \prod_\iota X_\iota, \lim$ is said to be a metric discrete limit space (cf. Stummel [21]) if, for every $\iota \in I$, the set X_ι is a metric space with distance $|\cdot,\cdot|_{X_\iota}$ and if the following condition holds.

(M) For every pair of sequences $(u_\iota),(v_\iota) \in \prod_\iota X_\iota$ such that (u_ι) or (v_ι) converges discretely, the following statements are equivalent:

$$\lim |u_\iota, v_\iota|_{X_\iota} = o \Longleftrightarrow \lim u_\iota = \lim v_\iota .$$

In particular, this condition (M) implies that, with every discretely convergent sequence (u_ι), all metrically equivalent sequences (v_ι), defined by $\lim|u_\iota, v_\iota| = o$, belong to the domain of definition of lim and have the same discrete limit $\lim u_\iota = \lim v_\iota$.

Now, let $X, \prod_\iota X_\iota, \lim^X$ and $Y, \prod_\iota Y_\iota, \lim^Y$ be two metric discrete limit spaces with the same index sequence I and let (A_ι) be a sequence of mappings $A_\iota : X_\iota \to Y_\iota, \iota \in I$. For the sake of notational simplicity, we will denote $\lim^X$, $\lim^Y$ by lim and the distances $|.,.|_{X_\iota}$, $|.,.|_{Y_\iota}$ by $|.,.|$. Evidently, in metric discrete limit spaces we have the equivalent characterization of the stability of (A_ι) by the following condition,

$$(b^M) \qquad \lim|u_\iota, v_\iota| = o \Longrightarrow \lim|A_\iota u_\iota, A_\iota v_\iota| = o$$

for every pair of discretely convergent sequences $(u_\iota), (v_\iota)$ such that $(A_\iota u_\iota)$ or $(A_\iota v_\iota)$ converges discretely. Hence a sequence (A_ι) is said to be <u>stable at the point</u> (u_ι) if, for every sequence (v_ι), the condition $\lim|u_\iota, v_\iota| = o$ implies $\lim |A_\iota u_\iota, A_\iota v_\iota| = o$. We call a sequence (u_ι) <u>discretely</u> (A_ι)-<u>convergent</u> if both (u_ι) and $(A_\iota u_\iota)$ converge discretely. Thus in metric discrete limit spaces, using this terminology, a sequence (A_ι) is stable if and only if (A_ι) is stable at every discretely (A_ι)-convergent sequence (u_ι).

To begin the study, we shall characterize the stability of a sequence of mappings by the asymptotic equicontinuity of the sequence. A sequence of mappings (A_ι) is said to be <u>asymptotically equicontinuous at the point</u> (u_ι) if, for every $\varepsilon > o$, there exist a $\nu \in I$ and a $\delta > o$ such that the relation $|u_\iota, v_\iota| \leq \delta$ implies $|A_\iota u_\iota, A_\iota v_\iota| \leq \varepsilon$ for every $\iota \geq \nu$ and every $v_\iota \in X_\iota$.

(1) <u>The sequence</u> (A_ι) <u>is asymptotically equicontinuous at the point</u> (u_ι) <u>if and only if</u> (A_ι) <u>is stable at this point.</u>

<u>Proof.</u> (i) Let (A_ι) be asymptotically equicontinuous at the point (u_ι) and let (v_ι) be any sequence such that $\lim|u_\iota, v_\iota| = o$. Then, for every $\varepsilon > o$, there exist a $\nu \in I$ and

a $\delta > o$ with the properties defined above. Next, there is an index $\nu' \geq \nu$ such that $|u_\iota, v_\iota| \leq \delta$ for all $\iota \geq \nu'$. Consequently, for every $\varepsilon > o$, one obtains $|A_\iota u_\iota, A_\iota v_\iota| \leq \varepsilon$ for all $\iota \geq \nu'$. This proves the convergence $\lim |A_\iota u_\iota, A_\iota v_\iota| = o$.

(ii) Conversely, if (A_ι) is stable at the point (u_ι), then (A_ι) is asymptotically equicontinuous at (u_ι). Otherwise, there would be an $\varepsilon_o > o$ such that, for every $\nu_t \in I$ and every $\delta = 1/t$, $t = 1,2,\ldots$, there exist a $\iota'_t \geq \nu_t$ and a $v_\iota^{(t)} \in X_\iota$ with the properties $|u_\iota, v_\iota^{(t)}| \leq 1/t$ and $|A_\iota u_\iota, A_\iota v_\iota^{(t)}| \geq \varepsilon_o$ for $\iota = \iota'_t$. The indices ν_t, $t = 1,2,\ldots$, can be chosen in such a way that $\iota'_1 < \iota'_2 < \iota'_3 < \ldots$. Hence, one has an infinite subsequence $I' = (\iota'_1, \iota'_2, \iota'_3, \ldots)$ of I. On setting $u'_\iota = u_\iota$ for $\iota \in I - I'$, $u'_\iota = v_\iota^{(t)}$ for $\iota = \iota'_t \in I'$, we obtain $|u_\iota, u'_\iota| \to o (\iota \in I)$. But, on the other hand, $|A_\iota u_\iota, A_\iota u'_\iota| \geq \varepsilon_o$, $\iota \in I'$, and this is a contradiction to the stability of (A_ι) at the point (u_ι).

We now turn to the characterization of the inverse stability of a sequence of mappings $A_\iota : X_\iota \to Y_\iota$, $\iota \in I$, in metric discrete limit spaces. In this case, the inverse stability of (A_ι) is obviously equivalent to the condition

$$(b_{-1}^M) \qquad \lim |A_\iota u_\iota, A_\iota v_\iota| = o \implies \lim |u_\iota, v_\iota| = o$$

for every pair of discretely convergent sequences $(A_\iota u_\iota), (A_\iota v_\iota)$ such that (u_ι) or (v_ι) converges discretely. Thus the sequence (A_ι) is said to be inversely stable at the point (u_ι) if, for every sequence (v_ι), the condition $\lim |A_\iota u_\iota, A_\iota v_\iota| = o$ implies $\lim |u_\iota, v_\iota| = o$. Hence (A_ι) is inversely stable if and only if (A_ι) is inversely stable at every discretely (A_ι)-convergent sequence (u_ι). Finally, the sequence (A_ι) is said to be <u>asymptotically inversely equicontinuous at the point</u> (u_ι) if, for every $\varepsilon > o$, there exist a $\nu \in I$ and a $\delta > o$ such that the relation $|A_\iota u_\iota, A_\iota v_\iota| \leq \delta$ implies $|u_\iota, v_\iota| \leq \varepsilon$ for every $\iota \geq \nu$ and every $v_\iota \in X_\iota$.

(2) <u>The sequence</u> (A_ι) <u>is asymptotically inversely equicontinuous at the point</u> (u_ι) <u>if and only if</u> (A_ι) <u>is inversely stable at this point.</u>

Proof. (i) Let (A_ι) be asymptotically inversely equicontinuous at (u_ι) and let (v_ι) be any sequence with the property $\lim|A_\iota u_\iota, A_\iota v_\iota| = o$. Then for every $\varepsilon > o$, there exist a $\nu \in I$ and a $\delta > o$ such that $|A_\iota u_\iota, A_\iota v_\iota| \leq \delta$ implies $|u_\iota, v_\iota| \leq \varepsilon$ for every $\iota \geq \nu$ and every $v_\iota \in X_\iota$. Further, there exists a $\nu' \geq \nu$ such that $|A_\iota u_\iota, A_\iota v_\iota| \leq \delta$ for every $\iota \geq \nu'$. Hence for every $\varepsilon > o$, we obtain $|u_\iota, v_\iota| \leq \varepsilon$ for every $\iota \geq \nu'$. This proves $\lim|u_\iota, v_\iota| = o$.

(ii) Conversely, let (A_ι) be inversely stable at the point (u_ι). Then (A_ι) is asymptotically inversely equicontinuous at (u_ι). If we suppose the contrary, there would be a number $\varepsilon_o > o$ such that, for every $\nu_t \in I$ and every $\delta = 1/t$, $t = 1,2,\ldots,$ there exist a $\iota_t' \geq \nu_t$ and a $v_\iota^{(t)} \in X_\iota$ with the property $|A_\iota u_\iota, A_\iota v_\iota^{(t)}| \leq 1/t$ and $|u_\iota, v_\iota^{(t)}| \geq \varepsilon_o$ for $\iota = \iota_t'$. Thus, one can find an infinite subsequence $I' = (\iota_1', \iota_2', \iota_3', \ldots)$ of I and a sequence of elements $u_\iota' = u_\iota$ for $\iota \in I - I'$, $u_\iota' = v_\iota^{(t)}$ for $\iota = \iota_t' \in I'$, such that $|A_\iota u_\iota, A_\iota u_\iota'| \to o(\iota \in I)$ and $|u_\iota, u_\iota'| \geq \varepsilon_o, \iota \in I'$, contrary to the inverse stability of (A_ι) at the point (u_ι).

A first important application of these theorems establishes the equivalence of discrete convergence and continuous discrete convergence for sequences of mappings in metric discrete limit spaces. A sequence of mappings $A_\iota : X_\iota \to Y_\iota, \iota \in I$, is said to be <u>continuously discretely convergent</u> to the mapping $A: X \to Y$, denoted by $A_\iota \xrightarrow[c]{} A(\iota \in I)$, if $\lim(\prod_\iota X_\iota, X)$, $\lim(\prod_\iota Y_\iota, AX)$ are surjective and if, for every $u \in X$, for every pair of sequences $u_\iota \to u$, $w_\iota \to Au(\iota \in I)$ and for every $\varepsilon > o$, there exist a $\nu \in I$ and a $\delta > o$ such that the condition $|u_\iota, v_\iota| \leq \delta$ implies $|w_\iota, A_\iota v_\iota| \leq \varepsilon$ for every $\iota \geq \nu$ and every $v_\iota \in X_\iota$.

(3) <u>The discrete convergence $A_\iota \to A(\iota \in I)$ is a necessary and sufficient condition for the continuous discrete convergence $A_\iota \xrightarrow[c]{} A(\iota \in I)$.</u>

Proof. (i) If $A_\iota \to A(\iota \in I)$, then (A_ι) is stable and $A, (A_\iota)$ is consistent. Thus we obtain from theorem (1) that (A_ι) is asymptotically equicontinuous at every discretely (A_ι)-convergent sequence (u_ι). The consistency condition implies that $\lim(\prod_\iota X_\iota, X)$ and

$\lim(\prod_\iota Y_\iota, AX)$ are surjective. For every $u \in X$ and every pair of sequences $u_\iota \to u$, $w_\iota \to Au(\iota \in I)$, the sequence (u_ι) is discretely (A_ι)-convergent, since $A_\iota \to A(\iota \in I)$. Hence for every $\varepsilon > o$, there exist a $\nu \in I$ and a $\delta > o$ such that the condition $|u_\iota, v_\iota| \leq \delta$ implies $|A_\iota u_\iota, A_\iota v_\iota| \leq \varepsilon/2$ for every $\iota \geq \nu$ and every $v_\iota \in X_\iota$. From $u_\iota \to u$, $A_\iota u_\iota \to Au$, it follows $|A_\iota u_\iota, w_\iota| \to o(\iota \in I)$. Thus there is an index $\nu' \geq \nu$ such that $|w_\iota, A_\iota u_\iota| \leq \varepsilon/2$ for all $\iota \geq \nu'$. Hence for every $\varepsilon > o$ and every $\iota \geq \nu'$, we have the relation $|w_\iota, A_\iota v_\iota| \leq |w_\iota, A_\iota u_\iota| + |A_\iota u_\iota, A_\iota v_\iota| \leq \varepsilon$.

(ii) If $A_\iota \underset{c}{\to} A(\iota \in I)$, then $\lim(\prod_\iota X_\iota, X)$ and $\lim(\prod_\iota Y_\iota, AX)$ are surjective. For every $u \in X$, for every pair of discretely convergent sequences $u_\iota \to u, w_\iota \to Au(\iota \in I)$ and for every $\varepsilon > o$, there exists a $\nu \in I$ such that, on setting $v_\iota = u_\iota$, the inequality $|A_\iota u_\iota, w_\iota| \leq \varepsilon$ holds for all $\iota \geq \nu$ so that the sequences $(w_\iota), (A_\iota u_\iota)$ are equivalent and $\lim A_\iota u_\iota = \lim w_\iota = Au$.

Using the above theorem, we obtain for sequences of bijective mappings $A: X \to Y$, $A_\iota: X_\iota \to Y_\iota, \iota \in I$, the equivalence of the discrete convergence and continuous discrete convergence of the sequence of inverse mappings

(4) $$A_\iota^{-1} \to A^{-1}(\iota \in I) \iff A_\iota^{-1} \underset{c}{\to} A^{-1}(\iota \in I).$$

Thus for every $w \in Y$, for every pair of sequences $w_\iota \to w, u_\iota \to A^{-1}w(\iota \in I)$ and for every $\varepsilon > o$, there exist a $\nu \in I$ and a $\delta > o$ such that the condition $|w_\iota, y_\iota| \leq \delta$ implies $|u_\iota, A_\iota^{-1} y_\iota| \leq \varepsilon$ for every $\iota \geq \nu$ and every $y_\iota \in Y_\iota$. Hence on setting $y_\iota = A_\iota v_\iota$, the condition $|w_\iota, A_\iota v_\iota| \leq \delta$ implies $|u_\iota, v_\iota| \leq \varepsilon$ for every $\iota \geq \nu$ and every $v_\iota \in X_\iota$.

3. Equicontinuity

Given a sequence of continuous mappings, stability is a necessary and sufficient condition for equicontinuity of the sequence. In this section we assume again that the index sequence I is denumerably infinite and linearly ordered, $I = \{\iota_1, \iota_2, \iota_3, \dots\}$, $\iota_1 < \iota_2 < \iota_3 < \dots$. Let $X, \prod_\iota X_\iota, \lim^X$ and $Y, \prod_\iota Y_\iota, \lim^Y$ be two metric discrete limit spaces with the index sequence I and let (A_ι) be a sequence of mappings $A_\iota : X_\iota \to Y_\iota$, $\iota \in I$. The sequence (A_ι) is said to be equicontinuous at the point $(u_\iota) \in \prod_\iota X_\iota$ if, for every $\varepsilon > 0$, there exists a $\delta > 0$ such that the condition $|u_\iota, v_\iota| \leq \delta$ implies $|A_\iota u_\iota, A_\iota v_\iota| \leq \varepsilon$ for every $\iota \in I$ and every $v_\iota \in X_\iota$. The sequence (A_ι) is called equicontinuous if (A_ι) is equicontinuous at every discretely (A_ι)-convergent sequence (u_ι). We are now in a position to establish the connection between stability and equicontinuity of the sequence (A_ι).

(1) Let $A_\iota : X_\iota \to Y_\iota$ be continuous for each $\iota \in I$. Then (A_ι) is equicontinuous at the point (u_ι) if and only if (A_ι) is stable at this point. The sequence (A_ι) is equicontinuous if and only if (A_ι) is stable.

Proof. (i) If the sequence (A_ι) is equicontinuous at (u_ι), then (A_ι) is obviously asymptotically equicontinuous and theorem 2.(1) implies that (A_ι) is stable at (u_ι).
(ii) Conversely, if (A_ι) is stable at (u_ι) then, for every $\varepsilon > 0$, there exist a $\nu \in I$ and a $\delta' > 0$ such that the condition $|u_\iota, v_\iota| \leq \delta'$ implies $|A_\iota u_\iota, A_\iota v_\iota| \leq \varepsilon$ for every $\iota \geq \nu$ and every $v_\iota \in X_\iota$. For $\nu \in I = (\iota_1, \iota_2, \ldots)$ there is a natural number t such that $\nu = \iota_t$. Since the mappings A_ι are continuous, there exists a $\delta'' > 0$ such that the condition $|u_\iota, v_\iota| \leq \delta''$ implies $|A_\iota u_\iota, A_\iota v_\iota| \leq \varepsilon$ for every $\iota \in \{\iota_1, \ldots, \iota_{t-1}\}$ and every $v_\iota \in X_\iota$. On setting $\delta = \min(\delta', \delta'')$, we obtain from $|u_\iota, v_\iota| \leq \delta$ the condition $|A_\iota u_\iota, A_\iota v_\iota| \leq \varepsilon$ for every $\iota \in I$ and every $v_\iota \in X_\iota$. Hence (A_ι) is equicontinuous at (u_ι).
(iii) Finally, the equivalence of pointwise stability and equicontinuity implies the equivalence of stability and equicontinuity at every discretely (A_ι)-convergent sequence.

A mapping $A_\iota : X_\iota \to Y_\iota$ is said to be continuously invertible at the point $u_\iota \in X_\iota$ if A_ι is injective and if the mapping $A_\iota^{-1} : Y'_\iota \to X_\iota$, $Y'_\iota = A_\iota X_\iota$, is continuous at the point $w_\iota = A_\iota u_\iota \in Y'_\iota$. Correspondingly, a sequence of mappings $A_\iota : X_\iota \to Y_\iota$, $\iota \in I$, is said to be equicontinuously invertible at the point $(u_\iota) \in \prod_\iota X_\iota$ if A_ι is injective for each $\iota \in I$ and if the sequence of mappings $A_\iota^{-1} : Y'_\iota \to X_\iota$, $Y'_\iota = A_\iota X_\iota$, $\iota \in I$, is equicontinuous at the point $(w_\iota) = (A_\iota u_\iota) \in \prod_\iota Y'_\iota$. Finally, a sequence (A_ι) is called <u>equicontinuously invertible</u> if (A_ι) is equicontinuously invertible at every discretely (A_ι)-convergent sequence (u_ι).

(2) <u>Let</u> $A_\iota : X_\iota \to Y_\iota$ <u>be continuously invertible for each</u> $\iota \in I$. <u>Then the sequence</u> (A_ι) <u>is equicontinuously invertible at the point</u> (u_ι) <u>if and only if</u> (A_ι) <u>is inversely stable at this point. The sequence</u> (A_ι) <u>is equicontinuously invertible if and only if</u> (A_ι) <u>is inversely stable.</u>

<u>Proof.</u> These statements follow at once from theorem (1), applied to the sequence of mappings $A_\iota^{-1} : Y'_\iota \to X_\iota, Y'_\iota = A_\iota X_\iota, \iota \in I$.

As an important application of these general theorems, one obtains easily stability criteria for sequences of continuous linear mappings in normed spaces. Let $X, \prod_\iota X_\iota, \lim^X$ and $Y, \prod_\iota Y_\iota, \lim^Y$ be two metric discrete limit spaces constituted of normed spaces $X_\iota, Y_\iota, \iota \in I$, over the same field of real or complex numbers. As usual, we define the distances on X_ι, Y_ι by means of the norms on X_ι, Y_ι by the equations $|u_\iota, v_\iota| = \|u_\iota - v_\iota\|$. Let us further assume that the sequences of null vectors in $\prod_\iota X_\iota$ resp. $\prod_\iota Y_\iota$ converge discretely to the null vector 0 in X resp. Y,

(3) $$\lim{}^X 0_\iota = 0, \qquad \lim{}^Y 0_\iota = 0.$$

Under these assumptions, we can prove the following theorem

(4) <u>A sequence of continuous linear mappings</u> $A_\iota : X_\iota \to Y_\iota, \iota \in I$, <u>is stable if and only if there exists a positive number</u> α_1 <u>such that the following inequalities hold</u>

(b^N) $$\|A_\iota u_\iota\|_{Y_\iota} \leq \alpha_1 \|u_\iota\|_{X_\iota}, \qquad u_\iota \in X_\iota, \iota \in I.$$

(5) <u>A sequence of continuously invertible linear mappings</u> $A_\iota : X_\iota \to Y_\iota, \iota \in I$, <u>is inversely stable if and only if there exists a positive number</u> α_0 <u>such that the following inequalities hold</u>

(b^N_{-1}) $$\alpha_0 \|u_\iota\|_{X_\iota} \leq \|A_\iota u_\iota\|_{Y_\iota}, \qquad u_\iota \in X_\iota, \iota \in I.$$

Proof. (i) Evidently, the inequalities (b^N) imply the stability condition (b^M). Conversely, suppose (A_ι) stable. Then we conclude from theorem (1) that the sequence (A_ι) is equicontinuous. Under the assumption (3), the sequence of null vectors $(0_\iota) \in \Pi_\iota X_\iota$ is discretely (A_ι)-convergent and thus (A_ι) is equicontinuous at the point (0_ι). Hence for $\varepsilon = 1$, there exists a $\delta > 0$ such that the condition $\|v_\iota\| \leq \delta$ implies $\|A_\iota v_\iota\| \leq 1$ for every $\iota \in I$ and every $v_\iota \in X_\iota$. Consequently, for every $\iota \in I$, for every $h_\iota \in X_\iota$ such that $\|h_\iota\| = 1$ and for every $v_\iota = \delta h_\iota$, we obtain the estimate $\|A_\iota h_\iota\| \leq 1/\delta$ so that $\|A_\iota\| \leq 1/\delta = \alpha_1$. This is equivalent to the inequalities (b^N).

(ii) We now consider a sequence of continuously invertible linear mappings $A_\iota : X_\iota \rightarrow Y_\iota$ and the associated inverse mappings $A_\iota^{-1} : Y'_\iota \rightarrow X_\iota, Y'_\iota = A_\iota X_\iota, \iota \in I$. Then the sequence (A_ι) is inversely stable if and only if the sequence (A_ι^{-1}) is stable. By theorem (4), applied to (A_ι^{-1}), a necessary and sufficient condition that (A_ι^{-1}) be stable is stated by the inequalities $\|A_\iota^{-1}\| \leq 1/\alpha_0$, uniformly for all $\iota \in I$. This condition is equivalent to the inequalities (b_{-1}^N).

4. Moduli of Continuity

The equicontinuity of a sequence of mappings may be characterized equivalently by inequalities using a modulus of continuity of the sequence. In this way we shall be able to establish fundamental error estimates, as will be shown in the next section. In the following, we call modulus of continuity every increasing real function $\mu : \mathbb{R}_+ \rightarrow \overline{\mathbb{R}}_+$ which is continuous at the point o such that $\mu(0) = 0$. A modulus of continuity is said to be strictly positive if $\mu(x) > 0$ for $x > 0$.

Let us consider two sequences of metric spaces $X_\iota, Y_\iota, \iota \in I$, and a sequence of mappings $A_\iota : X_\iota \rightarrow Y_\iota, \iota \in I$. Then we have the following basic theorem.

(1) The sequence (A_ι) is equicontinuous at the point (u_ι) if and only if there exists a modulus of continuity μ such that the following inequalities hold

(2) $$|A_\iota u_\iota, A_\iota v_\iota| \leq \mu(|u_\iota, v_\iota|), \qquad v_\iota \in X_\iota, \iota \in I.$$

Proof. (i) Suppose (A_ι) is equicontinuous at the point (u_ι). Then for every $\varepsilon > o$, there is a $\delta > o$ such that we have the relation

$$|(u_\iota),(v_\iota)| = \sup_\iota |u_\iota, v_\iota| \leq \delta \Longrightarrow |(A_\iota u_\iota),(A_\iota v_\iota)| = \sup_\iota |A_\iota u_\iota, A_\iota v_\iota| \leq \varepsilon$$

for every sequence (v_ι). Thus the modulus of continuity μ of the sequence (A_ι) at the point (u_ι) is defined by

$$\mu(s) = \sup \left\{ |(A_\iota u_\iota),(A_\iota v_\iota)| \;\middle|\; (v_\iota) \in \prod_\iota X_\iota, |(u_\iota),(v_\iota)| \leq s \right\}$$

for all $o \leq s < \infty$. Hence μ is an increasing real function, $\mu: \mathbb{R}_+ \longrightarrow \overline{\mathbb{R}}_+$ such that $\mu(o) = o$ and μ is continuous at $s = o$. Moreover, μ satisfies the following inequality

$$|(A_\iota u_\iota),(A_\iota v_\iota)| \leq \mu(|(u_\iota),(v_\iota)|) \quad (v_\iota) \in \prod_\iota X_\iota,$$

from which one obtains at once the inequalities (2).

(ii) Conversely, the inequalities (2) imply the equicontinuity of the sequence (A_ι) at the point (u_ι). Indeed, since μ is continuous at $s = o$ and $\mu(o) = o$, for every $\varepsilon > o$, there is a $\delta > o$ such that $\mu(s) \leq \varepsilon$ for all $o \leq s \leq \delta$. Thus for every $\iota \in I$ and every $v_\iota \in X_\iota$ the relation $s = |u_\iota, v_\iota| \leq \delta$ implies $|A_\iota u_\iota, A_\iota v_\iota| \leq \mu(s) \leq \varepsilon$.

We next consider sequences of continuously invertible mappings $A_\iota: X_\iota \longrightarrow Y_\iota, \iota \in I$. In this case, the above theorem may be applied to the sequence of inverse mappings $A_\iota^{-1}: Y'_\iota \rightarrow X_\iota, Y'_\iota = A_\iota X_\iota, \iota \in I$.

(3) The sequence (A_ι) is equicontinuously invertible at the point (u_ι) if and only if there exists a modulus of continuity μ_I such that the following inequalities are valid

(4) $$|u_\iota, v_\iota| \leq \mu_I(|A_\iota u_\iota, A_\iota v_\iota|), \qquad v_\iota \in X_\iota, \iota \in I.$$

Proof. The sequence (A_ι) is equicontinuously invertible at (u_ι) if and only if the sequence of inverse mappings $A_\iota^{-1}: Y'_\iota \rightarrow X_\iota, \iota \in I$, is equicontinuous at the point $(w_\iota) = (A_\iota u_\iota) \in \prod_\iota Y'_\iota$. By theorem (1), this is equivalent to the existence of a modulus of continuity

μ_1 such that the following inequalities hold,

$$|A_\iota^{-1} w_\iota, A_\iota^{-1} y_\iota| \leq \mu_1(|w_\iota, y_\iota|), \qquad y_\iota \in Y'_\iota, \iota \in I.$$

Hence, on setting $w_\iota = A_\iota u_\iota$, $y_\iota = A_\iota v_\iota$, $v_\iota \in X_\iota$, one obtains the inequalities (4).

The following important lemma shows that every strictly positive modulus of continuity μ has a generalized inverse function μ_{-1} which is also a modulus of continuity.

(5) For every strictly positive modulus of continuity μ there exists a strictly positive modulus of continuity μ_{-1} satisfying the condition $\mu_{-1}(\mu(s)) \leq s$, $o < s < \infty$.

Proof. For every modulus of continuity μ we define a generalized inverse function μ_{-1} by $\mu_{-1}(o) = o$ and

(6) $$\mu_{-1}(t) = \sup\{r \in \mathbb{R}_+ | \mu(r) < t\}, \qquad o < t < \infty.$$

As μ is assumed to be strictly positive, we have $\mu(s) > o$ for $s > o$ and hence

$$\mu_{-1}(\mu(s)) = \sup\{r \in \mathbb{R}_+ | \mu(r) < \mu(s)\} \leq \sup\{r \in \mathbb{R}_+ | r < s\} = s.$$

Since μ is continuous at $s = o$, there is, for every $t > o$, a $\delta' > o$ such that $\mu(\delta') < t$ and, consequently, $\mu_{-1}(t) \geq \delta' > o$. Evidently, μ_{-1} is an increasing real function. Finally, μ_{-1} is continuous at the point $o \in \mathbb{R}_+$ since, for every $\varepsilon > o$ and $\delta = \mu(\varepsilon) > o$, the inequality $\mu_{-1}(t) \leq \mu_{-1}(\delta) = \mu_{-1}(\mu(\varepsilon)) \leq \varepsilon$ holds for all $o \leq t \leq \delta$.

The following theorem establishes an interesting result which will be the basis of two-sided error estimates in the next section.

(7) The sequence (A_ι) is both equicontinuous and equicontinuously invertible at the point (u_ι) if and only if there exist two strictly positive moduli of continuity μ_1, μ_{-1} such that the following inequalities hold

(8) $$\mu_{-1}(|A_\iota u_\iota, A_\iota v_\iota|) \leq |u_\iota, v_\iota| \leq \mu_1(|A_\iota u_\iota, A_\iota v_\iota|), \qquad v_\iota \in X_\iota, \iota \in I.$$

Proof. (i) Suppose (A_ι) is equicontinuously invertible at (u_ι). Then, by theorem (3), we have the inequalities (4) and hence the inequalities on the right side of (8). Further suppose (A_ι) is equicontinuous at (u_ι). Then there exists a modulus of continuity μ such that the inequalities (2) are valid. We may assume that μ_1, μ are strictly positive. Otherwise, for example, one can replace μ by the new modulus of continuity $\mu(s)+\eta s$, $s \in \mathbb{R}_+$, for some positive number η. By lemma (5), the strictly positive modulus of continuity μ has a strictly positive generalized inverse function μ_{-1} satisfying the condition $\mu_{-1}(\mu(s)) \leq s, s \in \mathbb{R}_+$. Hence, the inequalities on the left side of (8) follow from (2) and the relation

$$\mu_{-1}(|A_\iota u_\iota, A_\iota v_\iota|) \leq \mu_{-1}(\mu(|u_\iota, v_\iota|)) \leq |u_\iota, v_\iota|, \qquad v_\iota \in X_\iota, \iota \in I.$$

(ii) Conversely, by theorem (3), the inequalities (8) imply that the sequence (A_ι) is equicontinuously invertible. Further, for every $\varepsilon > o$ and $\delta = \mu_{-1}(\varepsilon) > o$, for every $\iota \in I$ and every $v_\iota \in X_\iota$, the relation $|u_\iota, v_\iota| < \delta$ implies $\mu_{-1}(|A_\iota u_\iota, A_\iota v_\iota|) < \mu_{-1}(\varepsilon)$ and hence the inequality $|A_\iota u_\iota, A_\iota v_\iota| < \varepsilon$. This shows that (A_ι) is equicontinuous at the point (u_ι).

5. Characterization of Discrete Convergence

The results obtained in the preceeding sections permit a number of important characterizations of the concept of discrete convergence of continuous mappings in discrete limit spaces. These are summarized in the following theorems. Moreover, we shall establish error estimates for the discrete convergence of approximants and a-posteriori error estimates for the solutions of the approximating equations.

Let $X, \prod_\iota X_\iota, \lim^X$ and $Y, \prod_\iota Y_\iota, \lim^Y$ be two metric discrete limit spaces with the same index sequence I. Again we assume that I is denumerably infinite and linearly ordered.

(1) Let (A_ι) be a sequence of continuous mappings $A_\iota : X_\iota \to Y_\iota, \iota \in I$. Then the following statements are equivalent:

(i) $A_\iota \to A (\iota \in I)$. (ii) $A_\iota \xrightarrow[c]{} A (\iota \in I)$.

(iii) $A, (A_\iota)$ is consistent and (A_ι) is equicontinuous.

(iv) $A, (A_\iota)$ is consistent and, for every discretely (A_ι)-convergent sequence (u_ι), there exists a modulus of continuity μ such that the following inequalities hold,

$$|A_\iota u_\iota, A_\iota v_\iota| \leq \mu(|u_\iota, v_\iota|), \quad v_\iota \in X_\iota, \iota \in I.$$

Proof. The equivalence of discrete convergence and continuous discrete convergence follows from Theorem 2.(3). By theorem 1.(1), $A_\iota \longrightarrow A(\iota \in I)$ if and only if $A,(A_\iota)$ is consistent and (A_ι) is stable. Theorem 3.(1) implies the equivalence of stability and equicontinuity of the sequence (A_ι) of continuous mappings. Finally, by theorem 4.(1), this is the necessary and sufficient condition for the existence of a modulus of continuity for (A_ι) at every discretely (A_ι)-convergent sequence (u_ι).

In the following theorems, concerning the discrete convergence of inverse mappings, we assume the mappings to be bijective. Note, however, that these theorems are valid for sequences of continuously invertible mappings A_ι, which are not necessarily bijective from X_ι to Y_ι, when applied to the restricted mappings $A_\iota : X_\iota \longrightarrow Y'_\iota$, $Y'_\iota = A_\iota X_\iota$, $\iota \in I$. In this case, the discrete convergence of the associated inverse mappings $A_\iota^{-1} \rightarrow A^{-1}(\iota \in I)$ is equivalent to the inverse convergence relation (a_{-1}) and to the condition that $\lim(\prod_\iota A_\iota X_\iota, AX)$ is surjective.

(2) Let (A_ι) be a sequence of continuously invertible, bijective mappings $A_\iota : X_\iota \rightarrow Y_\iota, \iota \in I$, and let $A : X \rightarrow Y$ be bijective. Then the following statements are equivalent:

(i) $A_\iota^{-1} \rightarrow A^{-1}(\iota \in I)$. (ii) $A_\iota^{-1} \underset{c}{\rightarrow} A^{-1}(\iota \in I)$.

(iii) $A,(A_\iota)$ is consistent and (A_ι) is equicontinuously invertible.

(iv) $A,(A_\iota)$ is consistent and, for every discretely (A_ι)-convergent sequence (u_ι), there exists a modulus of continuity μ_1 such that the following inequalities are valid,

$$|u_\iota, v_\iota| \leq \mu_1(|A_\iota u_\iota, A_\iota v_\iota|), \quad v_\iota \in X_\iota, \iota \in I.$$

Proof. The equivalence of discrete convergence and continuous discrete convergence of inverse mappings has been shown in 2.(4). By theorem 1.(4), we have $A_\iota^{-1} \rightarrow A^{-1}(\iota \in I)$ if and only if $A,(A_\iota)$ is consistent and (A_ι) is inversely stable. Theorem 3.(2) states that the inverse stability of (A_ι) is equivalent to the equicontinuous invertibility of (A_ι). For bijective mappings, this means the equicontinuity of (A_ι^{-1}) at every discretely convergent sequence $(w_\iota) = (A_\iota u_\iota)$ such that $(u_\iota) = (A_\iota^{-1} w_\iota)$ converges discretely as well. Finally, theorem 4.(3) establishes the equivalence of this condition to the existence of a modulus of continuity μ_1 for every discretely (A_ι)-convergent sequence (u_ι).

The next theorem is particularly interesting with respect to error estimates since it establishes not only upper but also lower bounds for the errors.

(3) Let (A_ι) be a sequence of continuous, continuously invertible, bijective mappings $A_\iota : X_\iota \to Y_\iota, \iota \in I$, and let $A: X \to Y$ be bijective. Then the following conditions are equivalent:

(i) $A_\iota \to A$, $A_\iota^{-1} \to A^{-1} (\iota \in I)$. (ii) $A_\iota \underset{c}{\to} A$, $A_\iota^{-1} \underset{c}{\to} A^{-1} (\iota \in I)$.

(iii) $A, (A_\iota)$ is consistent, (A_ι) and (A_ι^{-1}) are equicontinuous.

(iv) $A, (A_\iota)$ is consistent and, for every discretely (A_ι)-convergent sequence (u_ι), there exist two strictly positive moduli of continuity μ_1, μ_{-1} such that the following two-sided inequalities are satisfied

$$\mu_{-1}(|A_\iota u_\iota, A_\iota v_\iota|) \leq |u_\iota, v_\iota| \leq \mu_1(|A_\iota u_\iota, A_\iota v_\iota|), \quad v_\iota \in X_\iota, \iota \in I.$$

Proof. Theorem (1) and Theorem (2) imply the equivalence of (i), (ii), (iii). The equivalence of (iii) and (iv) follows at once from theorem 4.(7).

In virtue of these results we can derive fundamental error estimates for the discrete convergence of solutions u_ι to the solution u of the equations

(4) $$Au = w, \qquad A_\iota u_\iota = w_\iota, \iota \in I.$$

In the following, let us assume $A: X \to Y$ to be bijective, (A_ι) to be a sequence of continuously invertible bijective mappings $A_\iota : X_\iota \to Y_\iota, \iota \in I$, and (w_ι) to be discretely convergent to w. Then the sequence $A, (A_\iota)$ is consistent if and only if, for every $u \in X$, there exists a pair of discretely convergent sequences $u'_\iota \to u$, $w'_\iota \to Au (\iota \in I)$ such that

$$\tau_\iota = |A_\iota u'_\iota, w'_\iota|, \quad \iota \in I,$$

is a null sequence. This sequence is said to be a discretization error or truncation error of $A,(A_\iota)$. When $A_\iota^{-1} \to A^{-1} (\iota \in I)$, theorem (2) implies the consistency of $A,(A_\iota)$ such that $\tau_\iota \to o(\iota \in I)$ and the existence of a modulus of continuity μ_1 at the point (u'_ι) such that the following inequalities are valid,

(5) $$|u'_\iota, A_\iota^{-1} w'_\iota| \leq \mu_1(\tau_\iota), \qquad \iota \in I.$$

When $A_\iota \to A$, $A_\iota^{-1} \to A^{-1} (\iota \in I)$ and A_ι is continuous for each $\iota \in I$, we obtain from theorem (3) the consistency of $A,(A_\iota)$ such that $\tau_\iota \to o(\iota \in I)$ and the existence of two moduli of continuity μ_1, μ_{-1} such that the following two-sided error estimates hold

(6) $$\mu_{-1}(\tau_\iota) \leq |u'_\iota, A_\iota^{-1} w'_\iota| \leq \mu_1(\tau_\iota), \qquad \iota \in I.$$

The sequence (w_ι) of the given data in (4) is discretely convergent to w if and only if the sequence

$$\sigma_\iota = |w_\iota, w'_\iota|, \qquad \iota \in I,$$

is a null sequence where (w'_ι) is a discretely convergent sequence $w'_\iota \to w = Au(\iota \in I)$. The sequence (σ_ι) is said to be a discretization error or truncation error of the data (w_ι). When $A_\iota^{-1} \to A^{-1} (\iota \in I)$, one has the consistency of $A,(A_\iota)$ and a modulus of continuity μ'_1 at the point $(A_\iota^{-1} w'_\iota)$ such that the following error estimates hold,

(7) $$|u_\iota, A_\iota^{-1} w'_\iota| \leq \mu'_1(\sigma_\iota), \qquad \iota \in I.$$

When $A_\iota \to A$, $A_\iota^{-1} \to A^{-1} (\iota \in I)$ and A_ι is continuous for each $\iota \in I$, then $A,(A_\iota)$ is consistent and there exist two moduli of continuity μ'_1, μ'_{-1} such that the following two-sided error estimates are valid

(8) $$\mu'_{-1}(\sigma_\iota) \leq |u_\iota, A_\iota^{-1} w'_\iota| \leq \mu'_1(\sigma_\iota), \qquad \iota \in I.$$

However, in most applications, the equations $A_\iota u_\iota = w_\iota$, $\iota \in I$, can only be solved approximately. For example, nonlinear equations are solved by iterative methods where only a finite number of steps can be effected. Moreover, numerical solutions have inevitably rounding errors. To estimate the error of approximations v_ι of the solutions $u_\iota = A_\iota^{-1} w_\iota$, we finally introduce the sequence

$$\rho_\iota = |w_\iota, A_\iota v_\iota|, \qquad \iota \in I,$$

which defines a measure for the defect or the residuum of the approximations v_ι, $\iota \in I$. By these means, we are able to establish a-posteriori error estimates for $|u_\iota, v_\iota|$ without knowing the solutions u_ι. Indeed, under the assumption $A_\iota^{-1} \to A^{-1} (\iota \in I)$, there exists a modulus of continuity μ_1'' at the point $(u_\iota) = (A_\iota^{-1} w_\iota)$ such that the error estimates

(9) $$|u_\iota, v_\iota| \leq \mu_1''(\rho_\iota), \qquad \iota \in I,$$

are valid. Under the assumption $A_\iota \to A, A_\iota^{-1} \to A^{-1} (\iota \in I)$ and A_ι continuous for each $\iota \in I$, we obtain from theorem (3) the existence of two strictly positive moduli of continuity μ_1'', μ_{-1}'' such that the following two-sided a-posteriori error estimates hold,

(10) $$\mu_{-1}''(\rho_\iota) \leq |u_\iota, v_\iota| \leq \mu_1''(\rho_\iota), \qquad \iota \in I.$$

Finally, a sequence $|u_\iota', v_\iota|$, $\iota \in I$, is said to be a total error of the approximations v_ι of the solutions $u_\iota = A_\iota^{-1} w_\iota$, $\iota \in I$, where (u_ι') is some discretely convergent sequence $u_\iota' \to u (\iota \in I)$. Obviously, the total error $|u_\iota', v_\iota|$, $\iota \in I$, and the associated sequence $|A_\iota u_\iota', A_\iota v_\iota|$, $\iota \in I$, can be estimated by the sum of the three error terms considered above,

$$|u_\iota', v_\iota| \leq |u_\iota', A_\iota^{-1} w_\iota'| + |A_\iota^{-1} w_\iota', u_\iota| + |u_\iota, v_\iota|,$$
$$|A_\iota u_\iota', A_\iota v_\iota| \leq \tau_\iota + \sigma_\iota + \rho_\iota, \qquad \iota \in I.$$

When $A_\iota^{-1} \to A^{-1} (\iota \in I)$, we obtain from theorem (3) a modulus of continuity μ_1 at the point (u_ι') such that the following error estimates hold,

(11) $$|u_\iota', v_\iota| \leq \mu_1(|A_\iota u_\iota', A_\iota v_\iota|) \leq \mu_1(\tau_\iota + \sigma_\iota + \rho_\iota), \qquad \iota \in I.$$

When $A_\iota \to A, A_\iota^{-1} \to A^{-1} (\iota \in I)$ and A_ι is continuous for each $\iota \in I$, one has two strictly positive moduli of continuity μ_1, μ_{-1} at the point (u_ι') such that the following two-sided estimates of the total errors are valid

(12) $$\mu_{-1}(|A_\iota u_\iota', A_\iota v_\iota|) \leq |u_\iota', v_\iota| \leq \mu_1(|A_\iota u_\iota', A_\iota v_\iota|) \leq \mu_1(\tau_\iota + \sigma_\iota + \rho_\iota), \qquad \iota \in I.$$

6. Further Results

Some further results can be established in metric discrete limit spaces with discretely convergent metrics. In this section, let $X, X_\iota, \iota \in I$, be metric spaces and let I be a denumerably infinite, linearly ordered index sequence. Then $X, \prod_\iota X_\iota$, lim is said to be a metric discrete limit space with discretely convergent metrics $|.,.|_{X_\iota} \rightarrow |.,.|_X, \iota \in I$, if the condition (M) holds and if the relation

$$\lim u_\iota = u, \ \lim v_\iota = v \Longrightarrow \lim |u_\iota, v_\iota|_{X_\iota} = |u,v|_X$$

is valid for every pair of discretely convergent sequences $(u_\iota), (v_\iota) \in \prod_\iota X_\iota$.

In many applications, the discrete convergence lim is defined in such a way that first, for every $u \in X$, a sequence of elements $(u'_\iota) \in \prod_\iota X_\iota$ is given with $\lim u'_\iota = u$. Then, every sequence $(u_\iota) \in \prod_\iota X_\iota$ being metrically equivalent to (u'_ι) is defined to be discretely convergent to u. In this way, for instance, a sequence of restriction operators $r_\iota : X \rightarrow X_\iota, \iota \in I$, constitutes a discrete convergence on setting $(u'_\iota) = (r_\iota u)$ for every $u \in X$. This procedure is based on the following theorem.

(1) The space $X, \prod_\iota X_\iota$, lim is a metric discrete limit space if and only if the following statement (i) holds. This space is a metric discrete limit space with discretely convergent metrics if and only if the following statements (i) and (ii) are valid:

(i) For every u in the range of lim in X, there exists a discretely convergent sequence (u'_ι) such that $\lim u'_\iota = u$ and, for every sequence (u_ι), the following conditions are equivalent

$$\lim u_\iota = u \Longleftrightarrow \lim |u_\iota, u'_\iota| = o.$$

(ii) For every u,v in the range of lim in X, the associated sequences $(u'_\iota), (v'_\iota)$ satisfy the relation

$$\lim u'_\iota = u, \quad \lim v'_\iota = v \Longrightarrow \lim |u'_\iota, v'_\iota| = |u,v|.$$

Proof. (i) Evidently, (i) is a necessary condition for $X, \prod_\iota X_\iota$, lim to be a metric discrete limit space. We now show that this condition is sufficient. Let $(u_\iota), (v_\iota)$ be any pair of sequences such that (u_ι) or (v_ι) converges discretely. We may assume that (v_ι) converges discretely and denote $\lim v_\iota = u$. Using condition (i), there is a discretely convergent sequence (u'_ι) such that $\lim u'_\iota = u$ and hence $\lim|v_\iota,u'_\iota| = o$. By means of the triangle inequality, one obtains

$$\left||u_\iota,u'_\iota| - |u_\iota,v_\iota|\right| \leq |u'_\iota,v_\iota| \to o \qquad (\iota\in I).$$

Then condition (i) implies

$$\lim|u_\iota,v_\iota| = o \Longleftrightarrow \lim|u_\iota,u'_\iota| = o \Longleftrightarrow \lim u_\iota = u.$$

(ii) Obviously, the conditions (i), (ii) are necessary for $X, \prod_\iota X_\iota, \lim$ to be a metric discrete limit space with discretely convergent metrics. Conversely, suppose (i) and (ii) are valid. For every pair of discretely convergent sequences $(u_\iota),(v_\iota)$ such that $\lim u_\iota = u$, $\lim v_\iota = v$, let $(u'_\iota),(v'_\iota)$ be the associated sequences defined by (i), (ii). Further, we have the following inequality

$$\left||u_\iota,v_\iota| - |u'_\iota,v'_\iota|\right| \leq \left||u_\iota,v_\iota| - |u'_\iota,v_\iota|\right| + \left||u'_\iota,v_\iota| - |u'_\iota,v'_\iota|\right| \leq$$
$$\leq |u_\iota,u'_\iota| + |v_\iota,v'_\iota|, \ \iota\in I.$$

By condition (i), the term on the right tends to zero. Finally, condition (ii) yields $|u'_\iota,v'_\iota| \to |u,v|$ and hence $|u_\iota,v_\iota| \to |u,v|$ $(\iota\in I)$.

Now let $X, \prod_\iota X_\iota, \lim^X$ and $Y, \prod_\iota Y_\iota, \lim^Y$ be two metric discrete limit spaces with discretely convergent metrics $|.,.|_{X_\iota} \to |.,.|_X$, $|.,.|_{Y_\iota} \to |.,.|_Y$ $(\iota\in I)$. Let $A,(A_\iota)$ be a sequence of mappings $A:X \to Y$, $A_\iota:X_\iota \to Y_\iota$, $\iota\in I$. Then we can state the following interesting theorem.

(2) <u>If the sequence</u> (A_ι) <u>converges discretely to</u> A, <u>then the mapping</u> A <u>is continuous.</u>

<u>Proof.</u> By theorem 2.(3), $A_\iota \to A(\iota\in I)$ implies $A \underset{c}{\to} A(\iota\in I)$. In this case, $\lim^X$ is surjective and, for every $u\in X$, there is a sequence $u'_\iota \to u$ so that $A_\iota u'_\iota \to Au(\iota\in I)$. For every $\varepsilon > o$ there exist a $\nu\in I$ and a $\delta > o$ such that the condition $|u'_\iota,v_\iota| \leq \delta$ implies $|A_\iota u'_\iota, A_\iota v_\iota| \leq \varepsilon$ for every $\iota \geq \nu$ and every $v_\iota \in X_\iota$. For $\delta' = \delta/2$ and for every $v\in X$ such that $|u,v| \leq \delta'$ there is a sequence $v'_\iota \to v$ so that $A_\iota v'_\iota \to Av(\iota\in I)$, $\lim|u'_\iota,v'_\iota| = |u,v|$ and hence $|u'_\iota,v'_\iota| \leq \delta$ for every $\iota \geq \nu'$ and a suitable $\nu' \geq \nu$. Thus we have $|A_\iota u'_\iota, A_\iota v'_\iota|$

$\leq \varepsilon$ for every $\iota \geq \nu'$ and therefore $\lim|A_\iota u'_\iota, A_\iota v'_\iota| = |Au, Av| \leq \varepsilon$. Consequently, for every $u \in X$ and every $\varepsilon > 0$, for $\delta' = \delta/2$ and every $v \in X$, the condition $|u,v| \leq \delta'$ implies $|Au, Av| \leq \varepsilon$, hence $A: X \to Y$ is continuous.

Finally, we are going to establish a useful criterion for the consistency of a sequence of mappings $A: X \to Y$, $A_\iota: X_\iota \to Y_\iota$, $\iota \in I$, in metric discrete limit spaces with discretely convergent metrics. This theorem generalizes a theorem of Grigorieff [7] for linear mappings.

(3) Let A be continuous, let $\lim(\prod_\iota X_\iota, X)$ and $\lim(\prod_\iota Y_\iota, AX)$ be surjective. Then $A, (A_\iota)$ is consistent if and only if there exists a dense subset Φ in X such that, for every $\varphi \in \Phi$, there is a sequence (φ_ι) with the property $\lim \varphi_\iota = \varphi$, $\lim A_\iota \varphi_\iota = A\varphi$.

Proof. (i) If $A, (A_\iota)$ is consistent, there exists, for every $u \in \Phi = X$, a discretely (A_ι)-convergent sequence $(u'_\iota) = (\varphi_\iota)$ having this property.

(ii) Conversely, for every $u \in X$ and every $t = 1,2,\ldots$, there exists an element $\varphi^{(t)} \in \Phi$ such that the inequalities $|u, \varphi^{(t)}| \leq 1/t$, $|Au, A\varphi^{(t)}| \leq 1/t$ hold, since A is continuous and Φ is dense in X. Next, there exist discretely convergent sequences $u_\iota \to u$, $w_\iota \to Au$ $(\iota \in I)$ and, for every $t = 1,2,\ldots$, discretely (A_ι)-convergent sequences $(\varphi_\iota^{(t)})$ so that $\varphi_\iota^{(t)} \to \varphi^{(t)}$, $A_\iota \varphi_\iota^{(t)} \to A\varphi^{(t)}$ $(\iota \in I)$. Thus we obtain

$$|u_\iota, \varphi_\iota^{(t)}| \to |u, \varphi^{(t)}|, \quad |w_\iota, A_\iota \varphi_\iota^{(t)}| \to |Au, A\varphi^{(t)}|, \quad \iota \in I.$$

Hence, for every $t = 1,2,\ldots$, there exists a $\nu_t \in I$ such that the inequalities $|u_\iota, \varphi_\iota^{(t)}| \leq 2/t$, $|w_\iota, A_\iota \varphi_\iota^{(t)}| \leq 2/t$ are valid for every $\iota \geq \nu_t$. The sequence $\nu_1, \nu_2, \nu_3, \ldots$ can be chosen in such a way that the relation $\nu_1 < \nu_2 < \nu_3 < \ldots$ holds. Finally, define $u'_\iota = \varphi_\iota^{(t)}$ for $\nu_t \leq \iota < \nu_{t+1}$, $t = 1,2,3,\ldots$ and $u'_\iota = u_\iota$ for $\iota < \nu_1$. Consequently, $|u_\iota, u'_\iota| \leq 2/t$ and $|w_\iota, A_\iota u'_\iota| \leq 2/t$ for every $t = 1,2,\ldots$ and every $\iota \geq \nu_t$. This implies $\lim|u_\iota, u'_\iota| = 0$, $\lim|w_\iota, A_\iota u'_\iota| = 0$ so that (u'_ι) is discretely (A_ι)-convergent and $\lim u'_\iota = u$, $\lim A_\iota u'_\iota = Au$.

References

1. Ansorge, R.: Problemorientierte Hierarchie von Konvergenzbegriffen bei der numerischen Lösung von Anfangswertaufgaben. Math. Z. 112, 13-22 (1969).

2. —— und Hass, R.: Konvergenz von Differenzenverfahren für lineare und nichtlineare Anfangswertaufgaben. Lecture Notes in Mathematics 159. Berlin-Heidelberg-New York: Springer 197o.

3. Aubin, J.P.: Approximation des espaces de distribution et des opérateurs différentiels. Bull. Soc. Math. France Mém. 12, 1-139 (1967).

4. —— Approximation of elliptic boundary-value problems. New York: Wiley 1972.

5. Browder, F.E.: Approximation-solvability of nonlinear functional equations in normed linear spaces. Arch. Rat. Mech. Anal. 26, 33-42 (1967).

6. Dieudonné, J.: Fundations of modern analysis. New York-London: Academic Press 196o.

7. Grigorieff, R.D.: Zur Theorie linearer approximationsregulärer Operatoren. I und II. To appear in Math. Nachr.

8. Henrici, P.: Discrete variable methods in ordinary differential equations. New York-London: Wiley 1962.

9. —— Error propagation for difference methods. New York-London: Wiley 1963.

1o. Krasnosel'skii, M.A., Vainikko, G.M., et al.: Approximate solution of operator equations. Groningen: Wolters-Noordhoff 1972.

11. Lax, P.D., and Richtmyer, R.D.: Survey of the stability of linear finite difference equations. Comm. Pure Appl. Math. 9, 267-293 (1956).

12. Pereyra, V.: Iterated deferred corrections for nonlinear operator equations. Numer. Math. 1o, 316-323 (1967).

13. Petryshyn, W.V.: Projection methods in nonlinear numerical functional analysis. J. Math. Mech. 17, 353-372 (1967).

14. —— On the approximation-solvability of nonlinear equations. Math. Ann. 177, 156-164 (1968).

15. Reinhardt, J.: Diskrete Konvergenz nichtlinearer Operatoren in metrischen Räumen. Frankfurt: Diplomarbeit 1971.

16. Rinow, W.: Die innere Geometrie der metrischen Räume. Berlin-Göttingen-Heidelberg: Springer 1961.

17. Spijker, M.N.: On the structure of error estimates for finite-difference methods. Numer. Math. 18, 73-1oo (1971).

18. —— Equivalence theorems for non-linear finite difference methods. Numerische Lösung nichtlinearer partieller Differential- und Integrodifferentialgleichungen. Lecture Notes in Mathematics 267, 233-264. Berlin-Heidelberg-New York: Springer 1972.

19. Stetter, H.J.: Stability of nonlinear discretization algorithms. In J.H. Bramble (Ed.): Numerical solution of partial differential equations, 111-123. New York: Academic Press 1966.

2o. Stummel, F.: Diskrete Konvergenz linearer Operatoren. I. Math. Ann. 19o, 45-92 (197o). II. Math. Z. 12o, 231-264 (1971). III. To appear, Proc. Oberwolfach Conference on Linear Operators and Approximation 1971. Int. Series of Numerical Mathematics, Vol. 2o. Basel: Birkhäuser.

21. —— Discrete convergence of mappings. To appear, Proc. Conference on Numerical Analysis, Dublin, August 1972. New York-London: Academic Press 1973.

22. —— Discrete convergence of differentiable mappings. To appear.

23. Vainikko, G.M.: Galerkin's perturbation method and the general theory of approximate methods for non-linear equations. USSR Comput. Math. and Math. Phys. 7, 1-41 (1967).

24. Watt, J.M.: Consistency, convergence and stability of general discretizations of the initial value problem. Numer. Math. 12, 11-22 (1968).

CONVERGENCE ESTIMATES FOR SEMI-DISCRETE GALERKIN METHODS FOR INITIAL-VALUE PROBLEMS

By Vidar Thomée

1. Introduction

In this paper we shall consider the approximate solution of the initial-value problem

(1.1) $$\frac{\partial u}{\partial t} = a\frac{\partial^q u}{\partial x^q}, \quad t > 0, \quad x \in R,$$

(1.2) $$u(x,0) = v(x),$$

where R denotes the real axis. We shall assume that this problem is correctly posed in Petrovskiǐ's sense so that

(1.3) $$\mathrm{Re}(a(i\xi)^q) \leq 0, \quad \xi \in R.$$

For q odd this holds if and only if a is real and for $q = 2m$ even if and only if $\mathrm{Re}(a(-1)^m)$ is non-positive. If the latter quantity is strictly negative the equation is parabolic in Petrovskiǐ's sense. Introducing the Fourier transform and its inverse,

$$(\mathcal{F}u)(\xi) = \hat{v}(\xi) = \int_R v(x)e^{-ix\xi}dx,$$

$$(\mathcal{F}^{-1}w)(x) = \frac{1}{2\pi}\int_R w(\xi)e^{ix\xi}d\xi,$$

the solution operator $E(t)$ of (1.1), (1.2) can be represented as

$$(E(t)v)(x) = \mathcal{F}^{-1}(\exp(ta(i\xi)^q)\hat{v}))(x);$$

by (1.3) $E(t)$ is a bounded operator in L_2.

We want to find by means of Galerkin's method an approximate solution of (1.1), (1.2) in terms of splines of order μ based on a regular mesh with mesh-width h. For this purpose let χ be the characteristic function of the interval $[-\frac{1}{2},\frac{1}{2}]$ and consider for $\mu \geq 1$ the function $\varphi = \chi^{*\mu}$ obtained by convolving χ with itself μ times. Then φ which has support in $[-\frac{1}{2}\mu,\frac{1}{2}\mu]$ is the B-spline of order μ. For h a (small) positive number and l in the set Z of integers, let $\varphi_l(x) = \varphi(h^{-1}x - l)$, with the dependence on μ and h suppressed in the notation and consider the set $\mathcal{S}_h = \{\sum_l c_l\varphi_l\}$ where c_l grows at most as a power of $|l|$ as $|l|$ tends to infinity. The elements of $\mathcal{S}_h$ the splines of order μ of at most power growth, are functions in $C^{\mu-2}$ which reduce to polynomials of degree at most $\mu-1$ in each interval $(jh,(j+1)h)$, $j \in Z$, for μ even and $((j-\frac{1}{2})h,(j+\frac{1}{2})h)$, $j \in Z$, for μ odd. For material on splines, cf. Schoenberg [7], [8], [9].

The Galerkin method consists in finding $u_h(\cdot,t) \in \mathcal{S}_h$ for $t \geq 0$ such that

(i) $u_h(jh,0) = v(jh)$ for $j \in Z$,

(ii) $\frac{\partial u_h}{\partial t} - a\,\frac{\partial^q u_h}{\partial x^q}$

is orthogonal with respect to the L_2 inner product $(\cdot,\cdot)$ to $\mathcal{S}_h$ for all $t > 0$.

In order not to demand excessive regularity of u_h we shall express (ii) in the weak form $(D = \frac{d}{dx})$

$$(\frac{\partial u_h}{\partial t}, \varphi_l) - a(-1)^m(D^{q-m}u_h, D^m\varphi_l) = 0 \quad \text{for} \quad l \in Z,$$

where $m = [q/2]$ so that $q = 2m$ or $2m+1$.

In terms of the coefficients of u_h with respect to the basis $\{\varphi_l\}$ in $\mathcal{S}_h$ this leads to an infinite system of ordinary differential equations in t, which under suitable assumptions on v can be solved to give a solution $u_h(\cdot,t) = G_h(t)v$.

For the purpose of studying the Galerkin solution operator $G_h(t)$ we introduce the spline interpolation operator S_h which takes a continuous function v of at most power growth into the element $v_h \in \mathcal{S}_h$ which agrees with v at the mesh-points hZ. We then associate with the Galerkin equations an operator $F_h(t)$ which can be thought of as a finite difference solution operator with the property that u_h agrees with $F_h(t)v$ on the mesh-lines so that the Galerkin solution operator can be represented as $G_h(t) = S_hF_h(t)$.

The finite difference solution operator $F_h(t)$ which has coefficients which vary with t/h^q is shown to be accurate of order $2\mu-q$ for q even and $2\mu-q+1$ for q odd, respectively, and using also simple properties of the spline interpolation operator S_h we shall therefore for smooth v arrive at the error estimates, as $h \to 0$,

$$(1.4) \qquad \|G_h(t)v - E(t)v\|_{L_2} = \begin{cases} O(h^{\min(\mu,2\mu-q)}), & q \text{ even}, \\ O(h^{\min(\mu,2\mu-q+1)}), & q \text{ odd}. \end{cases}$$

If we restrict the consideration to the mesh-points we obtain the sharper estimates, as $h \to 0$,

$$(1.5) \qquad \|G_h(t)v - E(t)v\|_{l_{2,h}} = \begin{cases} O(h^{2\mu-q}), & q \text{ even}, \\ O(h^{2\mu-q+1}), & q \text{ odd}, \end{cases}$$

where $\|\cdot\|_{l_{2,h}}$ denotes the discrete l_2-norm on hZ. This suggests that when $\mu > q$ for q even or $\mu > q+1$ for q odd, higher order splines could be used in the final interpolation to obtain more accurate results between the mesh-points.

Consider for example the simple hyperbolic equation (1.1) with a real and $q=1$, and let the Galerkin equations be based on piece-wise linear functions ($\mu = 2$). Then the error estimate at the mesh-points is fourth order whereas the global estimate is only second order.

Estimates of the type (1.4) have been proved by other authors in a variety of cases, cf. e.g. Douglas and Dupont [3], Dupont [4], Fix and Nassif [5], Price and Varga [6], Strang and Fix [10] and Swartz and Wendroff [11]; it is the observation that $G_h(t)$ can be represented as a finite difference solution operator at the mesh-points which is our basis for the proofs of the stronger estimates (1.5).

In our previous paper [12] where the program of the present paper was carried out for the special case of the heat equation, it was noticed that the finite difference operator $F_h(t)$ could then be written for $t = nk$ where $k = \text{const}.h^2$ as $F_h(nk) = F_k^n$ where F_k is a generalized finite difference operator with coefficients independent of h. This made it possible to apply existing theory for such finite difference operators. Since this theory also contains results in the maximum norm it was then possible to obtain maximum norm convergence results analogous to the L_2 estimates above. In this analysis the ordinary differential equations associated with the Galerkin equations were assumed to be solved exactly and the finite difference operator F_k based on a time step k is an auxiliary construction of the proof. We also showed in [12], however, that it is possible to first discretize the differential equation in time and to then apply Galerkin's method to generate genuine two-level finite difference approximations F_k matching the accuracy attained in the space variable.

It is clear that also in the more general situation described in this paper the representation of $F_h(t)$ for $t = nk$ as F_k^n is possible where now $k = \text{const}.h^q$, and that results from finite difference theory may again be utilized. In the hyperbolic case, in particular, the estimates in [2] can be applied to yield maximum norm estimates. We shall not here carry out the details of this nor discuss the possibility of discretization in time.

Throughout this paper C and c denote positive constants not necessarily the same at different occurrences.

2. Spline interpolation

We shall first state some well-known facts about Toeplitz operators on sequences. For this purpose consider sequences $y = (y_j)_{j=-\infty}^{\infty}$ in

$$\mathscr{S} = \{y; y_j = O(|j|^{-q}) \quad \text{as} \quad |j| \to \infty \quad \text{for all real} \quad q\},$$

$$\mathscr{S}' = \{y; y_j = O(|y|^{q}) \quad \text{as} \quad |j| \to \infty \quad \text{for some real} \quad q\}.$$

Let now for $\tau = (\tau_j) \in \mathscr{S}$ the Toeplitz operator T be defined by $z = Ty$ where

$$z_j = \sum_l \tau_{j-l} y_l. \tag{2.1}$$

It is easy to see that T maps $\mathscr{S}$ and $\mathscr{S}'$ into themselves.

For $y \in \mathscr{S}$ let $\mathscr{F}y = \tilde{y}$ denote its (discrete) Fourier transform,

$$(\mathscr{F}y)(\vartheta) = \tilde{y}(\vartheta) = \sum_{j \in Z} y_j e^{-ij\vartheta}.$$

In particular,

$$\tilde{\tau}(\vartheta) = \sum_j \tau_j e^{-ij\vartheta}$$

is the characteristic function of T. Notice that $\mathscr{F}\mathscr{S}$ is the set $C_{2\pi}^{\infty}$ of 2π-periodic functions in C^{∞} and that for $y \in \mathscr{S}$,

$$\mathscr{F}(Ty) = \tilde{\tau}\mathscr{F}y.$$

About the invertibility of T (or the solvability of (2.1)) we have the following well-known:

Lemma 2.1. Assume that $\tau(\vartheta) \neq 0$ for all real ϑ. Then T^{-1} exists on $\mathcal{S}$ and $\mathcal{S}'$ and is the Toeplitz operator defined by

$$(T^{-1}z)_j = \sum_l \gamma_{j-l} z_l, \text{ where } \frac{1}{\tilde{\tau}(\vartheta)} = \sum_j \gamma_j e^{-ij\vartheta}.$$

Let now as in the introduction χ be the characteristic function of $[-\frac{1}{2}, \frac{1}{2}]$, set $\varphi = \chi^{*\mu}$ and $\varphi_j(x) = \varphi(h^{-1}x-j)$ and consider the splines of order μ of at most power growth,

$$\mathcal{S}_h = \{v = \sum_{j \in Z} c_j \varphi_j;\ c = (c_j) \in \mathcal{S}'\}.$$

We associate with $\mathcal{S}_h$ the trigonometric polynomial

(2.2) $$g_\mu(\vartheta) = \sum_l \varphi(l) e^{-il\vartheta}.$$

Since φ is even and has its support in $[-\frac{1}{2}\mu, \frac{1}{2}\mu]$, g_μ is an even trigonometric polynomal of order $[\frac{1}{2}(\mu-1)]$. The polynomials g_μ can easily be determined recursively for $\mu = 1,2,\ldots$ (cf. [12]). We have:

Lemma 2.2. The trigonometric polynomial $g_\mu(\vartheta)$ is positive for $\mu \geq 1$. For $\mu \geq 2$ we have

$$g_\mu(\vartheta) = \sum_l \hat{\chi}(\vartheta+2\pi l)^\mu \quad \text{where} \quad \hat{\chi}(\vartheta) = \frac{2 \sin \frac{1}{2}\vartheta}{\vartheta}.$$

Proof. See [12].

Consider now functions v in the set C_* of continuous functions of at most power growth,

$$C_* = \{v \in C;\ v(x) = O(|x|^q) \text{ as } |x| \to \infty \text{ for some real } q\}.$$

Lemma 2.3. For $v \in C_*$ there is a unique $v_h = S_h v \in \mathcal{S}_h$ such that

(2.3) $$v_h(lh) = v(lh),\ l \in Z.$$

The coefficients (c_j) in $v_h = \sum_j c_j \varphi_j$ may be determined from

$$g_\mu(\vartheta)\tilde{c}(\vartheta) = \tilde{v}(\vartheta),$$

where $\tilde{c} = \mathcal{F}(c_j), \tilde{v} = \mathcal{F}(v(jh))$ and g_μ is defined by (2.2).

Proof. The equations (2.3) reduce to

$$\sum_j c_j \varphi(l-j) = v(lh),\ l \in Z,$$

and the result therefore follows by Lemmas 2.1 and 2.2.

Notice that without the requirement that $c \in \mathcal{S}'$ in the definition of $\mathcal{S}_h$ we would not have uniqueness (cf. [9]).

The linear operator $S_h: C_* \to \mathcal{S}_h$ defined in Lemma 2.3 is referred to as the spline interpolation operator. We shall need some estimates for this operator. We introduce the discrete l_2-norm on hZ,

(2.4) $$\|v\|_{l_{2,h}} = (h \sum_{j \in Z} |v(hj)|^2)^{1/2}.$$

We have:

Lemma 2.4. The spline interpolation operator S_h has the following properties:

(i) There is a constant C such that

$$\|S_h v\|_{L_2} \leq C\|v\|_{l_{2,h}} .$$

(ii) For $\frac{1}{2} < s \leq \mu$ there is a constant C such that

$$\|(S_h - I)v\|_{L_2} \leq C h^s \|v\|_{W_2^s}.$$

Proof. See [8], [1].

3. The Galerkin method

We now return to the continuous time Galerkin equations associated with the equation (1.1),

$$(3.1)\qquad \left(\frac{\partial u_h}{\partial t},\varphi_l\right) - a(-1)^m(D^{q-m}u_h,D^m\varphi_l) = 0,\ l\in Z,$$

where $m = \left[\frac{1}{2}q\right]$. Setting

$$u_h(x,t) = \sum_j c_j(t)\varphi_j(x),$$

the equations (3.1) may be written

$$(3.2)\qquad \sum_j \{c_j'(t)(\varphi_j,\varphi_l) - a(-1)^m c_j(t)(D^{q-m}\varphi_j,D^m\varphi_l)\} = 0,\ l\in Z.$$

We notice that (φ_j,φ_l) and $(D^{q-m}\varphi_j,D^m\varphi_l)$ only depend on $l-j$ so that this system can be considered as a convolution equation. We introduce for $\sigma \le 2\mu-2$ with $\nu = [\sigma/2]$ the trigonometric polynomial,

$$(3.3)\qquad g_{\mu,\sigma}(\theta) = h^{\sigma-1}(-i)^{\sigma-2\nu}\sum_l (D^{\sigma-\nu}\varphi_0,D^\nu\varphi_l)e^{-il\theta}.$$

The factor $h^{\sigma-1}$ makes $g_{\mu,\sigma}$ independent of h and we shall prove the following:

Lemma 3.1. We have for $\sigma \le 2\mu-2$,

$$g_{\mu,\sigma}(\theta) = \sum_l (\theta+2\pi l)^\sigma \hat{\chi}(\theta+2\pi l)^{2\mu} = \theta^\sigma\hat{\chi}(\theta)^{2\mu} + R_{\mu,\sigma}(\theta),$$

where as $\theta \to 0$,

$$R_{\mu,\sigma}(\theta) = \begin{cases} O(\theta^{2\mu}) & \text{for } \sigma \text{ even},\\ O(\theta^{2\mu+1}) & \text{for } \sigma \text{ odd}.\end{cases}$$

For σ even, $g_{\mu,\sigma}(\theta)$ is positive for $0 < |\theta| \le \pi$.

Proof. Since $g_{\mu,\sigma}$ is independent of h we may choose $h=1$. We then have by Parseval's relation

$$(D^{\sigma-\nu}\varphi_0, D^{\nu}\varphi_1) = \int D^{\sigma-\nu}\varphi(x+l)D^{\nu}\varphi(x)dx$$

$$= \frac{1}{2\pi}\int (i\xi)^{\sigma-\nu}\hat{\varphi}(\xi)e^{il\xi}\overline{(i\xi)^{\nu}\hat{\varphi}(\xi)}d\xi$$

$$= i^{\sigma-2\nu}\frac{1}{2\pi}\int \xi^{\sigma}\hat{\varphi}(\xi)^2 e^{il\xi}d\xi = i^{\sigma-2\nu}\frac{1}{2\pi}\mathcal{F}(\xi^{\sigma}\hat{\varphi}(\xi)^2)(-l),$$

and hence

$$(-i)^{\sigma-2\nu}(D^{\sigma-\nu}\varphi_0, D^{\nu}\varphi_1)e^{-il\vartheta} = \frac{1}{2\pi}\mathcal{F}_{\xi}((\xi+\vartheta)^{\sigma}\hat{\varphi}(\xi+\vartheta)^2)(-l)$$

$$= \mathcal{F}_{\xi}^{-1}((\xi+\vartheta)^{\sigma}\hat{\varphi}(\xi+\vartheta)^2)(l),$$

so that by Poisson's summation formula,

$$g_{\mu,\sigma}(\vartheta) = \sum_{l} (\vartheta+2\pi l)^{\sigma}\hat{\varphi}(\vartheta+2\pi l)^2,$$

which proves the first part of the lemma. We find at once

$$R_{\mu,\sigma}(\vartheta) = (2\sin\tfrac{1}{2}\vartheta)^{2\mu}\sum_{l\neq 0}(\vartheta+2\pi l)^{-(2\mu-\sigma)} = O(\vartheta^{2\mu}) \quad \text{as} \quad \vartheta\to 0.$$

When σ is odd the function $R_{\mu,\sigma}$ is in addition an odd function which proves the second part about $R_{\mu,\sigma}$. For σ even, $R_{\mu,\sigma}$ is non-negative which concludes the proof of the lemma.

Notice that by Lemma 2.2 we have $g_{\mu,0} = g_{2\mu}$.

Assuming that $(c_j(t))$ and $(c_j'(t))$ are in ℓ we may introduce their Fourier transforms $\tilde{c}(\vartheta,t)$ and $\frac{d}{dt}\tilde{c}(\vartheta,t)$, and conclude that with the notation (3.3) the equations (3.2) imply

$$hg_{\mu,0}(\vartheta)\frac{d}{dt}\tilde{c}(\vartheta,t) - ai^q h^{-(q-1)}g_{\mu,q}(\vartheta)\tilde{c}(\vartheta,t) = 0.$$

This ordinary differential equation with respect to t may be solved to yield

$$\tilde{c}(\theta,t) = \exp\left(\frac{t}{h^q}\, ai^q \frac{g_{\mu,q}(\theta)}{g_{\mu,0}(\theta)}\right) \tilde{c}(\theta,0).$$

It follows that

$$c_j(t) = \sum_l f_l(t/h^q) c_{j-l}(0),$$

where $f_l(t/h^q)$ are the Fourier coefficients of the exponential,

$$(3.4) \qquad \exp\left(\frac{t}{h^q}\, ai^q \frac{g_{\mu,q}(\theta)}{g_{\mu,0}(\theta)}\right) = \sum_l f_l(t/h^q) e^{-il\theta}.$$

By Lemma 2.3 this is equivalent to

$$(3.5) \qquad u_h(jh,t) = \sum_l f_l(t/h^q) v(jh-lh).$$

Since the exponential in (3.4) belongs to $C_{2\pi}^{\infty}$ we have $(f_l(t/h^q)) \in \mathcal{S}$ so that (3.5) defines $u_h(\cdot,t)$ not only for $(v(jh)) \in \mathcal{S}$ but more generally for $(v(jh)) \in \mathcal{S}'$. In particular, for $(v(jh)) \in l_2$ we let $u_h(\cdot,t) = G_h(t)v \in \mathcal{S}_h$ be the spline defined at the mesh-points by (3.5).

Let us introduce the finite difference operator $F_h(t)$ defined for all $x \in R$ by

$$(3.6) \qquad F_h(t)v(x) = \sum_l f_l(t/h^q) v(x-lh).$$

Setting

$$(3.7) \qquad p_h(\xi) = h^{-q} a i^q \frac{g_{\mu,q}(h\xi)}{g_{\mu,0}(h\xi)},$$

we have by (3.4) for the Fourier transform of $F_h(t)v$,

$$\mathcal{F}(F_h(t)v)(\xi) = \exp(tp_h(\xi))\hat{v}(\xi).$$

From (3.7) and Lemma 3.1 we obtain at once by considering the cases q even and q odd separately:

Lemma 3.2. If the initial-value problem (1.1), (1.2) is correctly posed in Petrovskiĭ's sense we have

$$\operatorname{Re} P_h(\xi) \leq 0 \quad \text{for} \quad \xi \in R.$$

If the equation (1.1) is parabolic there is a positive c such that

$$\operatorname{Re} P_h(\xi) \leq -c\xi^q \quad \text{for} \quad |h\xi| \leq \pi.$$

Since $F_h(t)v$ agrees with $G_h(t)v$ on hZ we have the following representation result.

Lemma 3.3. The solution operator $G_h(t)$ corresponding to the Galerkin equations (3.1) can be represented in the form $G_h(t) = S_h F_h(t)$ where S_h is the spline interpolation operator and $F_h(t)$ is the finite difference operator defined in (3.6).

4. Convergence estimates

With the notation of last section and with $P(\xi) = a(i\xi)^q$ we have the following consistency estimate.

Lemma 4.1. We have for $|h\xi| \leq \pi$,

$$|P_h(\xi) - P(\xi)| \leq \begin{cases} Ch^{2\mu-q}|\xi|^{2\mu} & \text{if } q \text{ even,} \\ Ch^{2\mu-q+1}|\xi|^{2\mu+1} & \text{if } q \text{ odd.} \end{cases}$$

Proof. By Lemma 3.1 we have for q even,

$$h^q P_h(h^{-1}\xi) = ai^q \frac{g_{\mu,q}(\xi)}{g_{\mu,0}(\xi)} = ai^q \frac{\xi^q \hat{\chi}(\xi)^{2\mu} + O(\xi^{2\mu})}{\hat{\chi}(\xi)^{2\mu} + O(\xi^{2\mu})}$$

$$= a(i\xi)^q + O(\xi^{2\mu}) \quad \text{as} \quad \xi \to 0,$$

and hence obviously

$$|h^q P_h(h^{-1}\xi) - P(\xi)| \leq C|\xi|^{2\mu} \quad \text{for} \quad |\xi| \leq \pi,$$

which proves the result for q even. The proof for odd q is analogous.

This result will be used in the following two lemmas on estimates for the solution operators on the Fourier transform side.

Lemma 4.2. Let the initial-value problem (1.1), (1.2) be correctly posed in Petrovskiĭ's sense and let $0 \leq s \leq 2\mu$ if q is even, $0 \leq s \leq 2\mu+1$ if q is odd. Then for $|h\xi| \leq \pi$, $0 \leq t \leq T$,

$$|\exp(tP_h(\xi)) - \exp(tP(\xi))| \leq \begin{cases} Ch^{\frac{2\mu-q}{2\mu}s}|\xi|^s & \text{if } q \text{ even} \\ Ch^{\frac{2\mu-q+1}{2\mu+1}s}|\xi|^s & \text{if } q \text{ odd.} \end{cases}$$

Proof. We have

$$\text{(4.1)} \qquad \exp(tP_h(\xi)) - \exp(tP(\xi))$$

$$= t(P_h(\xi) - P(\xi)) \int_0^1 \exp(stP_h(\xi) + (1-s)tP(\xi))ds$$

from which the results follow at once by (1.3) and Lemmas 3.2 and 4.1 for $s = 2\mu$ and $s = 2\mu+1$, respectively. Since also by (1.3) and Lemma 3.2,

$$|\exp(tP_h(\xi)) - \exp(tP(\xi))| \leq 2,$$

the result holds for general s.

Lemma 4.3. Let the equation (1.1) be parabolic and let $0 \leq s \leq 2\mu - q$. Then for $|h\xi| \leq \pi$, $t \geq 0$,

$$|\exp(tP_h(\xi)) - \exp(tP(\xi))| \leq Ch^s |\xi|^s.$$

Proof. In this case we obtain by Lemma 3.2,

$$|\exp(stP_h(\xi) + (1-s)tP(\xi))| \leq \exp(-ct\xi^q), \quad |h\xi| \leq \pi,$$

so that by (4.1) and Lemma 4.1,

$$|\exp(tP_h(\xi)) - \exp(tP(\xi))| \leq Cth^{2\mu-q}|\xi|^{2\mu}\exp(-ct\xi^q)$$

$$\leq C(h|\xi|)^{2\mu-q} \leq C(h|\xi|)^s,$$

which proves the result.

We shall need the following estimate for the discrete l_2-norm defined by (2.4):

Lemma 4.4. Let $s > \frac{1}{2}$. Then there is a constant C such that $w \in W_2^s$,

$$\|w\|^2_{l_{2,h}} \leq C\{\int_{-\pi/h}^{\pi/h} |\hat{w}(\xi)|^2 d\xi + h^{2s} \int_{-\infty}^{\infty} |\xi|^{2s} |\hat{w}(\xi)|^2 d\xi\}.$$

Proof. We have by Parseval's relation,

$$\|w\|^2_{l_{2,h}} = \frac{h}{2\pi} \int_{-\pi}^{\pi} |\tilde{w}_h(\xi)|^2 d\xi,$$

where

$$\tilde{w}_h(\xi) = \sum_j w(jh)e^{-ij\xi}.$$

Poisson's summation formula gives

$$\tilde{w}_h(\xi) = h^{-1} \sum_j \hat{w}(h^{-1}(\xi+2\pi j)),$$

and hence for $|\xi| \leq \pi$,

$$h^2|\tilde{w}_h(\xi)|^2 \leq 2|\hat{w}(\tfrac{\xi}{h})|^2 + 2|\sum_{j\neq 0} \hat{w}(\tfrac{\xi+2\pi j}{h})|^2$$

$$\leq 2|\hat{w}(\tfrac{\xi}{h})|^2 + 2\sum_{j\neq 0} |\xi+2\pi j|^{-2s} \sum_j |\xi+2\pi j|^{2s}|\hat{w}(\tfrac{\xi+2\pi j}{h})|^2$$

$$\leq C\{|\hat{w}(\tfrac{\xi}{h})|^2 + \sum_j |\xi+2\pi j|^{2s}|\hat{w}(\tfrac{\xi+2\pi j}{h})|^2\}.$$

Hence

$$\|w\|^2_{l_{2,h}} \leq Ch^{-1}\{\int_{-\pi}^{\pi} |\hat{w}(\tfrac{\xi}{h})|^2 d\xi + \sum_j \int_{-\pi}^{\pi} |\xi+2\pi j|^{2s} |\hat{w}(\tfrac{\xi+2\pi j}{h})|^2 d\xi\},$$

from which the result immediately follows.

We shall now state and prove the convergence estimates. We shall express these estimates in terms of the regularity of the initial data; in the parabolic case the smoothing property of the solution operator reduces the regularity assumptions compared to the general case of a correctly posed initial-value problem. We first give the results at the mesh-points.

Theorem 4.1. Let the initial-value problem (1.1), (1.2) be correctly posed in Petrovskiĭ's sense and let $\frac{1}{2} < s \leq 2\mu$ if q even, $\frac{1}{2} < s \leq 2\mu+1$ if q is odd. Then there is a C such that for $v \in W_2^s$, $0 \leq t \leq T$,

$$\|F_h(t)v - E(t)v\|_{l_{2,h}} \leq \begin{cases} Ch^{\frac{2\mu-q}{2\mu}s}\|v\|_{W_2^s} & \text{if } q \text{ even,} \\ Ch^{\frac{2\mu-q+1}{2\mu+1}s}\|v\|_{W_2^s} & \text{if } q \text{ odd.} \end{cases}$$

Proof. By Lemmas 4.4 and 4.2 we have for q even,

$$\|F_h(t)v - E(t)v\|^2_{l_{2,h}}$$

$$\leq C\{\int_{-\frac{\pi}{h}}^{\frac{\pi}{h}} |\exp(tP_h(\xi)) - \exp(tP(\xi))|^2|\hat{v}(\xi)|^2d\xi + h^{2s}\|v\|^2_{W_2^s}\}$$

$$\leq C\{\int_{-\frac{\pi}{h}}^{\frac{\pi}{h}} h^{\frac{2\mu-q}{2\mu}2s}|\xi|^{2s}|\hat{v}(\xi)|^2d\xi + h^{2s}\|v\|^2_{W_2^s}\},$$

from which the result follows at once. Odd q are treated analogously.

Theorem 4.2. Let the equation (1.1) be parabolic and let $\frac{1}{2} < s \leq 2\mu-q$. Then there is a C such that for $v \in W_2^s$, $t \geq 0$,

$$\|F_h(t)v - E(t)v\|_{l_{2,h}} \leq Ch^s \|v\|_{W_2^s}.$$

Proof. Follows analogously by Lemma 4.3.

It is now easy to obtain the global error estimates.

Theorem 4.3. Let the initial-value problem (1.1), (1.2) be correctly posed in Petrovskiǐ's sense and $s > \frac{1}{2}$. Then there is a C such that for $v \in W_2^s$, $0 \leq t \leq T$,

$$\|G_h(t)v - E(t)v\|_{L_2} \leq \begin{cases} Ch^{\min(\frac{2\mu-q}{2\mu}s,\ 2\mu-q,\ \mu)} \|v\|_{W_2^s}, & \text{if } q \text{ even}, \\ Ch^{\min(\frac{2\mu-q+1}{2\mu+1}s,\ 2\mu-q+1,\ \mu)} \|v\|_{W_2^s}, & \text{if } q \text{ odd}. \end{cases}$$

Theorem 4.4. Let the equation (1.1) be parabolic and $s > \frac{1}{2}$. Then there is a C such that for $v \in W_2^s$, $t \geq 0$,

$$\|G_h(t)v - E(t)v\|_{L_2} \leq Ch^{\min(s,2\mu-q,\mu)} \|v\|_{W_2^s}.$$

Proofs. By Lemmas 3.3 and 2.4 we have

$$\|G_h(t)v - E(t)v\|_{L_2} \leq \|S_h(F_h(t) - E(t))v\|_{L_2} + \|(S_h - I)E(t)v\|_{L_2}$$

$$\leq C\|(F_h(t) - E(t))v\|_{l_{2,h}} + Ch^{\min(s,\mu)}\|E(t)v\|_{W_2^s}$$

$$\leq C\|(F_h(t) - E(t))v\|_{l_{2,h}} + Ch^{\min(s,\mu)}\|v\|_{W_2^s}.$$

The results now follow from Theorems 4.1 and 4.2.

References

[1] J.H. Bramble and S.R. Hilbert, Estimation of linear functionals on Sobolev spaces with application to Fourier transforms and spline interpolation. SIAM J. Numer. Anal. 7(1970), 112-124.

[2] Ph. Brenner and V. Thomée, Stability and convergence rates in L_p for certain difference schemes. Math. Scand. 27(1970), 5-23.

[3] J. Douglas, Jr. and T. Dupont, Galerkin methods for parabolic equations. SIAM J. Numer. Anal. 7(1970), 575-626.

[4] T. Dupont, Galerkin methods for first order hyperbolics: an example. SIAM J. Numer. Anal. (to appear).

[5] G. Fix and N. Nassif, On finite element approximations to time dependent problems. Numer. Math. 19(1972), 127-135.

[6] H.S. Price and R.S. Varga, Error bounds for semi-discrete Galerkin approximations of parabolic problems with application to petroleum reservoir mechanics. Numerical Solution of Field Problems in Continuum Physics. AMS Providence R.I., 1970, 74-94.

[7] I.J. Schoenberg, Contributions to the problem of approximation of equidistant data by analytic functions, A and B. Quart. Appl. Math. 4(1946), 45-99, 112-141.

[8] I.J. Schoenberg, Cardinal interpolation and spline functions. J. Approximation Theory 2(1969), 335-374.

[9] I.J. Schoenberg, Cardinal interpolation and spline functions II. Interpolation of data of power growth. J. Approximation Theory. (to appear).

[10] G. Strang and G. Fix, A Fourier analysis of the finite element variational method. Mimeographed notes.

[11] B. Swartz and B. Wendroff, Generalized finite difference schemes, Math. Comp. 23(1969), 37-50.

[12] V. Thomée, Spline approximation and difference schemes for the heat equation. (to appear).

DIFFERENCE APPROXIMATIONS FOR SOME FUNCTIONAL DIFFERENTIAL EQUATIONS

By Robert J. Thompson

1. Introduction

The classic Lax-Richtmyer theory of finite difference approximations [4] treats equations of the form

$$\frac{du}{dt} = Au \quad 0 \leq t \leq T$$
$$u(0) = u_0 \,. \tag{1}$$

Here A is a constant linear operator mapping part of a Banach space $\mathcal{B}$ into $\mathcal{B}$. Although the theory is set in an abstract space, its importance derives from the fact that many important problems in mathematical physics are special cases of Eq. 1. Sometimes a more accurate mathematical model is obtained when a term which depends on past values of the solution is added to the differential equation. The equation then becomes a functional differential equation. Such equations arise in control theory, for example, and in the study of materials with memory.

In a recent paper Cryer and Tavernini [2] analyze several methods for the numerical solution of systems of ordinary functional differential equations. The paper also contains an extensive list of references to other works dealing with the numerical solution of functional differential equations. In [7] the author showed how to get convergent difference approximations for the particular functional equation which is obtained when the term $\int_{t-\tau}^{t} u(s)ds$ is added to the right side of Eq. 1. In what follows much more general functionals are considered. In particular, difference approximations are obtained for functional differential equations of the form:

This author was supported by the U. S. Atomic Energy Commission.

$$\frac{du}{dt} = Au + F(t,u) \qquad 0 \le t \le T$$
$$u(t) = \varphi(t) \qquad -\tau \le t \le 0 \; . \tag{2}$$

Here F denotes a functional which depends on t and the values of u on the interval $[-\tau,t]$. Since the differential equation is to be satisfied at $t = 0$, the values of u must be specified on the "initial interval" $[-\tau,0]$. That is, Eq. 2 is an initial function problem. The initial function φ is regarded as given, just as the initial value u_0 is regarded as given in Eq. 1.

In Sec. 2 an existence and uniqueness theorem is proved for Eq. 2. The theorem is a slight generalization of a theorem proved by the author in [7]. Sec. 3 deals with difference approximations for Eq. 2. It is shown that whenever the functional F can be approximated in an appropriate manner, then a convergent difference approximation for Eq. 1 can be modified to obtain a convergent difference approximation for Eq. 2.

2. Existence of Solutions

First some notation: Throughout the remainder of the paper $\mathcal{B}$ will denote a Banach space. T and τ are fixed positive numbers. If a and b are real numbers and $a < b$, $C[a,b]$ denotes the space of strongly continuous functions which map $[a,b]$ into $\mathcal{B}$. For a function u in $C[-\tau,T]$ and any t in $[0,T]$, $\|u\|_t$ denotes $\sup_{-\tau\le s\le t} \|u(s)\|$.

The equations that Lax and Richtmyer considered give rise to solution operators which form semigroups. Such semigroups provide the foundation for the present work. Suppose

> $\{E(t)\}$, $t \ge 0$, is a strongly continuous one-parameter semigroup of bounded linear operators from $\mathcal{B}$ into $\mathcal{B}$ and $E(0)$ is the identity. (3)

Let A be the infinitesimal generator of $\{E(t)\}$. Then Eq. 1 is what Lax and Richtmyer call a properly posed initial value problem, and for each u_0 in $\mathcal{B}$ the function defined by $E(t)u_0$, $0 \le t \le T$, is the generalized solution of Eq. 1. Also, there is a positive number K such that $\|E(t)\| \leq K$ for $0 \leq t \leq T$. (The definitions of these concepts and the proofs of these statements can be found in Chap. 3 of the book of Richtmyer and Morton [5].)

The functional F which appears in Eq. 2 is assumed to satisfy some continuity conditions. In particular:

F is a mapping from $[0,T] \times C[-\tau,T]$ into $\mathcal{B}$ for which

(i) There exists a positive number L such that $\|F(t,v)-F(t,w)\| \leq L\|v-w\|_t$ for $0 \leq t \leq T$ and v,w in $C[-\tau,T]$. (4)

(ii) For each v in $C[-\tau,T]$ the mapping from $[0,T]$ into $\mathcal{B}$ defined by $F(t,v)$ is in $C[0,T]$.

<u>Definition 1</u> A solution of Eq. 2 is a function u mapping $[-\tau,T]$ into $\mathcal{B}$ for which

(i) $u(t) = \varphi(t)$ for $-\tau \leq t \leq 0$, and

(ii) The strong derivative du/dt exists and equals $Au(t) + F(t,u)$ for $0 \leq t \leq T$.

As in the case of Eq. 1 it is often too much to ask for a solution of Eq. 2. For one thing, a solution must stay in the domain of A. On the other hand, as with Eq. 1, it will be shown that a "generalized solution" exists.

Before defining generalized solutions it is necessary to make several remarks about integrals in a Banach space: If $a < b$ and g is a function in $C[a,b]$, then $\int_a^b g(s)ds$ exists as a limit of Riemann sums (see Chap. I, Sec. 1.7 of Kato's book [3], for example). In a previous paper of the authors [7] it was noted that the following result can easily be proved:

<u>Lemma 1</u> Suppose $\{E(t)\}$, $t \geq 0$, satisfies (3) and suppose g is a function in $C[0,T]$. Then for each t in $[0,T]$ the function from $[0,t]$ into $\mathcal{B}$ defined by $E(t-s)g(s)$, $0 \leq s \leq t$, is in $C[0,t]$. (Note that

by the remarks in the preceding paragraph it follows that $\int_0^t E(t-s)g(s)ds$ exists.) In addition, the function from $[0,T]$ into $\mathcal{B}$ defined by $\int_0^t E(t-s)g(s)ds$ is in $C[0,T]$.

Now suppose u is a function in $C[-\tau,T]$. If $\{E(t)\}$, $t \geq 0$, satisfies (3) and F satisfies (4), then it follows from Lemma 1 that $\int_0^t E(t-s)F(s,u)ds$ exists and is a strongly continuous function of t. This permits the following definition:

Definition 2 A generalized solution of Eq. 2 is a function u in $C[-\tau,T]$ for which

(i) $u(t) = \varphi(t)$ for $-\tau \leq t \leq 0$, and

(ii) $u(t) = E(t)\varphi(0) + \int_0^t E(t-s)F(s,u)ds$ for $0 \leq t \leq T$.

The use of the term generalized solution is justified by this theorem:

Theorem 1 Suppose $\{E(t)\}$, $t \geq 0$, satisfies (3) and suppose F satisfies the conditions (4). Let A be the infinitesimal generator of $\{E(t)\}$ and suppose φ is a function in $C[-\tau,0]$. Then if a solution of Eq. 2 exists it must also be a generalized solution.

Proof Let u be a solution of Eq. 2. Then u is in $C[-\tau,T]$ and it need only be shown that u satisfies the integral equation $u(t) = E(t)\varphi(0) + \int_0^t E(t-s)F(s,u)ds$ for $0 \leq t \leq T$. For t in $[0,T]$ let $f(t) = F(t,u)$. By (ii) in (4) the function f is in $C[0,T]$. In addition, on $[0,T]$ u satisfies the differential equation $du/dt = Au + f(t)$, $u(0) = \varphi(0)$. By Theorem 3.2 in [6] it follows that $u(t) = E(t)\varphi(0) + \int_0^t E(t-s)f(s)ds$, $0 \leq t \leq T$. That is, $u(t) = E(t)\varphi(0) + \int_0^t E(t-s)F(s,u)ds$, $0 \leq t \leq T$.

Theorem 2 Suppose $\{E(t)\}$, $t \geq 0$, satisfies (3) and suppose F satisfies the conditions (4). Let A be the infinitesimal generator of $\{E(t)\}$ and suppose φ is a function in $C[-\tau,0]$. Then there exists a unique generalized solution of Eq. 2.

Proof $C[-\tau,T]$ with norm $\|v\|_T$ is a Banach space. It follows from the conditions (4) and Lemma 1 that for each function v in $C[-\tau,T]$ the function $\int_0^t E(t-s)F(s,v)ds$ is defined and continuous. So a transformation $\mathfrak{J}$ from $C[-\tau,T]$ into itself is defined as follows:

$$\mathfrak{J}[v](t) = \begin{cases} \varphi(t) & -\tau \leq t \leq 0 \\ E(t)\varphi(0) + \int_0^t E(t-s)F(s,v)ds & 0 \leq t \leq T \end{cases}$$

(Here $\mathfrak{J}[v]$ denotes the image of v under $\mathfrak{J}$ and $\mathfrak{J}[v](t)$ is the value of $\mathfrak{J}[v]$ at t.) The proof will be completed by showing that $\mathfrak{J}$ has a unique fixed point. Now the transformations $\mathfrak{J}^n$, n = 1,2,... are also defined and map $C[-\tau,T]$ into itself. Let v and w be in $C[-\tau,T]$. By induction it will be shown that

$$\|\mathfrak{J}^n[v](t)-\mathfrak{J}^n[w](t)\| \leq \begin{cases} 0 & -\tau \leq t \leq 0 \\ \frac{(KLt)^n}{n!}\|v-w\|_T & 0 \leq t \leq T \end{cases}$$

Here K is the bound on the operators E(t) and L is the Lipschitz constant from (4).

The inequality is trivial for $-\tau \leq t \leq 0$. For $0 \leq t \leq T$, $\|\mathfrak{J}[v](t)-\mathfrak{J}[w](t)\| = \|\int_0^t E(t-s)[F(s,v)-F(s,w)]ds\|$. The integrand is less than or equal to $K\|F(s,v)-F(s,w)\| \leq KL\|v-w\|_s \leq KL\|v-w\|_T$, and so $\|\mathfrak{J}[v](t)-\mathfrak{J}[w](t)\| \leq KLt\|v-w\|_T$. So the inequality holds for n = 1. Suppose then that it holds for n = 1,2,...,k. $\|\mathfrak{J}^{k+1}[v](t)-\mathfrak{J}^{k+1}[w](t)\| \leq \int_0^t K\|F(s,\mathfrak{J}^k[v])-F(s,\mathfrak{J}^k[w])\|ds$. The integrand is less than or equal to $KL\|\mathfrak{J}^k[v]-\mathfrak{J}^k[w]\|_s = KL \sup_{-\tau\leq r\leq s}\|\mathfrak{J}^k[v](r)-\mathfrak{J}^k[w](r)\|$. For $-\tau \leq r \leq 0$, $\|\mathfrak{J}^k[v](r)-\mathfrak{J}^k[w](r)\| = 0 \leq [(KLs)^k/k!]\|v-w\|_T$. For $0 \leq r \leq s$, $\|\mathfrak{J}^k[v](r)-\mathfrak{J}^k[w](r)\| \leq [(KLr)^k/k!]\|v-w\|_T \leq [(KLs)^k/k!]\|v-w\|_T$. Thus $KL\|\mathfrak{J}^k[v]-\mathfrak{J}^k[w]\|_s \leq [KL(KLs)^k/k!]\|v-w\|_T$ and so $\|\mathfrak{J}^{k+1}[v](t)-\mathfrak{J}^{k+1}[w](t)\| \leq \int_0^t [KL(KLs)^k/k!]\|v-w\|_T ds = [(KLt)^{k+1}/(k+1)!]\|v-w\|_T$. This completes the induction proof. Now $\|\mathfrak{J}^n[v]-\mathfrak{J}^n[w]\|_T \leq [(KLT)^n/n!]\|v-w\|_T$, and so $\mathfrak{J}$ is eventually contracting. By Theorem 3 in Chu and Diaz [1], $\mathfrak{J}$ has a unique fixed point.

3. Convergent Difference Approximations

In this section it is shown how a convergent difference approximation for Eq. 1 can be modified to obtain a convergent difference approximation for Eq. 2. Suppose then that a convergent approximation for Eq. 1 is available. That is,

> For each $\Delta t > 0$ let $C(\Delta t)$ be a linear operator mapping $\mathcal{B}$ into itself, and suppose that the family of operators $C(\Delta t)$ provides a convergent difference approximation for Eq. 1. (5)

In Chap. 3 of Richtmyer and Morton [5] it is shown that (5) implies that the operators $C(\Delta t)$ are stable. That is, there exist positive numbers B and η such that if $0 < \Delta t < \eta$ and n is a nonnegative integer for which $0 \leq n\Delta t \leq T$, then $\|C^n(\Delta t)\| \leq B$.

Roughly speaking, a convergent difference approximation for Eq. 2 is a means for approximating generalized solutions at certain discrete values of t. In particular,

Definition 3 For each $\Delta t > 0$ and for each integer n such that $-\tau \leq n\Delta t \leq T$, let $G_n(\Delta t)$ be an operator mapping $C[-\tau,0]$ into $\mathcal{B}$. The operators $G_n(\Delta t)$ are a convergent difference approximation for $du/dt = Au + F(t,u)$ if the following is true: let φ be in $C[-\tau,0]$ and let u be the corresponding generalized solution. Given $\epsilon > 0$, there exists a $\delta > 0$ such that if $0 < \Delta t < \delta$, $-\tau \leq t \leq T$ and n is an integer for which $-\tau \leq n\Delta t \leq T$ and $|t-n\Delta t| < \delta$, then $\|G_n(\Delta t)\varphi - u(t)\| \leq \epsilon$.

In order to get a difference approximation for Eq. 2 it may be necessary to approximate the functional F. For example, if F involves integrating u one might approximate F with a Riemann sum. More generally,

> Suppose F satisfies the conditions (4). For each $\Delta t > 0$ let $F_{\Delta t}$ be a mapping from $[0,T] \times C[-\tau,T]$ into $\mathcal{B}$ and suppose

(i) There exists a positive number L^* such that $\|F_{\Delta t}(t,v)-F_{\Delta t}(t,w)\| \leq L^*\|v-w\|_t$ for all $\Delta t > 0$, $0 \leq t \leq T$ and v,w in $C[-\tau,T]$. (6)

(ii) For each v in $C[-\tau,T]$ it is true that given $\epsilon > 0$ there exists a $\delta > 0$ such that if $0 < \Delta t < \delta$ and $0 \leq t \leq T$ then $\|F(t,v)-F_{\Delta t}(t,v)\| \leq \epsilon$.

The intent of the conditions (6) is that the functionals $F_{\Delta t}$ are readily computable while F may not be. Of course, if F can be computed all the $F_{\Delta t}$ may be taken equal to F.

Notation Although Δt is not fixed, the following notation should not be confusing: t_n will sometimes be used to denote $n\Delta t$ and $\|u\|_{t_n}$ will be denoted by $\|u\|_n$.

A difference approximation for Eq. 2 can now be defined. It looks complicated, but in essence it is just this: to advance an approximate solution to the next time step, operate on it with $C(\Delta t)$ and then add Δt times $F_{\Delta t}$.

Definition 4 Suppose $\Delta t > 0$. For φ in $C[-\tau,0]$, let

$$G_n(\Delta t)\varphi = \varphi(t_n) \qquad -\tau \leq t_n \leq 0$$

$$h_0(t) = \begin{cases} \varphi(t) & -\tau \leq t \leq 0 \\ \varphi(0) & 0 \leq t \leq T \end{cases}$$

For $n \geq 1$

$$G_n(\Delta t)\varphi = C(\Delta t)G_{n-1}(\Delta t)\varphi + \Delta t F_{\Delta t}(t_{n-1},h_{n-1})$$

$$h_n(t) = \begin{cases} h_{n-1}(t) & -\tau \leq t \leq t_{n-1} \\ G_{n-1}(\Delta t)\varphi + \dfrac{t-t_{n-1}}{\Delta t}\left[G_n(\Delta t)\varphi-G_{n-1}(\Delta t)\varphi\right] & t_{n-1} \leq t \leq t_n \\ h_n(t_n) & t_n \leq t \leq T\,. \end{cases}$$

Note that the functions h_n are all defined on the entire interval $[-\tau,T]$, but this is just a technical convenience. On the interval

$[0,t_n]$, h_n is just the piecewise linear function which interpolates $G_0(\Delta t)\varphi, G_1(\Delta t)\varphi, \ldots,$ and $G_n(\Delta t)\varphi$. A little reflection on the definition should make it apparent that if the operators $C(\Delta t)$ and the functionals $F_{\Delta t}$ can be evaluated, then the $G_n(\Delta t)\varphi$ can be computed.

For $n \geq 1$ it is easily verified that $G_n(\Delta t)\varphi$ can be written in the equivalent form

$$G_n(\Delta t)\varphi = C^n(\Delta t)\varphi(0) + \Delta t \sum_{k=0}^{n-1} C^{n-k-1}(\Delta t)F_{\Delta t}(t_k,h_k)$$

The following lemma is also easily verified and the proof is omitted.

<u>Lemma 2</u> For $t_{n-1} \leq t \leq t_n$

$$\left.\begin{array}{l} \|h_n(t)-G_{n-1}(\Delta t)\varphi\| \\ \|h_n(t)-G_n(\Delta t)\varphi\| \end{array}\right\} \leq \|G_n(\Delta t)\varphi-G_{n-1}(\Delta t)\varphi\|$$

<u>Theorem 3</u> Suppose $\{E(t)\}$, $t \geq 0$, satisfies (3), and let A be the infinitesimal generator of $\{E(t)\}$. Suppose F is a functional which satisfies (4) and the functionals $F_{\Delta t}$, $\Delta t > 0$ satisfy (6). Suppose $C(\Delta t)$, $\Delta t > 0$, are a family of operators which satisfy (5). Then the operators $G_n(\Delta t)$ of Definition 4 are a convergent difference approximation for Eq. 2.

<u>Proof</u> Clearly only positive values of t and t_n need be considered. Let φ be a function in $C[-\tau,0]$ and let u denote the corresponding generalized solution of Eq. 2. (The existence of u follows from Theorem 2.) Let f denote the function defined by $f(t) = F(t,u)$, $0 \leq t \leq T$. By (4), f is in $C[0,T]$ and by Theorem 2, $u(t) = E(t)\varphi(0) + \int_0^t E(t-s)f(s)ds$, $0 \leq t \leq T$. By Definition 2.5 in Thompson [6], u restricted to the interval $[0,T]$ is the generalized solution of $du/dt = Au + f(t)$, $0 \leq t \leq T$, $u(0) = \varphi(0)$. By Theorem 2.1 in [6] the operators

$$G_n(f,\Delta t) = \begin{cases} \varphi(0) & n = 0 \\ C^n(\Delta t)\varphi(0) + \Delta t \sum_{k=0}^{n-1} C^{n-k-1}(\Delta t) f(t_k) & n \geq 1 \end{cases}$$

are a convergent approximation for u.

Now, suppose $\varepsilon > 0$ is given and let $\varepsilon' = \varepsilon/(e^{3BL^*T}-2/3)$. Since $G_n(f,\Delta t)$ are convergent, there exists a $\delta > 0$ such that if $0 < \Delta t < \delta$, $0 \leq t \leq T$, $0 \leq t_n \leq T$ and $|t-t_n| < \delta$, then $\|u(t) - G_n(f,\Delta t)\| \leq \varepsilon'/6$. In addition, choose δ small enough so that for $0 < \Delta t < \delta$,

$$\|u(t_{k+1})-u(t_k)\| \leq 2\varepsilon' \text{ for } k = 0,1,\ldots \text{ as long as } t_{k+1} \leq T$$

and

$$\|F(t,u)-F_{\Delta t}(t,u)\| \leq \varepsilon'/(6BT) \qquad \text{for } 0 \leq t \leq T .$$

For $n \geq 1$, $0 < \Delta t < \delta$, $0 \leq t \leq T$, $0 < t_n \leq T$ and $|t-t_n| < \delta$, $\|u(t)-G_n(\Delta t)\varphi\| \leq \|u(t)-G_n(f,\Delta t)\| + \|G_n(f,\Delta t)-G_n(\Delta t)\varphi\| \leq \varepsilon'/6 + \Delta tB \sum_{k=0}^{n-1} \|f(t_k)-F_{\Delta t}(t_k,h_k)\|$. Now $\|f(t_k)-F_{\Delta t}(t_k,h_k)\| = \|F(t_k,u)-F_{\Delta t}(t_k,h_k)\| \leq \|F(t_k,u)-F_{\Delta t}(t_k,u)\| + \|F_{\Delta t}(t_k,u)-F_{\Delta t}(t_k,h_k)\| \leq \varepsilon'/(6BT) + L^*\|u-h_k\|_k$. Thus $\|u(t)-G_n(\Delta t)\varphi\| \leq \varepsilon'/6 + \Delta tB \sum_{k=0}^{n-1} \left[\varepsilon'/(6BT) + L^*\|u-h_k\|_k\right] = \varepsilon'/6 + n\Delta tB\varepsilon'/(6BT) + BL^*\Delta t \sum_{k=0}^{n-1} \|u-h_k\|_k \leq \varepsilon'/3 + BL^*\Delta t \sum_{k=0}^{n-1} \|u-h_k\|_k$. That is,

$$\|u(t)-G_n(\Delta t)\varphi\| \leq \varepsilon'/3 + BL^*\Delta t \sum_{k=0}^{n-1} \|u-h_k\|_k \tag{7}$$

Next it is shown by induction that for $n = 1,2,\ldots$, and as long as $t_n \leq T$, that

$$BL^*\Delta t \sum_{k=0}^{n-1} \|u-h_k\|_k \leq \left[(1+3BL^*\Delta t)^{n-1} - 1\right]\varepsilon' \tag{8}$$

and

$$\|u-h_n\|_n \leq 3(1+3BL^*\Delta t)^{n-1}\varepsilon' \tag{9}$$

$\|u-h_0\|_0 = 0$, so (8) holds for $n = 1$. For $0 \leq s \leq t_1$, $\|u(s)-h_1(s)\| \leq$

$\|u(s)-G_1(\Delta t)\varphi\| + \|G_1(\Delta t)\varphi-h_1(s)\|$ which, using (7), Lemma 2 and the fact that $\|u-h_0\|_0 = 0$, is less than or equal to $\varepsilon'/3 + \|G_1(\Delta t)\varphi-G_0(\Delta t)\varphi\| \leq \varepsilon'/3 + \|G_1(\Delta t)\varphi-u(t_1)\| + \|u(t_1)-u(t_0)\| + \|u(t_0)-G_0(\Delta t)\varphi\|$. Now, again using (7) and the fact that $\|u-h_0\|_0$, $\|G_1(\Delta t)\varphi-u(t_1)\| \leq \varepsilon'/3$. By the choice of δ, $\|u(t_1)-u(t_0)\| \leq 2\varepsilon'$. $u(t_0) = G_0(\Delta t)\varphi$, so $\|u(t_0)-G_0(\Delta t)\varphi\| = 0 < \varepsilon'/3$. Putting all this together yields: For $0 \leq s \leq t_1$, $\|u(s)-h_1(s)\| \leq 3\varepsilon'$. Thus $\sup_{-\tau\leq s\leq t_1} \|u(s)-h_1(s)\| \leq 3\varepsilon'$. That is, $\|u-h_1\|_1 \leq 3\varepsilon'$ which shows that (9) holds for $n = 1$. Now, suppose (8) and (9) hold for $n = 1,\ldots,m-1$ $(m \geq 2)$. $BL^*\Delta t \sum_{k=0}^{m-1} \|u-h_k\|_k = BL^*\Delta t \sum_{k=1}^{m-1} \|u-h_k\|_k$ which, using the induction hypothesis on (9), is less than or equal to $BL^*\Delta t \sum_{k=1}^{m-1} 3(1+3BL^*\Delta t)^{k-1}\varepsilon' = \left[(1+3BL^*\Delta t)^{m-1}-1\right]\varepsilon'$. That is, (8) holds for $n = m$. For $t_{m-1} \leq s \leq t_m$, $\|u(s)-h_m(s)\| \leq \|u(s)-G_m(\Delta t)\varphi\| + \|G_m(\Delta t)\varphi-h_m(s)\|$ which, using (7) and Lemma 2 is less than or equal to $\varepsilon'/3 + BL^*\Delta t \sum_{k=0}^{m-1} \|u-h_k\|_k + \|G_m(\Delta t)\varphi-G_{m-1}(\Delta t)\varphi\|$. Now $\|G_m(\Delta t)\varphi-G_{m-1}(\Delta t)\varphi\| \leq \|G_m(\Delta t)\varphi-u(t_m)\| + \|u(t_m)-u(t_{m-1})\| + \|u(t_{m-1})-G_{m-1}(\Delta t)\varphi\|$ which, using (7) and by the choice of δ, is less than or equal to $\varepsilon'/3 + BL^*\Delta t \sum_{k=0}^{m-1} \|u-h_k\|_k + 2\varepsilon' + \varepsilon'/3 + BL^*\Delta t \sum_{k=0}^{m-2} \|u-h_k\|_k \leq 2\varepsilon'/3 + 2BL^*\Delta t \sum_{k=0}^{m-1} \|u-h_k\|_k + 2\varepsilon'$. Thus $\|u(s)-h_m(s)\| \leq 3\varepsilon' + 3BL^*\Delta t \sum_{k=0}^{m-1} \| -h_k\|_k$. But it has been shown that (8) holds for $n = m$ and so $\|u(s)-h_m(s)\| \leq 3\varepsilon' + 3\left[(1+3BL^*\Delta t)^{m-1}-1\right]\varepsilon' = 3(1+3BL^*\Delta t)^{m-1}\varepsilon'$. Thus $\sup_{t_{m-1}\leq s\leq t_m} \|u(s)-h_m(s)\| \leq 3(1+3BL^*\Delta t)^{m-1}\varepsilon'$. Now $h_{m-1} = h_m$ on $[-\tau,t_{m-1}]$ so $\|u-h_m\|_{m-1} = \|u-h_{m-1}\|_{m-1}$ which, by the induction hypothesis on (9), is less than or equal to $3(1+3BL^*\Delta t)^{m-2}\varepsilon' \leq 3(1+3BL^*\Delta t)^{m-1}\varepsilon'$. Thus $\|u-h_m\|_m \leq 3(1+3BL^*\Delta t)^{m-1}\varepsilon'$. This completes the induction proof of (8) and (9). Now (7) and (8) imply that $\|u(t)-G_n(\Delta t)\varphi\| \leq \varepsilon'/3 + \left[(1+3BL^*\Delta t)^{n-1}-1\right]\varepsilon' = (1+3BL^*\Delta t)^{n-1}\varepsilon' - (2/3)\varepsilon' \leq e^{3BL^*T}\varepsilon' - (2/3)\varepsilon' = \varepsilon$. This completes the proof of Theorem 3.

References

[1] Chu, S. C., and Diaz, J. B., J. Math. Anal. Appl. 11, 440-446 (1965).

[2] Cryer, C. W., and Tavernini, L., SIAM J. Num. Anal. 9, 105-129 (1972).

[3] Kato, T., Perturbation Theory for Linear Operators, Springer, New York, 1966.

[4] Lax, P. D., and Richtmyer, R. D., Comm. Pure Appl. Math. 9, 267-293 (1956).

[5] Richtmyer, R. D., and Morton, K. W., Difference Methods for Initial-Value Problems, 2nd Edition, Interscience, New York, 1967.

[6] Thompson, R. J., J. SIAM 12, 189-199 (1964).

[7] Thompson, R. J., SIAM J. Num. Anal. 5, 475-487 (1968).

MONOTONIEEIGENSCHAFTEN VON DISKRETISIERUNGEN DES DIRICHLETPROBLEMS QUASILINEARER ELLIPTISCHER DIFFERENTIALGLEICHUNGEN

Von W. Törnig

1. Einführung

Löst man das Dirichletproblem nichtlinearer elliptischer Differentialgleichungen numerisch mit einem Diskretisierungsverfahren, so führt dies auf ein großes nichtlineares Differenzengleichungssystem. Von diesem ist zu fordern, daß es eine bzw. genau eine Lösung besitzt und iterativ lösbar ist. Erfüllt das Differentialgleichungsproblem ein Monotonieprinzip, etwa ein Randmaximumprinzip, so wird man in der Regel noch verlangen, daß das Differenzengleichungsproblem ein analoges diskretes Monotonieprinzip erfüllt.

Die damit zusammenhängenden Fragen sind bisher wenig untersucht worden. Mit der Frage der Existenz von Lösungen des Differenzengleichungsproblems als diskretes Analogon des Dirichletproblems einer allgemeinen quasilinearen gleichmäßig elliptischen Differentialgleichung im R^2 haben sich u. a. G. McAllister |1|, |1a|, T. Frank |4| und besonders eingehend R. S. Stepleman |8| befaßt. In |8| wird untersucht, teilweise basierend auf Ergebnissen von J. Bramble und B. Hubbard |2| bei linearen Problemen, wann ein Differenzenoperator vom "irreduzibel diagonaldominanten positiven Typ (IDDPT)" ist. (Vgl. Definition 1 in 2.) Es wird gezeigt, daß ein solches Differenzenproblem eine Lösung besitzt.

Hinreichend dafür, daß das Differenzengleichungsproblem einem diskreten Randmaximumprinzip genügt, ist, daß der Differenzenoperator vom "IDDP - Typ" ist. Im linearen Fall wurde dies von P. G. Ciarlet |3| gezeigt, für nichtlineare Differenzenoperatoren findet sich ein Beweis in |9|. Ähnlich wie bei linearen Problemen nehmen daher diese Differenzenoperatoren eine wichtige Sonderstellung ein.

Es erhebt sich daher die Frage, wie man IDDPT - Operatoren konstruieren kann. Sie ist, auch bei allgemeineren linearen zweidimensionalen Problemen, noch nicht befriedigend beantwortet worden, sofern die gemischte Ableitung $u_{x_1x_2}$ auftritt. In |2| wird für diesen Fall ein IDDPT-Operator konstruiert, in dem bei Benutzung von nicht notwendig benachbarten und unter Umständen weit auseinanderliegenden Gitterpunkten die Ableitung $u_{x_1 x_2}$ im allgemeinen auf komplizierte Art approximiert wird. Für die praktische Rechnung sind solche Approximationen oft nicht besonders geeignet. Das gilt in noch stärkerem Maße bei allgemeinen quasilinearen Problemen im R^2, für die in |8| notwendige und hinreichende Bedingungen angegeben werden, unter denen die Approximationen der 2. Ableitungen aus |2| auch hier zu einem Differenzenoperator vom IDDP-Typ führen. In 2. wird diese Bedingung zitiert.

Nun treten quasilineare elliptische Differentialgleichungen in den Anwendungen, etwa in der Physik, meist in spezieller Form auf: Sie lassen sich als Divergenzform darstellen oder noch häufiger als Eulersche Gleichung eines Variationsproblems. Mit der Konstruktion von IDDPT-Operatoren und IDDPT - ähnlichen Operatoren für das Dirichletproblem einer Klasse solcher Gleichungen, in denen die gemischten Ableitungen vorkommen, befaßt sich diese Arbeit.

Nach der Definition von IDDPT-Operatoren in 2. wird im 3. Abschnitt allgemein die Konstruktion solcher Operatoren untersucht und im wesentlichen auf das Vorgehen bei linearen Problemen zurückgeführt. Die Funktionalmatrix der entstehenden großen nichtlinearen Gleichungssysteme ist dabei überall symmetrisch und positiv definit, so daß verschiedene Iterationsverfahren, etwa nichtlineare SOR-Verfahren, Anwendung finden. In 4. werden für den Fall n = 2 zwei besonders einfache Operatoren angegeben, die von 2. Ordnung konsistent sind. Einer von ihnen ist ein IDDPT-Operator, der andere erfüllt das Monotonie-Prinzip, ist aber im strengen Sinne kein IDDPT-Operator.

2. Quasilineare Differenzenoperatoren vom IDDP-Typ

Es sei $G \subset R^n$ ein offenes, beschränktes und einfach zusammenhängendes Gebiet, und wir betrachten das Randwertproblem

$$
\begin{aligned}
Lu &= f(x, u, \operatorname{grad} u), && x \in G, \\
u &= g(x), && x \in \dot{G},
\end{aligned}
\tag{2.1}
$$

mit

$$\text{(2.2)} \qquad Lu = \sum_{i,j=1}^{n} \alpha_{ij}\,(x,\, u,\, \text{grad } u)\; u_{x_i x_j}.$$

Dabei wurde $x^T = (x_1, \ldots, x_n)$, $(\text{grad } u)^T = (u_{x_1}, \ldots, u_{x_n})$ gesetzt. Die Matrix (α_{ij}) sei ferner in $\bar{G} \times R^{n+1}$ positiv definit, d. h. Lu sei elliptisch.

Wie üblich überziehen wir nun $\bar{G}$ mit einem Gitter $\bar{G}_h = G_h + \dot{G}_h$, wobei $\dot{G}_h$ und $\dot{G}$ nicht notwendig gemeinsame Punkte haben. Weiterhin betrachten wir ein festes aber beliebiges h und kennzeichnen künftig Größen, die von h abhängen, nicht sämtlich als solche, wenn kein Mißverständnis zu befürchten ist. Bei festem h habe dann G_h genau N, $\dot{G}_h$ genau $\dot{N}$ Punkte. Setzt man

$$v^T = (v_1, \ldots, v_N),\quad \dot{v}^T = (\dot{v}_1, \ldots, \dot{v}_{\dot{N}}),\quad \bar{v}^T = (v^T, \dot{v}^T)$$

und diskretisiert das Randwertproblem (2.1), so erhält man ein nichtlineares Differenzengleichungssystem in den Werten der Gitterfunktion $\bar{v}$, das die Form hat

$$\text{(2.3)} \qquad \begin{aligned} (L^h\bar{v})_i &= \sum_{j=1}^{N} a_{ij}(\bar{v})v_j + \sum_{j=1}^{\dot{N}} \dot{a}_{ij}(\bar{v})\dot{v}_j = \zeta_j(\bar{v}),\quad i = 1,\ldots,N,\\ \dot{v}_k &= \dot{\gamma}_k\,,\quad k = 1,\ldots,\dot{N}. \end{aligned}$$

Gesucht sind die Werte v_i der Gitterfunktion v in den Punkten von G_h. Setzt man noch

$$A = (a_{ij}),\quad \dot{A} = (\dot{a}_{ij}),\quad \zeta^T = (\zeta_1, \ldots, \zeta_N),\quad \dot{\gamma}^T = (\dot{\gamma}_1, \ldots, \dot{\gamma}_{\dot{N}}),$$

so folgt das System

$$\text{(2.4)} \qquad \begin{aligned} L^h\bar{v} &\equiv A(\bar{v})v + \dot{A}(\bar{v})\dot{v} = \zeta(\bar{v}),\\ \dot{v} &= \dot{\gamma}. \end{aligned}$$

Mit

$$(2.5) \qquad \bar{A}(\bar{v}) = \begin{pmatrix} A(\bar{v}) & \dot{A}(\bar{v}) \\ O & I \end{pmatrix}, \quad \bar{\zeta}^T = (\zeta^T, \dot{\gamma}^T),$$

wobei O die $\dot{N} \times N$-Nullmatrix und I die $\dot{N} \times \dot{N}$-Einheitsmatrix ist, lautet es mit der $\bar{N} \times \bar{N}$-Matrix $\bar{A}$, $\bar{N} = N + \dot{N}$, noch kürzer

$$(2.6) \qquad \bar{L}^h\bar{v} \equiv \bar{A}(\bar{v})\bar{v} = \bar{\zeta}(\bar{v}).$$

<u>Definition 1</u> (vgl. |8|) $\bar{L}^h$ heißt vom "irreduzibel diagonaldominanten positiven Typ" (IDDPT), wenn für alle $\bar{x} \in R^{\bar{N}}$ gilt

1. $A(\bar{x})$ ist eine L-Matrix, d. h. $a_{ii}(\bar{x}) > 0$, $a_{ij}(\bar{x}) \leqq 0$, $i \neq j$,

2. $\dot{A}(\bar{x}) \leqq 0$,

3. $\bar{A}(\bar{x})$ ist diagonaldominant,

4. $A(\bar{x})$ ist strikt oder irreduzibel diagonaldominant.

Wie bereits erwähnt, ist diese Eigenschaft von $\bar{L}^h$ hinreichend dafür, daß ein Randmaximum-Prinzip gilt. Genauer wird in |9| gezeigt:

<u>Satz 1</u> $\bar{L}^h$ sei vom IDDP-Typ. Für jeden Vektor $\bar{v} \in R^{\bar{N}}$ mit

$$L^h\bar{v} \leqq 0$$

gilt dann

$$(2.7) \qquad \mathrm{Max}\,\{v_i;\ 1 \leqq i \leqq N\} \leqq \mathrm{Max}\,\{0;\ \mathrm{Max}\,\{\dot{v}_j;\ 1 \leqq j \leqq \dot{N}\}\}.$$

Für den Beweis vergleiche man |9|.

Die Frage ist nun, welche Diskretisierungen zum IDDP-Typ führen. Für den Fall, daß $u_{x_i x_j}$, $i \neq j$, nicht auftreten, kann man solche Verfahren im allgemeinen leicht aufstellen. Kommen die $u_{x_i x_j}$, $i \neq j$, jedoch vor, - und nur diesen Fall wollen wir hier zugrunde legen -, so führt diese Frage auch bei linearen Problemen schon zu Schwierigkeiten. Bramble und

Hubbard |2| haben eine solche Diskretisierung gefunden, die jedoch die eingangs beschriebenen Nachteile hinsichtlich ihrer praktischen Anwendbarkeit besitzt.

R. S. Stepleman |8| zeigt für das quasilineare Problem (2.1) mit (2.2), daß die Diskretisierung von Bramble und Hubbard für n = 2 bei Erfülltsein weiterer Bedingungen genau dann zu einem Differenzenoperator vom IDDP-Typ führt, wenn

$$(2.8) \qquad \inf_{(r,p,q)\in R^3} \frac{\alpha_{22}(x_1,x_2,r,p,q)}{|\alpha_{12}(x_1,x_2,r,p,q)|} > \sup_{(r,p,q)\in R^3} \frac{|\alpha_{12}(x_1,x_2,r,p,q)|}{\alpha_{11}(x_1,x_2,r,p,q)}, \qquad (x_1,x_2)\in \bar{G}$$

gilt.

Im linearen Fall ist (2.8) die Forderung der Elliptizität, im quasilinearen Fall ist sie für $\alpha_{12} \equiv 0$ stets erfüllt, d. h. wenn $u_{x_1x_2}$ nicht auftritt. Andernfalls ist die Forderung (2.8) sehr weitgehend. Eine Differentialgleichung, die über den mathematischen Bereich hinaus Bedeutung hat und (2.8) erfüllt, scheint bisher nicht bekannt zu sein.

Wir betrachten daher im folgenden speziellere Gleichungen, die jedoch in der Physik und verwandten Gebieten häufiger vorkommen.

3. Diskretisierung des Dirichletproblems quasilinearer Eulerscher Gleichungen und IDDPT-Operatoren

Mit $u_{x_i} = p_i$, $i = 1,\dots,n$, $p^T = (p_1,\dots,p_n)$ betrachten wir jetzt das Variationsproblem

$$(3.1) \qquad J[u] = \int_G f(\tfrac{1}{2}\, p^T\, S(x)p)\,dg = \text{Min}, \quad u = g \text{ auf } \dot{G}.$$

Dabei ist $S(x) = (S_{ij}(x))$ eine $n\times n$ - Matrix und $f(t) \in C^2([0,\infty))$. Die zugehörige Eulersche Differentialgleichung lautet

$$(3.2) \qquad Lu = -\sum_{i=1}^{n} \frac{\partial}{\partial x_i}\left[f'(\tfrac{1}{2}\, p^T\, S(x)p) \sum_{j=1}^{n} S_{ij}(x)u_{x_j}\right] = 0.$$

Wir setzen weiter voraus

(3.3)
1. $S(x)$ symmetrisch und positiv definit für $x \in \bar{G}$, $S(x) \in C^{0}(\bar{G})$.
2. $f'(t) > 0$ für alle $t \in [0,\infty)$.
3. Mit $F(p) = f(\frac{1}{2}\, p^T S(x) p)$ ist die Matrix $\left(F_{p_i p_j}\right)$ positiv definit für alle $x \in \bar{G}$.

Ist 2. erfüllt, so gilt z. B. 3., wenn $f''(t) \geq 0$, $t \in [0,\infty)$

Eine Reihe wichtiger Differentialgleichungen der Physik hat die Form (3.2), läßt sich also aus (3.1) mit einer bestimmten Funktion $f(t)$ herleiten. Dabei ist häufig sogar $S(x) \equiv I$. Es sei hier nur an einige Gleichungen erinnert:

1. Gleichung der Minimalflächen: $n = 2$, $f(t) = (1 + t)^{\frac{1}{2}}$, $S(x) \equiv I$.

2. Gleichung des Geschwindigkeitspotentials der Strömung idealer Gase:

   $n = 2$ bzw. $n = 3$, $f(t) = (v_{Max}^2 - t)^{\frac{\kappa}{\kappa-1}}$, $\kappa = c_p/c_v$, $S(x) \equiv I$.

3. Gleichung spezieller magnetostatischer Felder: $n = 2$, $f(t)=t-\alpha \log(\beta+t)$, $\alpha, \beta > 0$ konstant, $0 < \alpha < \beta$, $S(x) \equiv I$.

Wir ersetzen jetzt (3.1) durch ein diskretes Analogon $Q[\bar{v}] = \text{Min}$. Der leichteren Beschreibung wegen nehmen wir an, daß $\bar{G}_h$ gleichmäßig ist und alle Gitterkanten die Länge h haben. Es handelt sich also um ein n-dimensionales Würfelgitter, das aus M Teilwürfeln bestehe, die wir ebenso wie die N Punkte von G_h und die $\dot{N}$ Punkte von $\dot{G}_h$ nach einer bestimmten Ordnung numerieren. Weiter sei $x^{(\mu)}$ der Mittelpunkt des μ-ten Teilwürfels und $\bar{B}$ das abgeschlossene Gebiet, das aus allen Würfeln besteht und somit $\bar{G}$ approximiert: es gelte $\lim_{h \to 0} B = G$. Für eine beliebige Funktion $\phi(x) \in C^2(\bar{B})$ gilt dann die Integrationsformel

$$\int_B \phi(x)\,db = h^n \sum_{\mu=1}^{M} \phi(x^{(\mu)}) + O(h^2)$$

und mit $\alpha_\nu > 0$, $\nu = 1, \ldots, S$, $\sum_{\nu=1}^{S} \alpha_\nu = 1$ trivialerweise

$$\int_B \phi(x)\,db = h^n \sum_{\nu=1}^{S} \alpha_\nu \sum_{\mu=1}^{M} \phi(x^{(\mu)}) + O(h^2). \tag{3.4}$$

Die Formel wird auf (3.1) mit $\phi = f$ angewandt, woraufhin unter der Summe dann noch die Ableitungen $p_\rho(x^{(\mu)})$ auftreten, die durch Differenzapproximationen ersetzt werden. Diese können wiederum für $\nu = 1, 2, \ldots, S$ verschieden sein; im ν - ten Summanden sei etwa

$$(3.5) \qquad p_\rho(x^{(\mu)}) = \sum_{j=1}^{N} \alpha_j^{(\mu,\nu)\rho}\, u(x_j) + \sum_{j=1}^{\dot N} \dot\alpha_j^{(\mu,\nu)\rho}\, \dot u(\dot x_j) + O(h^\gamma),$$

$$\nu = 1, \ldots, S; \; \gamma \geqq 1.$$

Läßt man die Restglieder in (3.3) und (3.5) fort, so kann das gesuchte diskrete Analogon zum Variationsproblem in der Form

$$(3.6) \qquad \begin{aligned} Q[\bar v] &= h^n \sum_{\nu=1}^{S} \alpha_\nu \sum_{\mu=1}^{M} f(z_{\mu,\nu}(\bar v)) = \text{Min}, \\ \dot v &= \dot\gamma \quad \text{auf } \dot G_h \end{aligned}$$

geschrieben werden. Dabei ist $z_{\mu,\nu}$ ein Ausdruck in den Komponenten von $\bar v$, dessen genaue Darstellung hier und im folgenden nicht benötigt wird. Zur Bestimmung der v_i, $i = 1, \ldots, N$, erhält man dann das Gleichungssystem

$$(3.7) \qquad \begin{aligned} \frac{1}{h^n}\frac{\partial Q[\bar v]}{\partial v_i} &= (L^h\,\bar v)_i = \sum_{j=1}^{N} a_{ij}(\bar v) v_j + \sum_{j=1}^{\dot N} \dot a_{ij}(\bar v)\dot v_j = 0, \; i = 1,\ldots,N, \\ \dot v_k &= \dot\gamma_k, \; k = 1, \ldots, \dot N. \end{aligned}$$

Es hat daher wie (2.4) die Form

$$(3.8) \qquad \begin{aligned} L^h\,\bar v &= A(\bar v)v + \dot A(\bar v)\dot v = 0, \\ \dot v &= \dot\gamma, \end{aligned}$$

und wir definieren $\bar A(\bar v)$ wie (2.5). Sei $\bar u$ entsprechend $\bar v$ durch die Werte $u(x_j)$ bzw. $\dot u(\dot x_j)$ der exakten Lösung des Variationsproblems (3.1), deren Existenz wir voraussetzen, definiert, so gilt unter den Voraussetzungen über $f(t)$

$$(3.9)\qquad I[u] - Q[\bar{u}] = O(h^n),\ n = 1,\ 2;\quad h \to 0.$$

Die Matrizen A, $\dot{A}$ bzw. $\bar{A}$ in (3.8) haben nun eine spezielle Form. Wir definieren die Matrizen

$$(3.10)\qquad B^{(\mu,\nu)} = (\beta_{ij}^{(\mu,\nu)}),\ \dot{B}^{(\mu,\nu)} = (\dot{\beta}_{ij}^{(\mu,\nu)}),\ \bar{B}^{(\mu,\nu)} = \begin{pmatrix} B^{(\mu,\nu)} & \dot{B}^{(\mu,\nu)} \\ O & I \end{pmatrix}$$

mit den Elementen

$$(3.11)\qquad \beta_{ij}^{(\mu,\nu)} = \sum_{\rho,\sigma=1}^{n} S_{\rho\sigma}(x^{(\mu)})\ \alpha_i^{(\mu,\nu)\rho}\alpha_j^{(\mu,\nu)\sigma}$$

$$= (\alpha_i^{(\mu,\nu)})^T S(x^{(\mu)})\alpha_j^{(\mu,\nu)},$$

$$(3.12)\qquad \dot{\beta}_{ij}^{(\mu,\nu)} = \sum_{\rho,\sigma=1}^{n} S_{\rho\sigma}(x^{(\mu)})\ \alpha_i^{(\mu,\nu)\rho}\dot{\alpha}_j^{(\mu,\nu)\sigma}$$

$$= (\alpha_i^{(\mu,\nu)})^T S(x^{(\mu)})\ \dot{\alpha}_j^{(\mu,\nu)},$$

mit

$$(\alpha_i^{(\mu,\nu)})^T = (\alpha_i^{(\mu,\nu)1},\ \ldots,\ \alpha_i^{(\mu,\nu)n}),$$

$$(\dot{\alpha}_j^{(\mu,\nu)})^T = (\dot{\alpha}_j^{(\mu,\nu)1},\ \ldots,\ \dot{\alpha}_j^{(\mu,\nu)n}).$$

Dann rechnet man aus:

$$(3.13)\qquad A(\bar{v}) = \sum_{\mu=1}^{M}\{\sum_{\nu=1}^{S}\alpha_\nu\ f'(z_{\mu,\nu}(\bar{v}))B^{(\mu,\nu)}\},$$

$$(3.14)\qquad \dot{A}(\bar{v}) = \sum_{\mu=1}^{M}\{\sum_{\nu=1}^{S}\alpha_\nu\ f'(z_{\mu,\nu}(\bar{v}))\dot{B}^{(\mu,\nu)}\}.$$

In |6| wird gezeigt: Die stets symmetrische Funktionalmatrix

$$\left(\frac{1}{h^n}\ \frac{\partial^2 Q}{\partial v_i\ \partial v_j}\right),\ i,\ j = 1,\ \ldots,\ N,$$

des Systems (3.7) ist unter den oben angegebenen Voraussetzungen über f(t) genau dann überall positiv definit, wenn die Matrix des aus (3.7) für $f(t) \equiv t$, $S(x) \equiv I$ entstehenden linearen Systems positiv definit ist. Wir setzen dies jetzt voraus. Dann kann (3.7) iterativ gelöst werden, etwa mit einem nichtlinearen SOR-Verfahren für $0 < \omega < 2$, vgl. |7| . Für $f(t) \equiv t$, $S(x) \equiv I$ entsteht aus (3.2) die Laplace-Gleichung $\Delta_n u = 0$.

Hinreichende Bedingungen dafür, daß $\bar{L}^h$ vom IDDP-Typ ist, können nun leicht angegeben werden; es gilt der

<u>Satz 2</u> Unter den Voraussetzungen (3.3) gelte

1. $\beta_{ij}^{(\mu,\nu)} \leqq 0,\ i \neq j,\ i, j = 1,\ldots,N;\ \dot{\beta}_{ik}^{(\mu,\nu)} \leqq 0,\ i = 1,\ldots,N,$

   $k = 1, \ldots, \dot{N};\ \mu = 1, \ldots, M;\ \nu = 1, \ldots, S.$

2. $B = \sum_{\mu=1}^{M} \sum_{\nu=1}^{S} B^{(\mu,\nu)}$ ist irreduzibel.

3. $\dot{B} = \sum_{\mu=1}^{M} \sum_{\nu=1}^{S} \dot{B}^{(\mu,\nu)}$ besitzt mindestens ein nichtverschwindendes Element.

Dann ist $\bar{L}^h$ vom IDDP-Typ.

<u>Beweis</u> Aus 1. folgt mit (3.13), (3.14) unmittelbar

$$(3.15)\qquad a_{ij}(\bar{v}) \leqq 0,\ i \neq j,\ i, j = 1,\ldots,N;\ \dot{a}_{ik}(\bar{v}) \leqq 0, \text{ also } \dot{A}(\bar{v}) \leqq 0.$$

Weiter gilt mit

$$\sigma_i^{(\mu,\nu)} = \sum_{\substack{j=1\\ j\neq i}}^{N} \beta_{ij}^{(\mu,\nu)} + \sum_{k=1}^{\dot{N}} \dot{\beta}_{ik}^{(\mu,\nu)}$$

$$(3.16)\qquad s_i(\bar{v}) = \sum_{\substack{j=1\\ j\neq i}}^{N} a_{ij}(\bar{v}) + \sum_{k=1}^{\dot{N}} \dot{a}_{ik}(\bar{v}) = \sum_{\mu=1}^{M} \sum_{\nu=1}^{S} \alpha_\nu f'(z_{\mu,\nu}(\bar{v}))\sigma_i^{(\mu,\nu)} < 0,$$

$$i = 1, \ldots, N.$$

Denn zunächst ist wegen $\sigma_i^{(\mu,\nu)} \leqq 0$ auch $s_i(\bar{v}) \leqq 0$. Die Konsistenz in (3.5) erfordert

$$\sum_{j=1}^{N} \alpha_j^{(\mu,\nu)} + \sum_{k=1}^{\dot{N}} \dot{\alpha}_k^{(\mu,\nu)} = 0,$$

und somit

$$(3.17)\quad a_{ii}(\bar{v}) + s_i(\bar{v}) = \sum_{\mu=1}^{M} \sum_{\nu=1}^{S} \alpha_\nu f'(z_{\mu,\nu}(\bar{v}))\,(\alpha_i^{(\mu,\nu)})^T S(x^{(\mu)}) \left[\sum_{j=1}^{N} \alpha_j^{(\mu,\nu)} + \sum_{k=1}^{\dot{N}} \dot{\alpha}_k^{(\mu,\nu)}\right] = 0,$$

$$i = 1, \ldots, N.$$

Wäre für ein $i = i_o$ nun $s_{i_o}(\bar{v}) = 0$, so folgte nach (3.17) $a_{i_o i_o}(\bar{v}) = 0$ und wegen der Relation (3.16) zwischen $s_i(\bar{v})$ und den $\sigma_i^{(\mu,\nu)}$ auch $\sigma_{i_o}^{(\mu,\nu)} = 0$, $\mu = 1, \ldots, M$; $\nu = 1, \ldots, S$. Dann enthielte aber B in der i_o-ten Zeile, und wegen der Symmetrie auch in der i_o-ten Spalte, kein nichtverschwindendes Element, im Widerspruch zur Irreduzibilität von B. Daher gilt wegen (3.17)

$$(3.18)\qquad s_i(\bar{v}) < 0, \quad a_{ii}(\bar{v}) > 0, \quad i = 1, 2, \ldots, N.$$

Aus (2.5), (3.15), (3.17) folgt die Diagonaldominanz von $\bar{A}(\bar{v})$, weiter gilt

$$(3.19)\qquad \sum_{j=1}^{N} a_{ij}(\bar{v}) = - \sum_{k=1}^{\dot{N}} \dot{a}_{ik}(\bar{v}) \geqq 0.$$

Nach 3. hat $\dot{B}$ ein nichtverschwindendes Element, es gibt daher mindestens ein $\dot{\beta}_{i_o k_o}^{(\mu_o,\nu_o)} < 0$. Nach (3.14) ist dann $\dot{a}_{i_o k_o}(\bar{v}) < 0$ und nach (3.19)

$$(3.20)\qquad \sum_{j=1}^{N} a_{i_o j}(\bar{v}) > 0.$$

Daher ist $A(\bar{v})$ diagonaldominant, und zwar strikt, wenn $\dot{A}(\bar{v})$ in jeder Zeile mindestens ein nichtverschwindendes Element besitzt. Mit B ist weiter auch $A(\bar{v})$ irreduzibel. Daher ist $\bar{L}^h$ vom IDDP-Typ.

Es sei $\bar{L}_o^h$ der Operator, der sich aus $\bar{L}^h$ im Spezialfall $f(t) \equiv t$ ergibt. Dann gilt offenbar der

Satz 3 $\bar{L}^h$ ist genau dann für jede Funktion $f(t) \in C^1 [0,\infty)$ mit $f'(t) > 0$ vom IDDP-Typ, wenn $\bar{L}^h_o$ vom IDDP-Typ ist.

Der Beweis dieses Satzes folgt unmittelbar aus der Darstellung (3.13), (3.14).

4. Zwei Differenzenoperatoren im Fall n = 2

Um für den Fall n = 2 einfache Differenzenoperatoren genauer untersuchen zu können, ist es zweckmäßig, die Werte der Gitterfunktionen und andere darauf bezogene Größen durch 2 Indizes zu kennzeichnen. Dazu bezeichnen wir die Gitterpunkte selbst mit $x_{\lambda,\mu}$, $\dot{x}_{\lambda,\mu}$, die Werte der Gitterfunktion entsprechend mit $v_{\lambda,\mu}$, $\dot{v}_{\lambda,\mu}$. Für die entsprechenden Werte der exakten Lösung u des Differentialgleichungsproblems schreiben wir $u(x_{\lambda,\mu})$, $\dot{u}(\dot{x}_{\lambda,\mu})$. Weiter sei

$$P = \{ (\lambda,\mu) \mid x_{\lambda+l,\mu+m} \in \bar{G}_h,\ l,\ m = 0,\ 1 \},$$

$$(4.1) \qquad Q = \{ (\lambda,\mu) \mid x_{\lambda,\mu} \in G_h \},$$

$$\dot{Q} = \{ (\lambda,\mu) \quad \dot{x}_{\lambda,\mu} \in \dot{G}_h \}.$$

Wie bisher bezeichnen wir mit v die Gitterfunktion mit den Komponenten $v_{\lambda,\mu}$, $(\lambda,\mu) \in Q$, mit $\dot{v}$ die mit den Elementen $\dot{v}_{\lambda,\mu}$, $(\lambda,\mu) \in \dot{Q}$.

Mit $\bar{\lambda} = \lambda + \frac{1}{2}$, $\bar{\mu} = \mu + \frac{1}{2}$ nimmt dann (3.4) die Form an

$$(4.2) \qquad \int_B \phi\ (x)db = h^2 \sum_{\nu=1}^{s} \alpha_\nu \sum_{(\lambda,\mu)\in P} \phi(x_{\bar{\lambda},\bar{\mu}}) + O(h^2).$$

In Analogie zu (3.5) gelte

$$(4.3) \qquad p_\rho(x_{\bar{\lambda},\bar{\mu}}) = \sum_{(k,l)\in Q} \alpha_{k,l}^{(\bar{\lambda},\bar{\mu})\nu,\rho} u(x_{k,l}) + \sum_{(k,l)\in \dot{Q}} \dot{\alpha}_{k,l}^{(\bar{\lambda},\bar{\mu})\nu,\rho} \dot{u}(\dot{x}_{k,l}) + O(h^\gamma),$$

$$\rho = 1,\ 2;\ \nu = 1,\ \ldots,\ s;\ \gamma \geq 1,$$

ferner in Analogie zu (3.7)

$$\frac{1}{h^2}\frac{\partial Q[\bar{v}]}{\partial v_{k,l}} = (L^h\bar{v})_{k,l} = \sum_{(m,n)\in Q} a_{k,l;m,n}(\bar{v})v_{m,n}$$

$$(4.4)\qquad + \sum_{(m,n)\in \dot{Q}} \dot{a}_{k,l;m,n}(\bar{v})\dot{v}_{m,n} = 0,\ (k,l)\in Q,$$

$$\dot{v}_{\lambda,\mu} = \dot{\gamma}_{\lambda,\mu},\ (\lambda,\mu)\in \dot{Q}.$$

Mit

$$A = (a_{k,l;m,n}),\ \dot{A} = (\dot{a}_{k,l;m,n}),\ \zeta^T = 0^T,\ \dot{\gamma}^T = (\dot{\gamma}_{k,l})$$

entsteht dann wieder das Gleichungssystem (3.8). Die Größen (3.11), (3.12) gehen über in

$$(4.5)\qquad \beta^{(\bar{\lambda},\bar{\mu})\nu}_{k,l;m,n} = \sum_{\rho,\sigma=1}^{2} s_{\rho\sigma}(x_{\bar{\lambda},\bar{\mu}})\alpha^{(\bar{\lambda},\bar{\mu})\nu\rho}_{k,l}\alpha^{(\bar{\lambda},\bar{\mu})\nu\sigma}_{m,n}$$

$$= (\alpha^{(\bar{\lambda},\bar{\mu})\nu}_{k,l})^T S(x_{\bar{\lambda},\bar{\mu}})\alpha^{(\bar{\lambda},\bar{\mu})\nu}_{m,n},$$

$$(4.6)\qquad \dot{\beta}^{(\bar{\lambda},\bar{\mu})\nu}_{k,l;m,n} = \sum_{\rho,\sigma=1}^{2} s_{\rho\sigma}(x_{\bar{\lambda},\bar{\mu}})\alpha^{(\bar{\lambda},\bar{\mu})\nu\rho}_{k,l}\dot{\alpha}^{(\bar{\lambda},\bar{\mu})\nu\sigma}_{m,n}$$

$$= (\alpha^{(\bar{\lambda},\bar{\mu})\nu}_{k,l})^T S(x_{\bar{\lambda},\bar{\mu}})\dot{\alpha}^{(\bar{\lambda},\bar{\mu})\nu}_{m,n}.$$

Mit

$$(4.7)\qquad B^{(\bar{\lambda},\bar{\mu})\nu} = (\beta^{(\bar{\lambda},\bar{\mu})\nu}_{k,l;m,n}),\ \dot{B}^{(\bar{\lambda},\bar{\mu})\nu} = (\dot{\beta}^{(\bar{\lambda},\bar{\mu})\nu}_{k,l;m,n})$$

gilt dann schließlich in Analogie zu (3.13), (3.14)

$$(4.8)\qquad A(\bar{v}) = \sum_{(\lambda,\mu)\in P}\ \sum_{\nu=1}^{s} \alpha_\nu f'(z^\nu_{\bar{\lambda},\bar{\mu}}(\bar{v}))B^{(\bar{\lambda},\bar{\mu})\nu}$$

$$(4.9) \qquad \dot{A}(\bar{v}) = \sum_{(\lambda,\mu)\in P} \sum_{\nu=1}^{s} \alpha_\nu f'(z^\nu_{\bar{\lambda},\bar{\mu}}(\bar{v}))\, \dot{B}^{(\bar{\lambda},\bar{\mu})\nu} .$$

Wir betrachten jetzt gemäß (4.3) die beiden Diskretisierungen

1. $s = 1$, also $\alpha_1 = 1$.

$$p_1(x_{\bar{\lambda},\bar{\mu}}) = \frac{1}{2h}\left[u(x_{\lambda+1,\mu}) - u(x_{\lambda,\mu}) + u(x_{\lambda+1,\mu+1}) - u(x_{\lambda,\mu+1})\right] + O(h^2),$$

$$p_2(x_{\bar{\lambda},\bar{\mu}}) = \frac{1}{2h}\left[u(x_{\lambda,\mu+1}) - u(x_{\lambda,\mu}) + u(x_{\lambda+1,\mu+1}) - u(x_{\lambda+1,\mu})\right] + O(h^2).$$

2. $s = 4$, $\alpha_\nu = \frac{1}{4}$, $\nu = 1, \ldots, 4$.

$$p_1(x_{\bar{\lambda},\bar{\mu}}) = \frac{1}{h}\left[u(x_{\lambda+1,\mu}) - u(x_{\lambda,\mu})\right] + O(h),\ \nu = 1, 2,$$

$$p_1(x_{\bar{\lambda},\bar{\mu}}) = \frac{1}{h}\left[u(x_{\lambda+1,\mu+1}) - u(x_{\lambda,\mu+1})\right] + O(h),\ \nu = 3, 4,$$

$$p_2(x_{\bar{\lambda},\bar{\mu}}) = \frac{1}{h}\left[u(x_{\lambda,\mu+1}) - u(x_{\lambda,\mu})\right] + O(h),\ \nu = 1, 3,$$

$$p_2(x_{\bar{\lambda},\bar{\mu}}) = \frac{1}{h}\left[u(x_{\lambda+1,\mu+1}) - u(x_{\lambda+1,\mu})\right] + O(h),\ \nu = 2, 4.$$

Das Gitter G_h sei ferner im Sinne von |8|

A. Zusammenhängend, d. h. je zwei Punkte aus G_h lassen sich durch einen Polygonzug verbinden, der aus Verbindungsgeraden der Länge h zwischen zwei benachbarten Punkten aus G_h besteht.

B. Regulär, d. h. aus $x_{\lambda,\mu} \in G_h$ folgt $x_{\lambda\pm\alpha,\mu\pm\beta} \in \bar{G}_h$, $\alpha, \beta = 0, 1$.

Nach (4.3) ergeben sich dann unmittelbar die Werte für

$$\alpha_{k,l}^{(\bar{\lambda},\bar{\mu})\nu,\rho} \quad , \quad \dot{\alpha}_{k,l}^{(\bar{\lambda},\bar{\mu})\nu,\rho} .$$

Für beide Diskretisierungen gilt bei hinreichend glattem f(t) $\eta = 2$ in (3.9), obwohl bei der zweiten Diskretisierung jeweils $\gamma = 1$ ist. Darüber hinaus sind beide durch die Diskretisierungen gegebenen Verfahren zur numerischen Lösung des Randwertproblems

(4.1o) $$Lu = 0 \text{ in } G,\ u = g \text{ auf } \dot{G}$$

konsistent von der Ordnung 2, es gilt also mit dem durch (3.2) gegebenen Operator L für jedes $u \in C^4(\bar{B})$

(4.11) $$(L^h\bar{u})_{k,l} - Lu(x_{k,l}) = O(h^2),\ x_{k,l} \in G_h.$$

Man bestätigt dies a posteriori durch Entwicklung der linken Seite von (4.11) nach Potenzen von h.

Wir bezeichnen die Operatoren (2.6), die aus den oben definierten Diskretisierungen 1. 2. entstehen, mit $\bar{L}_1^h$, $\bar{L}_2^h$. Um zu untersuchen, wann diese vom IDDP-Typ oder ähnlicher Art sind, wenden wir Satz 2 an. Dann gilt zunächst der

Satz 4 Die Voraussetzungen (3.3) über die Matrix S(x) und die Funktion f(t) seien erfüllt, das Gitter G_h sei zusammenhängend und regulär. Genau dann erfüllt $\bar{L}_1^h$ bzw. $\bar{L}_2^h$ die Bedingung 1. des Satzes 2, wenn $S_{11}(x) \equiv S_{22}(x)$ bzw. $S_{12}(x) \equiv 0$, $x \in B$, gilt.

Beweis A. Wir betrachten zunächst die 1. Diskretisierung und setzen $B^{(\bar{\lambda},\bar{\mu})1} = B^{(\bar{\lambda},\bar{\mu})}$, $\dot{B}^{(\bar{\lambda},\bar{\mu})1} = \dot{B}^{(\bar{\lambda},\bar{\mu})}$. Diese Matrizen können nach (4.5), (4.6) berechnet werden. Schreibt man kürzer S_{ij} statt $S_{ij}(x_{\bar{\lambda},\bar{\mu}})$, so rechnet man aus, daß in der Hauptdiagonalen von $B^{(\bar{\lambda},\bar{\mu})}$ jeweils nur Ausdrücke der Form

(4.12) $$\frac{1}{4h^2}(S_{11} \pm 2S_{12} + S_{22}),$$

außerhalb der Hauptdiagonalen von $B^{(\bar{\lambda},\bar{\mu})}$ und in $\dot{B}^{(\bar{\lambda},\bar{\mu})}$ außer 0 nur Ausdrücke der Form

(4.13) $$\pm\frac{1}{4h^2}(S_{11} - S_{22}),$$

(4.14) $$-\frac{1}{4h^2}(S_{11} \pm 2S_{12} + S_{22})$$

auftreten. Nach Voraussetzung ist S(x) positiv definit, also (4.12) positiv und demnach (4.14) negativ. Bedingung 1. des Satzes 2 ist somit genau dann erfüllt, wenn $(S_{11} - S_{22}) \leqq 0$, $-(S_{11} - S_{22}) \leqq 0$, d. h.

$S_{11} = S_{22}$ gilt. Dies ist aber für alle $(\lambda,\mu) \in P$, alle $h > 0$ und wegen $S(x) \in C^0(\bar{G})$ genau für $S_{11}(x) \equiv S_{22}(x)$, $x \in B$, der Fall.

B. Bei der 2. Diskretisierung treten in der Hauptdiagonalen der Matrizen $B^{(\bar{\lambda},\, \bar{\mu})\nu}$, $\nu = 1, \ldots, 4$, nur Ausdrücke der Form

$$(4.15)\qquad \begin{array}{l} \frac{1}{h^2}(S_{11} \pm 2S_{12} + S_{22}) \\ \frac{1}{h^2} S_{11}, \\ \frac{1}{h^2} S_{22}, \\ 0, \end{array}$$

außerhalb der Hauptdiagonalen von $B^{(\bar{\lambda},\, \bar{\mu})\nu}$ und in $\dot{B}^{(\bar{\lambda},\, \bar{\mu})\nu}$, $\nu = 1, \ldots, 4$, nur Ausdrücke der Form

$$(4.16)\qquad \begin{array}{l} \pm \frac{1}{h^2}(S_{12} - S_{22}), \\ - \frac{1}{h^2}(S_{11} \pm S_{12}) \\ \pm \frac{1}{h^2} S_{12}, \\ 0 \end{array}$$

auf. Nach Voraussetzung sind (4.15) nicht negativ und (4.16) genau dann nicht positiv, wenn $S_{12} = 0$. Derselbe Schluß wie unter A. führt dann auch hier zur Behauptung $S(x) \equiv 0$, $x \in B$. Damit ist der Satz bewiesen.

Da die Diagonalelemente der Matrizen $\sum_{\nu=1}^{s} B^{(\bar{\lambda},\, \bar{\mu})\nu}$ im 1. Fall die Form (4.12) haben, also positiv sind, im 2. Fall sich zu $\frac{2}{h^2}(S_{11} + S_{22}) > 0$ errechnen, so enthält

$$(4.17)\qquad B = \sum_{(\lambda,\mu) \in P} \sum_{\nu=1}^{s} B^{(\bar{\lambda},\, \bar{\mu})\nu}$$

in jeder Zeile ein nichtverschwindendes Element. Da das Gitter regulär ist, kann ferner in beiden Fällen

$$(4.18)\qquad \dot{B} = \sum_{(\lambda,\mu) \in P} \sum_{\nu=1}^{s} \dot{B}^{(\bar{\lambda},\, \bar{\mu})\nu}$$

nicht nur Nullen enthalten.

Bei beiden Diskretisierungen treten dann in jeder Gleichung des Systems (3.8) genau 5 Werte der Gitterfunktion $\bar{v}$ in benachbarten Punkten auf, und zwar bei der

Diskretisierung 1. in den Punkten $x_{k,l}$, $x_{k \pm 1, l \pm 1}$, $(k,l) \in Q$, bei der Diskretisierung 2. in den Punkten $x_{k,l}$, $x_{k \pm 1,l}$, $x_{k,l \pm 1}$, $(k,l) \in Q$.

Bei der Diskretisierung 2. ist für ein zusammenhängendes und reguläres Gitter dann bekannt, daß für hinreichend kleines h die Matrix B irreduzibel ist. (Vgl. etwa |10|, S. 18 ff, 181 ff). Man beachte dabei, daß mit $f(t) \equiv t$ nach (4.8)

(4.19) $$A = \frac{1}{4} B = \frac{1}{4} \sum_{(\lambda,\mu) \in P} \sum_{\nu=1}^{4} B^{(\bar{\lambda}, \bar{\mu})\nu}$$

gilt. Wir haben daher den

<u>Satz 5</u> Unter den Voraussetzungen des Satzes 4 erfüllt die Diskretisierung 2. die Bedingungen 2., 3. des Satzes 2, liefert unter Berücksichtigung von Satz 4 also einen IDDPT-Operator.

Wir betrachten jetzt die Vektoren v' bzw. v" mit den Komponenten $v_{k,l}$, $(k,l) \in Q$, $k + l$ gerade bzw. $v_{k,l}$, $(k,l) \in Q$, $k + l$ ungerade. Entsprechend definieren wir $\dot{v}'$, $\dot{v}''$.

Bei der Diskretisierung 1. ist B reduzibel, das Gleichungssystem

(4.2o) $$Bv = - \dot{B}\dot{v}$$

zerfällt in die beiden Gleichungssysteme

(4.21) $$B'v' = - \dot{B}'\dot{v}, \qquad B''v'' = - \dot{B}''\dot{v}'',$$

wobei B', B" für hinreichend kleines h irreduzibel sind (Vgl. |5|). Denn in jeder Gleichung von $Bv + \dot{B}\dot{v} = 0$ treten, wie oben erwähnt, genau die 5 Werte $v_{k,l}$, $v_{k \pm 1, l \pm 1}$ auf, wie man durch Berechnen der Matrizen $B^{(\bar{\lambda}, \bar{\mu})}$, $\dot{B}^{(\bar{\lambda}, \bar{\mu})}$ leicht bestätigt. Daher genügen B', $\dot{B}'$ und

B'' , $\dot{B}''$ jeweils den Voraussetzungen des Satzes 2.

Setzen wir $f(t) \equiv t$ und

$$(\bar{v}')^T = ((v')^T, (\dot{v}')^T), \quad (\bar{v}'')^T = ((v'')^T, (\dot{v}'')^T),$$

$$A' = \sum_{(\lambda,\mu) \in P} B^{(\bar{\lambda}, \bar{\mu})} = B',$$

entsprechend A'' , $\dot{A}'$, $\dot{A}''$ sowie $\bar{A}'$, $\bar{A}''$, ferner

$$(4.22) \qquad \bar{L}'^h \bar{v}' \equiv \bar{A}' \bar{v}', \quad \bar{L}''^h \bar{v}'' \equiv \bar{A}'' \bar{v}'',$$

so liefert die Diskretisierung 1. jeweils Operatoren $\bar{L}'^h$, $\bar{L}''^h$ vom IDDP-Typ. Dies gilt jedoch im allgemeinen nur für lineare Probleme, d. h. mit $f(t) = c_1 t + c_2$. Bei nichtlinearen Problemen treten in dem Term $f'(z_{\bar{\lambda},\bar{\mu}}(\bar{v}))$ Komponenten $v_{\lambda,\mu}$, $\dot{v}_{\lambda,\mu}$ auf, deren Indizessumme sowohl gerade als auch ungerade ist.

Obwohl $\bar{L}^h$ bei der Diskretisierung 1. kein IDDPT-Operator im strengen Sinn ist, gilt das in Satz 1 formulierte Monotonieprinzip, d. h. auch das Randmaximumprinzip. Die Diskretisierung 1. ist daher der Diskretisierung 2. im allgemeinen gleichwertig. Durch Linearkombination der beiden betrachteten Diskretisierungen lassen sich auf verschiedene Art weitere IDDPT-Operatoren gewinnen, wobei die $S_{ij}(x)$ nur geringen Einschränkungen unterworfen sind.

Literatur

|1| McAllister, G.: Some nonlinear elliptic partial differential equations and difference equations. J. Soc. Industr. Appl. Math. 12 (1964) 772 - 777.

|1a| McAllister, G.: Quasilinear uniformly elliptic partial differential equations and difference equations. SIAM J. Numer. Anal. 3 (1966) 13 - 33

|2| Bramble, J. u. Hubbard, B.: A theorem on error estimation for finite difference analogues of the Dirichlet problem for elliptic equations. Contributions to differential equations, 2 (1963) 319 - 34o.

|3| Ciarlet, P. G.: Discrete maximum principle for finite difference operators. Aeq. Math. 4 (197o) 338 - 352.

|4| Frank, T.: Error bounds on numerical solutions of Dirichlet problems for quasilinear elliptic equations. Thesis, University of Texas, Austin 1967.

|5| Meis, Th.: Zur Diskretisierung nichtlinearer elliptischer Differentialgleichungen. Computing 7 (1971) 344 - 352.

|6| Meis, Th. u. Törnig, W.: Diskretisierung des Dirichletproblems nichtlinearer elliptischer Differentialgleichungen. Erscheint in "Methoden und Verfahren der Mathematischen Physik", Bd 8, 1973.

|7| Ortega, J. M. u. Rheinboldt, W. C.: Iterative solution of nonlinear equations in several variables. Academic Press, New York, London 197o.

|8| Stepleman, R. S.: Difference analogues of quasi-linear elliptic Dirichletproblems with mixed derivatives. Math. of Comp. 25 (1971) 257 - 269.

|9| Törnig, W.: Monotonie- und Randmaximumsätze bei Diskretisierungen des Dirichletproblems allgemeiner nichtlinearer elliptischer Differentialgleichungen. Erscheint 1974.

|1o| Varga, R. S.: Matrix iterative analysis. Prentice Hall, Englewood Cliffs, N. J. 1962.

ANSCHRIFTEN DER AUTOREN

Allgower, E. Prof. Dr.
Forschungsinstitut für Mathematik
der Eidg. Technischen Hochschule
CH 8oo6 Zürich

Ansorge, R. Prof. Dr.
Institut für Angewandte Mathematik
der Universität Hamburg
D 2ooo Hamburg 13
Rothenbaumchaussee 41

Van de Craats, J. Dr.
Mathematisch Instituut
Rijksuniversiteit Leiden
NL Leiden
Wassenaarseweg 8o

Engels, H. Dr.
Zentralinstitut für Angewandte Mathematik
der Kernforschungsanlage Jülich GmbH
D 517o Jülich 1
Postfach 365

Esser, H. Dr.
Institut für Geometrie und Praktische Mathematik
der Technischen Hochschule
D 51oo Aachen
Templergraben 55

Graf Finck von Finckenstein, K. Dr.
Max-Planck-Institut für Plasmaphysik
D 8o46 Garching

Frehse, J. Dr.
D 6o79 Buchschlag
Im Birkeneck 1

Geiger, C. Dr.
Institut für Angewandte Mathematik
der Universität Hamburg
D 2ooo Hamburg 13
Rothenbaumchaussee 41

Gekeler, E. Dr.
Universität Mannheim (WH)
D 68oo Mannheim 1
Postfach 2428

Gorenflo, R. Prof. Dr.
Institut für Geometrie und Praktische Mathematik
der Technischen Hochschule
D 51oo Aachen
Templergraben 55

Henry, M. S. Prof. Dr.
Department of Mathematics
Montana State University
USA Bozeman, Montana 59715

Hertling, J. Dr.
Institut für Numerische Mathematik
der Technischen Hochschule Wien
A 1o4o Wien
Karlsplatz 13

Van der Houwen, P. J. Dr.
Mathem. Center
Universität Amsterdam
NL Amsterdam

Locher, F. Dr.
Mathematisches Institut
der Universität Tübingen
D 74oo Tübingen
Brunnenstr. 27

Meuer, H. W. Dr.
Zentralinstitut für Angewandte Mathematik
der Kernforschungsanlage Jülich GmbH
D 517o Jülich 1
Postfach 365

Pittnauer, F. Prof. Dr.
Universität Dortmund
Abteilung Mathematik
D 46oo Dortmund
August-Schmidt-Straße

Reinhardt, J. Dr.
Johann Wolfgang Goethe Universität
Fachbereich Mathematik
D 6ooo Frankfurt
Robert-Mayer-Straße 6 - 1o

Schober, G. Prof. Dr.
Department of Mathematics
Indiana University
USA Bloominton, Indiana 474o1

Stummel, F. Prof. Dr.
Johann Wolfgang Goethe Universität
Fachbereich Mathematik
D 6ooo Frankfurt
Robert-Mayer-Straße 6 - 1o

Thomée, V. Prof. Dr.
Department of Mathematics
University of Göteborg
Göteborg, Sweden

Thompson, R. J. Prof. Dr.
Sandia Laboratories
USA Albuquerque, New Mexico 87115

Törnig, W. Prof. Dr.
Fachbereich Mathematik
der Technischen Hochschule
D 61oo Darmstadt
Hochschulstr. 1

Zeller, K. Prof. Dr.
Mathematisches Institut
der Universität Tübingen
D 74oo Tübingen
Brunnenstr. 27

.ecture Notes in Mathematics

omprehensive leaflet on request

ol. 146: A. B. Altman and S. Kleiman, Introduction to Grothendieck uality Theory. II, 192 pages. 1970. DM 18,–

ol. 147: D. E. Dobbs, Cech Cohomological Dimensions for Comutative Rings. VI, 176 pages. 1970. DM 16,–

'ol. 148: R. Azencott, Espaces de Poisson des Groupes Localement ompacts. IX, 141 pages. 1970. DM 16,–

ol. 149: R. G. Swan and E. G. Evans, K-Theory of Finite Groups and)rders. IV, 237 pages. 1970. DM 20,–

'ol. 150: Heyer, Dualität lokalkompakter Gruppen. XIII, 372 Seiten. 970. DM 20,–

'ol. 151: M. Demazure et A. Grothendieck, Schémas en Groupes I. SGA 3). XV, 562 pages. 1970. DM 24,–

'ol. 152: M. Demazure et A. Grothendieck, Schémas en Groupes II. GA 3). IX, 654 pages. 1970. DM 24,–

ol. 153: M. Demazure et A. Grothendieck, Schémas en Groupes III. GA 3). VIII, 529 pages. 1970. DM 24,–

ol. 154: A. Lascoux et M. Berger, Variétés Kähleriennes Compactes. 'II, 83 pages. 1970. DM 16,–

ol. 155: Several Complex Variables I, Maryland 1970. Edited by Horváth. IV, 214 pages. 1970. DM 18,–

ol. 156: R. Hartshorne, Ample Subvarieties of Algebraic Varieties. V, 256 pages. 1970. DM 20,–

ol. 157: T. tom Dieck, K. H. Kamps und D. Puppe, Homotopietheorie. , 265 Seiten. 1970. DM 20,–

ol. 158: T. G. Ostrom, Finite Translation Planes. IV. 112 pages. 1970. M 16,–

ol. 159: R. Ansorge und R. Hass. Konvergenz von Differenzenverhren für lineare und nichtlineare Anfangswertaufgaben. VIII, 145 eiten. 1970. DM 16,–

ol. 160: L. Sucheston, Constributions to Ergodic Theory and Probaility. VII, 277 pages. 1970. DM 20,–

ol. 161: J. Stasheff, H-Spaces from a Homotopy Point of View. , 95 pages. 1970. DM 16,–

ol. 162: Harish-Chandra and van Dijk, Harmonic Analysis on Reduce p-adic Groups. IV, 125 pages. 1970. DM 16,–

ol. 163: P. Deligne, Equations Différentielles à Points Singuliers eguliers. III, 133 pages. 1970. DM 16,–

ol. 164: J. P. Ferrier, Seminaire sur les Algebres Complètes. II, 69 pas. 1970. DM 16,–

ol. 165: J. M. Cohen, Stable Homotopy. V, 194 pages. 1970. DM 16,–

ol. 166: A. J. Silberger, PGL_2 over the p-adics: its Representations, pherical Functions, and Fourier Analysis. VII, 202 pages. 1970. M 18,–

ol. 167: Lavrentiev, Romanov and Vasiliev, Multidimensional Inverse oblems for Differential Equations. V, 59 pages. 1970. DM 16,–

ol. 168: F. P. Peterson, The Steenrod Algebra and its Applications: conference to Celebrate N. E. Steenrod's Sixtieth Birthday. VII, 17 pages. 1970. DM 22,–

ol. 169: M. Raynaud, Anneaux Locaux Henséliens. V, 129 pages. 1970. M 16,–

ol. 170: Lectures in Modern Analysis and Applications III. Edited by T. Taam. VI, 213 pages. 1970. DM 18,–

ol. 171: Set-Valued Mappings, Selections and Topological Properties 2^X. Edited by W. M. Fleischman. X, 110 pages. 1970. DM 16,–

ol. 172: Y.-T. Siu and G. Trautmann, Gap-Sheaves and Extension Coherent Analytic Subsheaves. V, 172 pages. 1971. DM 16,–

ol. 173: J. N. Mordeson and B. Vinograde, Structure of Arbitrary urely Inseparable Extension Fields. IV, 138 pages. 1970. DM 16,–

ol. 174: B. Iversen, Linear Determinants with Applications to the icard Scheme of a Family of Algebraic Curves. VI, 69 pages. 1970. M 16,–

ol. 175: M. Brelot, On Topologies and Boundaries in Potential Theory. , 176 pages. 1971. DM 18,–

'ol. 176: H. Popp, Fundamentalgruppen algebraischer Mannigfaltigeiten. IV, 154 Seiten. 1970. DM 16,–

ol. 177: J. Lambek, Torsion Theories, Additive Semantics and Rings Quotients. VI, 94 pages. 1971. DM 16,–

Vol. 178: Th. Bröcker und T. tom Dieck, Kobordismentheorie. XVI, 191 Seiten. 1970. DM 18,–

Vol. 179: Seminaire Bourbaki – vol. 1968/69. Exposés 347-363. IV. 295 pages. 1971. DM 22,–

Vol. 180: Séminaire Bourbaki – vol. 1969/70. Exposés 364-381. IV, 310 pages. 1971. DM 22,–

Vol. 181: F. DeMeyer and E. Ingraham, Separable Algebras over Commutative Rings. V, 157 pages. 1971. DM 16,–

Vol. 182: L. D. Baumert. Cyclic Difference Sets. VI, 166 pages. 1971. DM 16,–

Vol. 183: Analytic Theory of Differential Equations. Edited by P. F. Hsieh and A. W. J. Stoddart. VI, 225 pages. 1971. DM 20,–

Vol. 184: Symposium on Several Complex Variables, Park City, Utah, 1970. Edited by R. M. Brooks. V, 234 pages. 1971. DM 20,–

Vol. 185: Several Complex Variables II, Maryland 1970. Edited by J. Horváth. III, 287 pages. 1971. DM 24,–

Vol. 186: Recent Trends in Graph Theory. Edited by M. Capobianco/ J. B. Frechen/M. Krolik. VI, 219 pages. 1971. DM 18.–

Vol. 187: H. S. Shapiro, Topics in Approximation Theory. VIII, 275 pages. 1971. DM 22,–

Vol. 188: Symposium on Semantics of Algorithmic Languages. Edited by E. Engeler. VI, 372 pages. 1971. DM 26,–

Vol. 189: A. Weil, Dirichlet Series and Automorphic Forms. V, 164 pages. 1971. DM 16,–

Vol. 190: Martingales. A Report on a Meeting at Oberwolfach, May 17-23, 1970. Edited by H. Dinges. V, 75 pages. 1971. DM 16,–

Vol. 191: Séminaire de Probabilités V. Edited by P. A. Meyer. IV, 372 pages. 1971. DM 26,–

Vol. 192: Proceedings of Liverpool Singularities – Symposium I. Edited by C. T. C. Wall. V, 319 pages. 1971. DM 24,–

Vol. 193: Symposium on the Theory of Numerical Analysis. Edited by J. Ll. Morris. VI, 152 pages. 1971. DM 16,–

Vol. 194: M. Berger, P. Gauduchon et E. Mazet. Le Spectre d'une Variété Riemannienne. VII, 251 pages. 1971. DM 22,–

Vol. 195: Reports of the Midwest Category Seminar V. Edited by J. W. Gray and S. Mac Lane. III, 255 pages. 1971. DM 22,–

Vol. 196: H-spaces – Neuchâtel (Suisse)- Août 1970. Edited by F. Sigrist, V, 156 pages. 1971. DM 16,–

Vol. 197: Manifolds – Amsterdam 1970. Edited by N. H. Kuiper. V, 231 pages. 1971. DM 20,–

Vol. 198: M. Hervé, Analytic and Plurisubharmonic Functions in Finite and Infinite Dimensional Spaces. VI, 90 pages. 1971. DM 16.–

Vol. 199: Ch. J. Mozzochi, On the Pointwise Convergence of Fourier Series. VII, 87 pages. 1971. DM 16,–

Vol. 200: U. Neri, Singular Integrals. VII, 272 pages. 1971. DM 22,–

Vol. 201: J. H. van Lint, Coding Theory. VII, 136 pages. 1971. DM 16,–

Vol. 202: J. Benedetto, Harmonic Analysis on Totally Disconnected Sets. VIII, 261 pages. 1971. DM 22,–

Vol. 203: D. Knutson, Algebraic Spaces. VI, 261 pages. 1971. DM 22,–

Vol. 204: A. Zygmund, Intégrales Singulières. IV, 53 pages. 1971. DM 16,–

Vol. 205: Séminaire Pierre Lelong (Analyse) Année 1970. VI, 243 pages. 1971. DM 20,–

Vol. 206: Symposium on Differential Equations and Dynamical Systems. Edited by D. Chillingworth. XI, 173 pages. 1971. DM 16,–

Vol. 207: L. Bernstein, The Jacobi-Perron Algorithm – Its Theory and Application. IV, 161 pages. 1971. DM 16,–

Vol. 208: A. Grothendieck and J. P. Murre, The Tame Fundamental Group of a Formal Neighbourhood of a Divisor with Normal Crossings on a Scheme. VIII, 133 pages. 1971. DM 16,–

Vol. 209: Proceedings of Liverpool Singularities Symposium II. Edited by C. T. C. Wall. V, 280 pages. 1971. DM 22,–

Vol. 210: M. Eichler, Projective Varieties and Modular Forms. III, 118 pages. 1971. DM 16,–

Vol. 211: Théorie des Matroïdes. Edité par C. P. Bruter. III, 108 pages. 1971. DM 16,–

Please turn over

Vol. 212: B. Scarpellini, Proof Theory and Intuitionistic Systems. VII, 291 pages. 1971. DM 24,-

Vol. 213: H. Hogbe-Nlend, Théorie des Bornologies et Applications. V, 168 pages. 1971. DM 18,-

Vol. 214: M. Smorodinsky, Ergodic Theory, Entropy. V, 64 pages. 1971. DM 16,-

Vol. 215: P. Antonelli, D. Burghelea and P. J. Kahn, The Concordance-Homotopy Groups of Geometric Automorphism Groups. X, 140 pages. 1971. DM 16,-

Vol. 216: H. Maaß, Siegel's Modular Forms and Dirichlet Series. VII, 328 pages. 1971. DM 20,-

Vol. 217: T. J. Jech, Lectures in Set Theory with Particular Emphasis on the Method of Forcing. V, 137 pages. 1971. DM 16,-

Vol. 218: C. P. Schnorr, Zufälligkeit und Wahrscheinlichkeit. IV, 212 Seiten 1971. DM 20,-

Vol. 219: N. L. Alling and N. Greenleaf, Foundations of the Theory of Klein Surfaces. IX, 117 pages. 1971. DM 16,-

Vol. 220: W. A. Coppel, Disconjugacy. V, 148 pages. 1971. DM 16,-

Vol. 221: P. Gabriel und F. Ulmer, Lokal präsentierbare Kategorien. V, 200 Seiten. 1971. DM 18,-

Vol. 222: C. Meghea, Compactification des Espaces Harmoniques. III, 108 pages. 1971. DM 16,-

Vol. 223: U. Felgner, Models of ZF-Set Theory. VI, 173 pages. 1971. DM 16,-

Vol. 224: Revêtements Etales et Groupe Fondamental. (SGA 1). Dirigé par A. Grothendieck XXII, 447 pages. 1971. DM 30,-

Vol. 225: Théorie des Intersections et Théorème de Riemann-Roch. (SGA 6). Dirigé par P. Berthelot, A. Grothendieck et L. Illusie. XII, 700 pages. 1971. DM 40,-

Vol. 226: Seminar on Potential Theory, II. Edited by H. Bauer. IV, 170 pages. 1971. DM 18,-

Vol. 227: H. L. Montgomery, Topics in Multiplicative Number Theory. IX, 178 pages. 1971. DM 18,-

Vol. 228: Conference on Applications of Numerical Analysis. Edited by J. Ll. Morris. X, 358 pages. 1971. DM 26,-

Vol. 229: J. Väisälä, Lectures on n-Dimensional Quasiconformal Mappings. XIV, 144 pages. 1971. DM 16,-

Vol. 230: L. Waelbroeck, Topological Vector Spaces and Algebras. VII, 158 pages. 1971. DM 16,-

Vol. 231: H. Reiter, L^1-Algebras and Segal Algebras. XI, 113 pages. 1971. DM 16,-

Vol. 232: T. H. Ganelius, Tauberian Remainder Theorems. VI, 75 pages. 1971. DM 16,-

Vol. 233: C. P. Tsokos and W. J. Padgett. Random Integral Equations with Applications to Stochastic Systems. VII, 174 pages. 1971. DM 18,-

Vol. 234: A. Andreotti and W. Stoll. Analytic and Algebraic Dependence of Meromorphic Functions. III, 390 pages. 1971. DM 26,-

Vol. 235: Global Differentiable Dynamics. Edited by O. Hájek, A. J. Lohwater, and R. McCann. X, 140 pages. 1971. DM 16,-

Vol. 236: M. Barr, P. A. Grillet, and D. H. van Osdol. Exact Categories and Categories of Sheaves. VII, 239 pages. 1971, DM 20,-

Vol. 237: B. Stenström. Rings and Modules of Quotients. VII, 136 pages. 1971. DM 16,-

Vol. 238: Der kanonische Modul eines Cohen-Macaulay-Rings. Herausgegeben von Jürgen Herzog und Ernst Kunz. VI, 103 Seiten. 1971. DM 16,-

Vol. 239: L. Illusie, Complexe Cotangent et Déformations I. XV, 355 pages. 1971. DM 26,-

Vol. 240: A. Kerber, Representations of Permutation Groups I. VII, 192 pages. 1971. DM 18,-

Vol. 241: S. Kaneyuki, Homogeneous Bounded Domains and Siegel Domains. V, 89 pages. 1971. DM 16,-

Vol. 242: R. R. Coifman et G. Weiss, Analyse Harmonique Non-Commutative sur Certains Espaces. V, 160 pages. 1971. DM 16,-

Vol. 243: Japan-United States Seminar on Ordinary Differential and Functional Equations. Edited by M. Urabe. VIII, 332 pages. 1971. DM 26,-

Vol. 244: Séminaire Bourbaki - vol. 1970/71. Exposés 382-399. IV, 356 pages. 1971. DM 26,-

Vol. 245: D. E. Cohen, Groups of Cohomological Dimension One. V, 99 pages. 1972. DM 16,-

Vol. 246: Lectures on Rings and Modules. Tulane University Ring and Operator Theory Year, 1970-1971. Volume I. X, 661 pages. 1972. DM 40,-

Vol. 247: Lectures on Operator Algebras. Tulane University Ring and Operator Theory Year, 1970-1971. Volume II. XI, 786 pages. 1972. DM 40,-

Vol. 248: Lectures on the Applications of Sheaves to Ring Theory. Tulane University Ring and Operator Theory Year, 1970-1971. Volume III. VIII, 315 pages. 1971. DM 26,-

Vol. 249: Symposium on Algebraic Topology. Edited by P. J. Hilton. VII, 111 pages. 1971. DM 16,-

Vol. 250: B. Jónsson, Topics in Universal Algebra. VI, 220 pages. 1972. DM 20,-

Vol. 251: The Theory of Arithmetic Functions. Edited by A. A. Gioia and D. L. Goldsmith VI, 287 pages. 1972. DM 24,-

Vol. 252: D. A. Stone, Stratified Polyhedra. IX, 193 pages. 1972. DM 18,-

Vol. 253: V. Komkov, Optimal Control Theory for the Damping of Vibrations of Simple Elastic Systems. V, 240 pages. 1972. DM 20,-

Vol. 254: C. U. Jensen, Les Foncteurs Dérivés de $\varprojlim$ et leurs Applications en Théorie des Modules. V, 103 pages. 1972. DM 16,-

Vol. 255: Conference in Mathematical Logic - London '70. Edited by W. Hodges. VIII, 351 pages. 1972. DM 26,-

Vol. 256: C. A. Berenstein and M. A. Dostal, Analytically Uniform Spaces and their Applications to Convolution Equations. VII, 130 pages. 1972. DM 16,-

Vol. 257: R. B. Holmes, A Course on Optimization and Best Approximation. VIII, 233 pages. 1972. DM 20,-

Vol. 258: Séminaire de Probabilités VI. Edited by P. A. Meyer. VI, 253 pages. 1972. DM 22,-

Vol. 259: N. Moulis, Structures de Fredholm sur les Variétés Hilbertiennes. V, 123 pages. 1972. DM 16,-

Vol. 260: R. Godement and H. Jacquet, Zeta Functions of Simple Algebras. IX, 188 pages. 1972. DM 18,-

Vol. 261: A. Guichardet, Symmetric Hilbert Spaces and Related Topics. V, 197 pages. 1972. DM 18,-

Vol. 262: H. G. Zimmer, Computational Problems, Methods, and Results in Algebraic Number Theory. V, 103 pages. 1972. DM 16,-

Vol. 263: T. Parthasarathy, Selection Theorems and their Applications. VII, 101 pages. 1972. DM 16,-

Vol. 264: W. Messing, The Crystals Associated to Barsotti-Tate Groups: with Applications to Abelian Schemes. III, 190 pages. 1972. DM 18,-

Vol. 265: N. Saavedra Rivano, Catégories Tannakiennes. II, 418 pages. 1972. DM 26,-

Vol. 266: Conference on Harmonic Analysis. Edited by D. Gulick and R. L. Lipsman. VI, 323 pages. 1972. DM 24,-

Vol. 267: Numerische Lösung nichtlinearer partieller Differential- und Integro-Differentialgleichungen. Herausgegeben von R. Ansorge und W. Törnig, VI, 339 Seiten. 1972. DM 26,-

Vol. 268: C. G. Simader, On Dirichlet's Boundary Value Problem. IV, 238 pages. 1972. DM 20,-

Vol. 269: Théorie des Topos et Cohomologie Etale des Schémas. (SGA 4). Dirigé par M. Artin, A. Grothendieck et J. L. Verdier. XIX, 525 pages. 1972. DM 50,-

Vol. 270: Théorie des Topos et Cohomologie Etle des Schémas. Tome 2. (SGA 4). Dirige par M. Artin, A. Grothendieck et J. L. Verdier. V, 418 pages. 1972. DM 50,-

Vol. 271: J. P. May, The Geometry of Iterated Loop Spaces. IX, 175 pages. 1972. DM 18,-

Vol. 272: K. R. Parthasarathy and K. Schmidt, Positive Definite Kernels, Continuous Tensor Products, and Central Limit Theorems of Probability Theory. VI, 107 pages. 1972. DM 16,-

Vol. 273: U. Seip, Kompakt erzeugte Vektorräume und Analysis. IX, 119 Seiten. 1972. DM 16,-

Vol. 274: Toposes, Algebraic Geometry and Logic. Edited by. F. W. Lawvere. VI, 189 pages. 1972. DM 18,-

Vol. 275: Séminaire Pierre Lelong (Analyse) Année 1970-1971. VI, 181 pages. 1972. DM 18,-

Vol. 276: A. Borel, Représentations de Groupes Localement Compacts. V, 98 pages. 1972. DM 16,-

Vol. 277: Séminaire Banach. Edité par C. Houzel. VII, 229 pages. 1972. DM 20,-